AF334452

Springer Series in Reliability Engineering

Series Editor

Professor Hoang Pham
Department of Industrial and Systems Engineering
Rutgers, The State University of New Jersey
96 Frelinghuysen Road
Piscataway, NJ 08854-8018
USA

Other titles in this series

The Universal Generating Function in Reliability Analysis and Optimization
Gregory Levitin

Warranty Management and Product Manufacture
D.N.P. Murthy and Wallace R. Blischke

Maintenance Theory of Reliability
Toshio Nakagawa

System Software Reliability
Hoang Pham

Reliability and Optimal Maintenance
Hongzhou Wang and Hoang Pham

Applied Reliability and Quality
B.S. Dhillon

Shock and Damage Models in Reliability Theory
Toshio Nakagawa

Risk Management
Terje Aven and Jan Erik Vinnem

Satisfying Safety Goals by Probabilistic Risk Assessment
Hiromitsu Kumamoto

Offshore Risk Assessment (2nd Edition)
Jan Erik Vinnem

The Maintenance Management Framework
Adolfo Crespo Márquez

Human Reliability and Error in Transportation Systems
B.S. Dhillon

Complex System Maintenance Handbook
D.N.P. Murthy and Khairy A.H. Kobbacy

Recent Advances in Reliability and Quality in Design
Hoang Pham

Product Reliability
D.N.P. Murthy, Marvin Rausand and Trond Østerås

Mining Equipment Reliability, Maintainability, and Safety
B.S. Dhillon

Advanced Reliability Models and Maintenance Policies
Toshio Nakagawa

Pierre-Jacques Courtois

Justifying the Dependability of Computer-based Systems

With Applications in Nuclear Engineering

 Springer

Pierre-Jacques Courtois, Pr.Dr.Ir.
Université Catholique de Louvain (UCL)
Louvain-la-Neuve
Belgium

and

BEL V
Federal Agency for Nuclear Safety Subsidiary
Brussels
Belgium

ISBN 978-1-84800-371-2   e-ISBN 978-1-84800-372-9

DOI 10.1007/978-1-84800-372-9

Springer Series in Reliability Engineering ISSN 1614-7839

British Library Cataloguing in Publication Data
Courtois, Pierre Jacques, 1937-
  Justifying the dependability of computer-based systems :
  with applications in nuclear engineering. - (Springer
  series in reliability engineering)
  1. Computer systems - Reliability 2. Nuclear power plants -
  Safety measures 3. Nuclear engineering - Safety measures
  I. Title
  005
ISBN-13: 9781848003712

Library of Congress Control Number: 2008930488

Some parts of this material were prepared as an account of work sponsored by AV Nuclear, Brussels and the European Commission. Neither AVN nor the EC make any warranty or assume any legal responsibility for any third party's use of any information, product or process disclosed in this book.

Cover design: deblik, Berlin, Germany

Printed on acid-free paper

9 8 7 6 5 4 3 2 1

springer.com

Expansion de la luz, centrifuga y centripeta, 1913–1914
Gino Severini
Fundación Colección Thyssen-Bornemisza, Madrid

To Dorinda

Preface

Safety is a paradoxical system property. It remains immaterial, intangible and invisible until a failure, an accident or a catastrophy occurs and, too late, reveals its absence. And yet, a system cannot be relied upon unless its safety can be explained, demonstrated and certified.

The practical and difficult questions which motivate this study concern the evidence and the arguments needed to justify the safety of a computer based system, or more generally its dependability. Dependability is a broad concept integrating properties such as safety, reliability, availability, maintainability and other related characteristics of the behaviour of a system in operation. How can we give the users the assurance that the system enjoys the required dependability? How should evidence be presented to certification bodies or regulatory authorities? What best practices should be applied? How should we decide whether there is enough evidence to justify the release of the system?

To help answer these daunting questions, a method and a framework are proposed for the justification of the dependability of a computer-based system. The approach specifically aims at dealing with the difficulties raised by the validation of software. Hence, it should be of wide applicability despite being mainly based on the experience of assessing Nuclear Power Plant instrumentation and control systems important to safety.

To be viable, a method must rest on a sound theoretical background. Hence, the book explores some of the most fundamental aspects of the evaluation of dependability: the nature of claims, arguments and evidence, the ways to deal with their interpretation and valuation and the different types of uncertainty. The approach emphasizes the necessity – for the demonstration of dependability – of adequate models, representations and interpretations of the system at the plant, architecture, design, and operation levels.

Finally, we propose a set of practices that should be adopted to produce a convincing justification of the dependability of a computer-based system. These practices may be considered as the first constituents of a discipline, two tropisms of which appear to be the necessity of maintaining structured levels of evidence and well-defined interpretation structures.

An obvious objective is to rationalize dependability arguments. However, whatever degree of rationalization may be achieved – a consensus among stakeholders based on expert judgment will remain necessary. An essential benefit of the framework is to identify the nature and the scope of the uncertainty that needs to be resolved by consensus, and to reduce this scope to what is strictly required by an axiomatic approach.

Justifying dependability turns out to be based on system prescriptions, on system descriptions and on analysis. Accordingly, after having set up the context, the book is organized into three main parts: a prescriptive, descriptive, and methodological part. The prescriptive part deals with the structure and syntax of the justification in terms of arguments. The descriptive part investigates and formalizes its complex semantics: the required evidence. The last part infers practical recommendations on how to make a dependability case under different real industrial constraints.

Nevertherless, the justification of computer system dependability remains extremely difficult. With a singular strength, the analysis requires an obstinate desire for understanding and strenuous work to make concepts precise, establish their interdependencies and test their validity. This work, needless to say, has not the last word; hopefully it throws some light on the issues and indicates ways to attack them.

In retrospect, one of its more specific contributions perhaps is the identification of an iterative expansion-reduction structuring process for the evaluation of dependability. Probably no painting could better illustrate and symbolize this expansion-reduction than Severini's representation of the centrifugal-centripetal dance of light – represented on the first page, unfortunately in black and white only.

Luxor, Egypt
February 2008

Pierre-Jacques Courtois

Acknowledgements

This work owes a lot to my collaboration with Professor David Lorge Parnas, now living in Ottawa, Canada. The countless lively and inspiring discussions I have had with Dave over the last thirty five years have strongly influenced my research, directly and indirectly, in contents and forms; the models of Chapter 9, in particular, are in part built with his tools.

The research is also based on experience gained over the last fifteen years at the Association Vinçotte Nuclear (AVN), which was then the nuclear authorized inspection and licensing body of Belgium[1]. I am deeply indebted to my collaborators in AVN, to its former director, Pierre Govaerts, and also to architect engineers, nuclear operators and I&C designers whom I met over these years, and who generously shared their know-how in nuclear safety control and instrumentation in Belgium and abroad. I am especially grateful to my AVN colleagues Claude Baudelet and Alain Geens; they contributed with many technical ideas and advice; Alain read and contributed to the improvement of an early draft of the manuscript.

Professors Jean-Claude Laprie, Jean Arlat and Mohamed Kaâniche, LAAS-CNRS laboratory, Toulouse, Jean-Pierre Auclair, SNCF, France, Philippe David, ESA-ESTEC, Noordwijk, The Netherlands, Dr. J. Hyvarinen from STUK, Finland, Pr. Peter B. Ladkin, University of Bielefeld, Germany, Pr. B. Littlewood, City University, London, Dr. Reddersen, Colenco, Switzerland must also be thanked for helpful discussions and comments over the last few years.

This research received the support of the European Commission in the context of the 5th framework programme research project CEMSIS (Cost Effective Modernization of Systems Important to Safety; environment project FIS5-1999-0035). Thanks are due to colleagues of CEMSIS for many useful and provocative interactions during the three years of the project; in particular to P. Bishop and R.

---

[1] In 2008, part of AVN became a subsidiary, now called BEL V, of the Belgian Federal Agency for Nuclear Control (FANC).

Bloomfield, from Adelard, UK, to J.-B. Chabannes and Thuy Nguyen, from EdF, France, Paul Tooley and D. Pavey from British Electric, UK.

Regulator fellow experts from the task force on safety critical software of the European Commission Nuclear Regulator Working Group (NRWG) also contributed with useful advice. Some ideas behind the assessment approach proposed here find their origin in the difficult issues we met, as a group of regulators, and tried hard to solve during the last fifteen years and more that we have been working together.

Last but not least, the Springer editors, Simon Rees and Anthony Doyle are warmly thanked for their numerous useful and professional comments, their boundless patience and their encouragement; Cornelia Kresser and Sorina Moosdorf from le-tex publishing services oHG, Leipzig, were also very helpful in the last stages of the editing and production of the book.

In spite of friends' and colleagues' advices, the work may still contain errors and be clumsy in places. Plagiarising Walt Whitman[2], I can only say: "Faulty as it is, it is by far my special and entire self-chosen utterance".

Grateful acknowledgement is made to the following for permission to use or reproduce previously published material:
Editions Gallimard, Paris
Electrabel, Belgium
Fundación Colección Thyssen-Bornemisza, Madrid
Springer-Verlag, London

---

[2] In his announcement for the press of the 1892 edition, now dubbed "deathbed", of his book *"Leaves of Grass"*.

Contents

PART III Descriptions

PART IV   Methodological Implications

PART I

The Context

1

Introduction

Despite its huge importance for industry, the process of approving software-based equipment for executing safety critical functions remains highly complex, poorly understood and, as a result, often not efficiently mastered by regulators, licensees and suppliers; "Understanding Computer System Dependability" might have been a more appropriate title for this work.

The difficulties – some are devilishly complex – are both practical and theoretical. In practice, several engineering specialties need to be applied, in the field of computer hardware and software design and engineering, and also in the application domain in which the computer is embedded. To add to the difficulty, several industrial sectors still think that today complex software systems can be developed in the same ways analogue instrumentation and control systems were manufactured thirty five years ago, with more or less the same requirement specification approach, production tools an test strategies; a state of affairs which does not contribute to produce justifiably dependable software.

Testing is another issue. Digital systems are more sensitive to "small" defects than analogous systems. A single erroneous bit can cause havoc in the system and its environment. Interpolation and continuity arguments are therefore not valid and, for a same level of test coverage, a digital system requires comparatively more test cases and more insight into its behaviour than an analogous counterpart.[3]

The approval process also raises great theoretical challenges. The justification of computer dependability lies at the intersection of several disciplines: computer science, mathematical logic, applied probabilities, and statistics. At first sight, the contents of this book may therefore appear as an unusual – perhaps even somewhat disparate – mixture of pragmatic issues and academic background. The reason, however, for this apparent heterogeneity is none other than the desire to let no important facet of the problem aside.

---

[3] This problem is in particular well illustrated by the issue of testing analogue/digital converters, discussed in Section 9.5.16.

A review of licensing approaches and of guidelines issued by nuclear regulators and operators was carried out in 2000 by the CEMSIS project [85]; it showed that except for procedures that formalize negotiations between licensee and licensor, seldom a systematic method is defined or in use for demonstrating the safety of software based systems. Most practitioners attack the problem in the best possible *ad hoc* way.

If a systematic and well-planned approach is not followed, licensing costs, resources and delays may outweigh the benefits expected from a system upgrade or a modernization. The temptation to reduce costs at the expense of safety then augments.

The most significant contribution of this book is a pragmatic framework to allow a cost-effective justification of dependability. A method is proposed for the assessment of computer/software equipment for "Systems Important to Safety" (SIS), as they are coined by the International Atomic Energy Agency (IAEA) (*cf.* [46], for instance). The framework helps to justify safety and to license embedded software and hardware newly developed, replaced or upgraded in such systems.

The framework provides a *hierarchical structure for constructing arguments* relying on a variety of disparate sources of evidence; the structure provides modularity and allows the integration of arguments from previous subsystem safety cases. Questions addressed here are: what evidence and arguments are needed to convince an assessor that a system is safe enough? How should the certification authorities handle evidence? What best practices should be followed? How should we decide whether there is enough evidence to justify the release of the system or not?

This approach may depart from certain current licensing practices, which basically require compliance with rules, design principles or standards. Because they often fail to demonstrate convincingly by themselves that a system has the specific dependability properties required by a given application, such practices often entail licensing delays and costs. In this respect, an approach focusing on specific dependability properties opens perspectives for more efficiency and flexibility. Besides, it may remain applicable in situations where standards may not apply. A standard, for instance, may not accept certain practices or alternative sorts of evidence while these may be perfectly adequate or even necessary to implement or to justify specific dependability properties.

A founding principle of the work is the recognition that the justification of dependability is made up of three ingredients: system descriptions, prescriptions, and analysis. Accordingly, after a contextual introduction, the book is organized into three main parts: prescriptive, descriptive, and methodological.

The *prescriptive part* gives recommendations on how to start the justification with initial dependability claims, and on how to organize claims, evidence and arguments.

These notions are of course quite general and have been in use under various forms and in different contexts of human activity. They have been formally studied in models of causality and counterfactual inference, in both deterministic and probabilistic contexts. They were more recently applied to the safety justification of computer-based systems. When they are not precisely studied, however, these concepts often carry with them imprecise notions related to the rationale and modalities for the choice of claims, the validation of the arguments, the backing of the evidence, and the assumptions behind the models used.

The present framework can be seen as a proposal for new mechanisms, which should help engineers, regulators and licensees to follow a claim-based approach under industrial constraints and for engineering purposes. The claim-evidence-argument multi-level structure which is proposed here rests on the precise concepts of *claim conjunction, inference, expansion and delegation*. These concepts allow dependability requirements made at the application level to be supported by different types of evidence at the plant-computer interface, computer architecture, hardware and software design and control of operation levels.

These levels of evidence correspond to levels of causality. To each level are associated specific system functions and undesired events likely to cause these functions to fail. A function may fail as a consequence of undesired events occurring at the same level or at lower implementation levels. Proximate and immediate causes of failures are dealt with at each level, assuming that lower implementation causes of failures are properly dealt with at lower levels. The consequence is that the arguments constructed on these levels of evidence are *hierarchical* (by their level structure), *backward* (by their proofs), *hologramatic* (each level "containing" the lower levels) *and recursive* (each level requiring new claims).

This conceptual framework can accommodate deterministic logic as well as probabilistic claims when uncertainty is present. Its claim-based structure makes it particularly suitable to a so-called risk informed approach to safety; that is when safety needs to be assessed against the consequences of specific undesired events, deterministic or random. The relations between the framework and three basic principles, or "theories", of safety are also investigated: prevention, precaution and enlightened catastrophism.

The *descriptive part* studies an as yet unexplored aspect, albeit essential, of dependability cases: the essential roles played by models. Models play a critical role in the demonstration of dependability, in two different ways. First, dependability like other attributes of a system can only be apprehended by means of observations and by models, the latter being necessary to give meaning and purpose to the former. Even more importantly, models are indispensable for the construction of arguments and the delivery of proof obligations.

Models must be chosen for their usefulness and effectiveness in interpreting, in explaining, in predicting and proving, balanced against the cost of designing and using them. The approach followed here is based on Model Theory. Models consist of structures, states (or classes of), relations between state variables, time and event classes. Models are indispensable to *valuate* the evidence and to provide precise descriptions of the *documentation* needed for the safety case. The use of precise models has also side-benefits. Some new aspects of the nature of claims on diversity, hazard mitigation and controllability, and on requirement properties are revealed.

The *methodological part* identifies the inter-relations between models at the different levels of evidence (plant interface, architecture, design, operation). Conditions for re-using arguments and for the composability of safety cases result from these model inter-relations. Different instantiations of the framework have been considered, in particular for systems of different safety criticality, and for systems using pre-existing platforms or Components-off-the-Shelf (COTS) software. Dependability inter-relations between subsystems are in this case of a *"guaranteed service – rely condition"* type.

The essential characteristics of the framework are to be *product- and claim-oriented* and to be *model-driven*. These characteristics are exploited to derive *a set of practices*, or a discipline that can be followed by a licensee and a regulator to efficiently produce a justification of the safety of a computer based system. Recommendations are given for building a safety case. Basic rules for the expansion of claims and the *construction of dependability arguments* are given. A method is also proposed to systematically elicitate claims and the evidence they require.

The main innovative features of the approach are (i) the use of layered models for the representation of the system and the organization of the evidence, and (ii) the mechanisms of inductive claim expansion and delegation of evidence onto lower levels for the construction of arguments. The major benefits are:

- Better structuring of arguments to back up and to limit subjective expert judgment;
- More natural and systematic transformation of dependability properties into functional system requirements and claims;
- Possibilities of assessing the weight of particular components of evidence,
- Modular re-use of arguments of sub-safety cases.

This dependability justification approach gives precedence to – and focuses on – the dependability properties of the system, and by the same token is cost effective in terms of efforts and resources spent on the justification: two advantages which meet the respective priorities of the regulator and the licensee and should therefore make their negotiations easier and more efficient.

Not all aspects of the framework have been thoroughly explored yet. Some particular aspects of the inter-relations between levels of evidence would need further investigations and experiments. Because claims and arguments should dictate which evidence is required, and not the contrary, these further experiments should be carried out by research projects intimately associated with real industrial system design and safety cases developments. This would allow a realistic assessment of the framework and would provide the freedom to determine the best evidence that needs to be produced, instead of being constrained by pre-existing evidence.

Notes for the Reader

A basic understanding of Chapters 1 to 7 does not require the mathematics of Chapter 8. As for Chapters 9 to 11, the reader not interested in detailed mathematical developments may skip Sections 8.7 to 8.11 included. Besides, a basic understanding of the analysis carried out in Chapter 9 does not require the details of the claim expansions of Sections 9.5 to 9.7.

The table of contents is intentionally rather detailed so as to help follow the cross-references across chapters that remain inevitable. Concepts, definitions and notations that are not explicit in the table are listed in a complementary index at the end of the book.

2

Current Practices

Several possibilities are offered to Assessors and Regulators to approve the commissioning and the use of a computer-based system important to safety. They may condition the approval on the provision of evidence that a set of rules, laws, standards, design criteria, (or even sometimes "beliefs"! [13]) is complied with. They may condition their approval on the success of a pre-defined method, such as the so-called "Three Leg" approach advocated in Britain for nuclear instrumentation [34, 43]; or on the provision of evidence that certain specific residual risks are acceptable or that safety goals are attained or dependability properties are present.

Any combination of these approaches is possible, of course. The framework proposed here, however, gives preference to the latter option and aims at providing a practical and convincing justification that some well-defined safety properties are satisfied. *The approach is therefore goal- or claim- oriented and not rule- or standard-oriented.*

The purpose of the framework is to organize arguments and evidence so as to justify specific requirements – identified upfront – on the dependability of the use of a computer-based system. We do not study in detail the so-called requirement engineering aspects of capturing and defining these dependability requirements. This is the role of experts of the application domain. We assume, as it is often the case, that these requirements are given and documented in some natural or application related language. We address instead the problem of matching these requirements onto claims, arguments and evidence on the dependability of the architecture, the software and hardware design and the modes of operation of the system. Tracing the application requirements into the design is in practice a major difficulty for designers, licensees and assessors. The necessary documentation must cover different levels of design, from the most abstract to the most detailed descriptions; its volume can be enormous for the implementation of the most simply stated safety requirements. Nevertheless it must remain readable, understandable and traceable to all protagonists, including users and independent assessors who did not take part in the design. These qualities require structure and discipline. Those are the focal points of our work.

Hence, the primary goal here is not to demonstrate compliance of a system with a set of laws, rules or standards. Rule or standard compliance is of course not excluded; but we hold the viewpoint that it is more appropriately invoked as evidence to support a particular claim or subclaim, and not as a primary objective of a dependability justification not even of a regulatory assessment.

This approach may therefore depart from certain legal frameworks and licensing practices. It is however believed that compliance with laws, rules, design principles or standards often fails to demonstrate convincingly *on its own* that a system is safe enough for a given application, thereby entailing licensing delays and costs. The three-leg approach suffers somewhat from the same shortcomings. By collecting evidence in three orthogonal directions (process, product and staff quality) that are often not clearly mutually related, one can fail to establish a convincing system property.

Another reason for giving preference to a safety property oriented justification is the wide gap between the semantic domain at the high level of application at which application safety requirements are formulated – *e.g.* those addressing the operation of a nuclear plant – and the semantics of the deep low levels of the computer software and hardware implementation. No general single and structured set of rules or guidelines or standards is actually able to coherently cover this entire gap without leaving many aspects uncovered, thereby giving rise to many problems of interpretation.

Standards[4] raise other problems. They are often behind the state of the art. They lack flexibility and do not let the licensee focus on the most specific safety properties that are required from the system. They state what has to be achieved, but say little on how to achieve it.

As an example, an influential standard on software for computers embedded in the safety systems of nuclear power plants claimed in a 1986 version which remained applicable until 2006: "The following principles for development (design and coding) are based on experience in producing error-free and understandable software". Consistent with itself in not admitting that despite the highest possible level of design quality, software defects may still defeat safety functions, this standard said almost nothing about software fault tolerance, defect and failure consequence analysis, design diversification, *etc.* Worse, perhaps, it did not provide recommendations to deal with real software, that is with software likely to contain undetected bugs; hence, all those who were committed to follow the standard had no other choice than to somehow fool themselves and others, losing enormous amounts of time and resources in trying to produce convincing evidence that this standard, although based on false and unreal premises, had actually been

---

[4] One can distinguish two kinds of standards. Those of the first kind define precise methods or benchmarks to calibrate, measure, or test a product in a given environment. These standards are effective and indispensable. Those of the second kind specify requirements on system design, development or assessment processes. The above discussion addresses this second type.

duly complied with. Other issues raised by computer and software standards are discussed in [104].

We already mentioned that a claim-oriented approach is also more general and flexible than a standard or rule based approach: it remains applicable in situations where standards may not. A standard, for instance, may not accept certain designs (*e.g.* computer processor interrupts as the standard mentioned above) or alternative types of evidence (*e.g.* formal proofs as a replacement for tests [25]), while these may be perfectly adequate or even necessary in specific circumstances.

A claim-oriented approach may give the impression of focusing on the qualities of the product and neglecting the necessary attributes required from the development process. This view is, however, incorrect. It will be made clear, in particular in Section 7.5, that evidence on both the product and the process are needed and exploited by the justification scheme. This is not to say that a dependability argumentation along the lines proposed in this book is necessarily a substitute for a legal and licensing framework in place; but it should at least be a useful complement or alternative, for no other reason than the simple fact that justifying the safety of a software based system will always entail more than merely showing compliance with a guide or a standard.

3

Axiomatic Justification and Uncertainty

Il y a une faille dans toute chose;
c'est par là qu'entre la lumière.
Leonard Cohen

The design of an engineered system and the analysis of the dependability of its behaviour can seldom exhaustively consider in detail, check and rule out all possible chains of occurrences of undesired events. The safety analysis of a real system almost always leaves open certain issues.

Uncertainties can be of two types: those that directly affect the behaviour of the system and those that affect the safety analysis itself. These latter uncertainties are inherent to the methods we are forced to use to assess safety and risks: they can affect the validity of the assumptions on which rests the analysis, or the correctness of the arguments or the plausibility of evidence. When such uncertainties imply risks, both uncertainties and risks have to be minimized (this is the so-called precautionary principle, *cf.* Section 7.5). But the crucial point is that uncertainties can never be totally and demonstrably excluded. Ideally, at least, they should not be left *implicit*. Possible sources of uncertainty should be identified and documented; to some extent, as we shall see, this is possible.

On the other hand, nothing is absolute and certain enough when safety, and in particular human life, is at stake. In such conditions, very understandably, one expects a justification of safety to be as rational, complete, definitive and deterministic as possible. Under such a view, probabilities are often considered as a mathematical concept which is not more than a convenient measure of our ignorance. Use of statistical data is allowed but for limited purposes only: for instance, to anticipate the failure behaviour of simple standard components for which large enough and statistically valid samples of data are available.

Such deterministic views, however, whatever their apparent absoluteness, are necessarily also based on uncertain grounds, essentially because they cannot question and demonstrate everything. A system dependability evaluation must start somewhere, in a given pre-existing context and on premises that are unquestionably accepted in this context. It is therefore necessarily founded on axioms that are taken for granted without demonstration.

Hence, there exist two predominant trends in the analysis of dependability: a Cartesian one which tends to exclude analyses based on probabilities and a pragmatic one which considers probabilities as being necessary to deal with inherent uncertainties. These views are sometimes recognised as complementary, but can also be in strong opposition, especially when design or software failures have to be taken into account.

This opposition is not new in Science. At the end of the nineteenth century, Max Planck believed so strongly in the fundamental aspect of the second principle of thermodynamics that he resisted adopting the statistical mechanical interpretation proposed in 1872 by Ludwig Boltzmann. Although, later in life, he envisaged the possibility of using Boltzmann's microscopic concept of entropy, he remained deeply convinced that "the law on entropy was absolute and thereby its foundation was non-probabilistic" [56].

Nevertheless, pursuing the entropy analogy further, one can conjecture that, in the context of dependability, the two antagonistic views, deterministic and probabilistic, should find complementary niches, deterministic analysis being dedicated to macroscopic system behaviours, stochastic analyses to local behaviours of hardware and possibly software components and to risk and sensitivity analyses of specific design options. Clearly, when safety is at stake, the functionalities expected from a system must be understood and specified in deterministic terms. It would be meaningless to start the design of a railway crossing control system with the requirement specification that, upon arrival of a train, the barrier shall be closed with a probability higher than a given value. On the other hand, it would be foolish to ignore that hardware components can fail, some with smaller failure rates than others, and that only stochastic models are able to evaluate these failure rates and the reliability of redundant or diverse arrangements.

More pragmatically, the work presented here can be viewed as an attempt to reconcile the two views in the context of dependability assessment. We define the dependability of a system important to safety in terms of specific properties and qualities that can be formulated as deterministic or probabilistic assertions. The existence of these properties has to be justified by means of arguments that relate them to evidence. An argument is a set of inference rules, which shows – within well-defined models – how one property, or a set of them, is the logical consequence of evidence components.

In this approach, the arguments and their inference rules are considered as being deterministic: the domains and the ranges of logical rules are invariable and the effects of the rules are not random. The method does not allow random use or random selection of inference rules in arguments. Chapter 6 explores the nature of arguments and justifies this restriction. Thus, the argumentation backbone of a claim is unique and fully deterministic: a claim is not supported by alternative arguments, or by some argument A with some probability a, and by some other argument B with probability $(1 - a)$.

The correctness of the arguments, once it has been agreed upon by a consensus among stakeholders, is considered as being undisputable. If doubts or uncertainty remain on the axiomatic correctness of an argument, it means that evidence is missing. Claims requesting additional evidence must then be explicitly added and justified so as to resolve these uncertainties.

The models – *i.e.* the assumptions, languages and interpretations used to describe the system and formulate dependability properties and evidence – are also considered as being precisely defined and agreed upon by a consensus among stakeholders. All uncertainties must also be removed from these models. An assumption that would appear uncertain or questionable cannot be accepted as such. If it cannot be removed, it must be replaced by a claim and by evidence that this uncertainty is acceptable on the basis of different and unquestionable assumptions. Assumptions must be restricted to undisputable and commonly agreed laws, properties and practices of the contextual universe in which the dependable system is to be implemented.

Likewise, evidence components are facts or data that are must be taken for granted and agreed upon by all parties involved in the safety case. There must be confidence in *the axiomatic truth* of these facts and data. Should this confidence turn out to be insufficient, explicit claims must then be added to request evidence that removes uncertainty beyond any reasonable doubt. If reasonable certainty cannot be attained by justifiable claims, the system design must be considered as being inappropriate to satisfy the dependability requirements.

These constraints may appear as being the limitations of a method unable to deal with uncertainties. On the contrary, we believe that whatever methods and models are used, the assessment of dependability always rests on the acceptance of given facts, assumptions and models. Our approach recognizes the existence of sources of uncertainty but requires them to be dealt *explicitly*, with appropriate claims and evidence. Claims can be deterministic or probabilistic assertions. But all claims must be supported by deterministic (mathematical or logical) arguments, based on undisputable axiomatic models and having recourse to statistical or operational plausible unquestionable evidence. To be undisputable, assumptions, arguments and facts of evidence must ultimately be the result of an underlying *consensus* among stakeholders. The method is based on the recognition that such a consensus is unavoidable. The justification scheme can be viewed as an attempt at reducing this indispensable consensus to a minimum and to make it easier and more cost-effective to achieve by maximising the part of the justification that can be rationalized by logical arguments.

Any consensus based on expert judgment of course implies subjectivity and thus some uncertainty about the correctness of the choices made by the experts. Here, this inevitable uncertainty – sometimes called *epistemic* – concerns the choice of model assumptions, the completeness of the initial claims, the correctness of the logical arguments, and the plausibility of the evidence. After a consensus has been established, this residual *epistemic* uncertainty must somehow be felt as being "logically unreasonable", *i.e.* addressing concerns that are beyond any reasonable doubt. In this sense, *epistemic* uncertainty is to be opposed to the *random* uncertainty that can be well understood, evaluated by physical and probability models, or measured by statistical methods. The framework presented here clearly differentiates these two kinds of uncertainty, and circumscribes the sources of epistemic uncertainty by placing maximum reliance and emphasis on explicit models, assumptions and logics in a demonstrable manner (more detail on this is given in Section 7.4).

4

Justification and Dependability Case

We use the term "justification" and not "demonstration" because a demonstration of the dependability of any real, industrial or commercial system in the true mathematical sense is likely to be beyond reach.

A further distinction is necessary between *dependability justification* and *dependability case*. The justification framework presented here provides a method to organize and structure technical claims supported by technical arguments using technical factual evidence. The framework does not propose a method to lead possibly controversial negotiations between parties with different interests (*e.g.* between a licensee and the regulator). These negotiable and controversial aspects, as well as the legal and juridical issues, albeit equally important, are outside the scope of this work and belong to what is usually called the safety case.

The UK Nuclear Safety Directorate gives the following definition [43]: "A safety case is the totality of documented information and arguments which substantiates the safety of the plant, activity, operation or modification in question. It provides a written demonstration that relevant standards have been met and that risks have been reduced as low as reasonably practicable (ALARP)".

In practice and in addition to a justification of the use of the system hardware and software, a dependability case must achieve a contractual agreement between stakeholders on the safety issues and the risks to be covered, the feasibility and timing aspects of the project, the resources involved, the methods deployed for design and analysis, the responsibilities, required qualifications and independency of the companies and experts involved (*cf. e.g.* [43]).

The dependability justification addressed here is thus one of the technical bases of the dependability case that has to be supplied by a licensee to the licensor in support of a license application. Other aspects require knowledge and expertise beyond that of I&C and software engineers. However, as noted in [51], it is vital that these engineers feel responsible not only for ensuring that their part is sound but also that it fits and is used appropriately in the wider scope of the *dependability case*.

4.1 Cost Minimization and the Proportionality Principle

Despite its limited context, such a technical basis remains of course an essential part of a safety case. In particular, a sound, complete and convincing justification of a well-defined set of dependability properties can help reduce the costs of licensing and safety cases in several ways:

- By allowing the safety case protagonists to focus and restrict their attention to only those claims that are dictated by the application, the replacement or upgrade of the system, or by the use of a new technology;
- By allowing protagonists to focus attention on the critical safety issues specifically raised by the system to be developed, replaced or upgraded, thereby avoiding endless debates on the applicability of – and compliance with – rules or standards that are not necessarily circumscribed and/or adapted to a given project, environment or system;
- By protecting manufacturers – when a technical aspect is clearly not the key to the safety argument – from being required by so-called "applicable" laws, rules, standards, or inexperienced experts, to document it anyway, diverting their attention away from areas that are genuinely important;
- By allowing safety claims to be prioritised, and resources to be allocated on their justification according to these priorities;
- By organizing claims and arguments so that the required evidence is limited to what is arguably necessary and sufficient;
- By helping a regulator or an independent assessor to identify and to scrutinize the evidence, and to assess its most important attributes: *plausibility*, *epistemic uncertainty* and *weight*. (*cf.* Section 7.2).

These cost reductions directly contribute to what could be called a *proportionality principle*, which demands that efforts to improve dependability be apportioned in function of the criticality of the safety issues at stake.

A proper dependability justification should also help reduce costs by allowing:

- Claims to be re-used in other dependability justifications, with their supporting arguments and evidence;
- Designers to evaluate the risks associated with their design decisions before these decisions are actually implemented; this facility requires that the dependability justification be elaborated along with the system design; as we shall see this method of design offers in fact many advantages.

Some may argue that cost reduction should not be an issue when safety is at stake, on the assumption that safety always brings extra costs and is necessarily expensive. Besides being unrealistic, this argument ignores that the cost reductions mentioned above concern the justification of dependability and not its implementation.

4.2 Risk-based Assessment

The approach presented in this book shares some common aspects with the so-called risk-based or risk-informed safety assessment more recently proposed in the literature (see *e.g.* [99]) and by regulators. Our initial step is to claim certain specific dependability properties of the system and then justify the existence of these properties, by opposition to a systematic safety evaluation of all possible modes of behaviours of the system. This process is analogous to starting "from failure modes and their consequences to be prevented and systematically seek to eliminate all physically meaningful causal paths that could lead to it, in opposition to probabilistic safety assessments (PSA) that systematically evaluate all potential outcomes from given set of initiators" [99]. Somewhat similarly, our approach starts with specific initial requirements on hazard *prevention* and seek all logically meaningful implication paths – or chains of subclaims – needed to support these initial requirements.

It is advocated in [99] that it is necessary for a risk-based approach to be based on "a perfectly rational concept to address hazards". Whether such a rational concept can exist or is inconsistent carrying in itself a contradiction of terms between hazards and rationality is unclear. In contrast, the approach that we propose is based on a rational argument structure to address requirements on the dependability of a system, a concept believed to be more practical and productive.

A claim-oriented justification is indeed suited to address risks, independently of the interpretation of the concept of risk, deterministic or – more classically – probabilistic. This is of special advantage, not only for safety critical systems, but also for safety related systems [49], *i.e.* for systems that have an indirect impact on safety only, but sometimes more subtle and difficult to apprehend. These systems often have also more complex designs. Their safety assessment is often hard, and an exhaustive evaluation often is impossible or too costly. There is a consensus of the European Nuclear Regulators (see common position 1.10.3.3 in [28]) that for these systems "in order to evaluate the possibility of relaxing certain requirements of the safety demonstration, as a minimum, the consequences of the potential modes of failures of the computer system shall be evaluated". To address the consequences of well-identified failure modes, a claim-based approach is often very appropriate.

4.3 Two Illustrative Case Studies: A Process Instrumentation (SIP) and a Radioactive Materials Handling System

Two dependability justification examples of system modernization are used throughout the book to introduce and illustrate basic concepts and their applications.

The first example is taken from a real nuclear safety system. The SIP (process instrumentation system) is a typical example of an upgrade project. It was renovated in the French palier N4 and has also been recently renovated in Belgium. The SIP concept includes a rather wide functionality. It is used here to illustrate the global aspects of the justification approach, especially the elicitation of the initial dependability claims made at the system-plant interface level.

The description of the SIP, given in Appendix A, summarizes the requirements and the constraints of the modernization, as they are typically known at the beginning of a modernization project. They include functional and non-functional requirements, as well as constraints on the plant – system interface, on the system implementation and on the operation and operators' interface. They result from prior plant safety analyses, and from the requirements on the existing system to be replaced. These latter requirements usually suffer little modifications, and if and when they do (due to the new technology or to needed additional functionality), those also are approved by a prior safety analysis.

These constraints and requirements typically are – as we shall see in Chapter 5 – the initial material from which are formulated the initial dependability claims of the safety justification.

The second example (Appendix B) addresses a nuclear material handling system – similar to the CEMSIS public domain case study [85]. It has a more specific functionality and is used to illustrate the application of the method at the implementation and operation levels.

PART II

Prescriptions

5

Requirements, Claims and Evidence

5.1 Where to Start? The First Foundation of Dependability Justification

"The only way to get somewhere, you know, is to figure out where you are going before you go there" says the farmer to Rabbit in John Updike's "*Rabbit, Run*" [101]. This common sense advice takes on a special meaning in the context of dependability, because the point of departure of the justification is also its destination. We regard a *dependability justification* as the set of arguments and evidence components which support a carefully selected set of requirements addressing the dependability of the operation of a computer-based system in a given environment, for instance an industrial plant. These requirements need careful attention; they are the raw material we start off with and also the properties we have to justify in the end.

For reasons that will become clearer later, these requirements will be referred to in this book as being the *initial dependability requirements or initial safety, availability, security... requirements:*

Definition 5.1 *Initial Dependability Requirements* are specific to the application supported by the computer-based system and to the environment of the system. However, and this is important, they do not assume the use of a computer system. They specify *what* has to be achieved in the environment under specific constraints, not *how* it must be achieved. Basically, they require that some properties be maintained in the environment (*e.g.* an industrial plant) under specified constraints, whatever system is to be implemented.

The *specification* of the *initial dependability requirements* is the role of application, plant and/or safety experts; these experts, and only they can and must establish and validate the properties and the relations that must be maintained in the environment. No computer expert need be involved at this stage. Contrary to certain current practices, one could perhaps even argue that they should preferably take no part in this specification so as to keep a neat separation between the specification of what an ideal system is expected to do and the constraints inherent to a real implementation. This separation of concerns guarantees that the ideal system specification remains a non-ambiguous reference against which to validate the actual system.

5.2 The Initial Dependability Requirements (CLM0)

Two types of *initial dependability requirements* can be distinguished. *Functional requirements* specify – among other functions – the detection and control of hazards, the necessary protection and prevention functions, safety and security actions, inventories, alarm generation, *etc.*

Non-functional requirements specify the properties that any implementation of these functions should satisfy; accuracy, stability, testability, maintainability are typical attributes that may be required. Targets on properties of the system behaviour within its environment and during operation such as timeliness, reliability, and availability are also possible requirements.

Initial Dependability requirements may therefore be expressed as deterministic or probabilistic assertions. But, contrary to what often is unfortunately the case, the *dependability requirements* must not be confused with the qualities required – *e.g.* by the state of the art or by standards – from the design or the project development processes. Such expected qualities are not of lesser importance but they are not part of *what* has to be justified; they are part of the justification. They are a *means* to provide evidence. They should be stated in separate quality assurance and V&V plans. The quality of the development and the V&V processes achieved by observing these plans contribute to producing or improving the evidence needed to support the claims into which the dependability requirements have to be expanded at the various levels of the implementation.

Initial dependability requirements may be formulated in a natural, informal or application-oriented language. In order to make things concrete, let us give two examples drawn from the case study in appendix B:

no_displacmt.clm0: Horizontal vehicle and vertical lift displacements are prevented when pallet is being inserted or extracted from a cabinet (*cf.* requirement (B1), page 297)

pfd_clm0: The probability of failure per demand shall be less than 10^{-4} per handling operation

In the first example, a specific functionality is required from the system, namely a protection function ensuring that the displacements are functionally impossible while handling a pallet of nuclear material. The second requirement is *non-functional*. It requires that the system enjoy a certain property, in this case the achievement of a reliability level. The first requirement is of a deterministic nature; the second one is probabilistic. None of these requirements refers to a computer. A more detailed elaboration of initial dependability requirements will be given by the embedded system analysis of Chapter 9 (specifically in Section 9.3).

Initial dependability requirements are the foundation of the justification and considered as constituting its level-0. Their documented specifications shall be conveniently referred to as being level-0 statements, collectively denoted **CLM0**. Although these initial requirement specifications are data that are given as input to the justification, and not claims, this notation is chosen for reasons of simplicity and homogeneity that will become clear later. Each requirement should be identified by a mnemonic name and by its level, its statement being labeled in this book by a symbolic short-hand label <**name.clm0**>.

5.2.1 PIE's, Constraints, Safe States

In addition, the **CLM0** specifications must be based on and include a precise description of the environment; precisely:

- All *events* that can be generated by the environment and can affect the system behaviour;
- The *constraints* and restrictions imposed by the environment on the system behaviour;
- The *safe* environment *states.*

These *events*, often called *postulated initiating events*, must be described with their consequences and effects. By definition, these events and only those are the events that will have to be controlled by the system or that are assumed to affect its behaviour. The monitoring, control and mitigation of these events are primary – and often the only – functions of a protection or safety system. They correspond to hazards, causes of damage and failures that may occur in the environment; they may be generated by the failure of a motor, by a pipe break, the malfunctioning of a valve, or a loss of power. They may also be natural events such as earthquakes, floods, tornadoes causing electromagnetic or seismic interferences; also human–induced events such as plane crashes, ship collisions and terrorist attacks.

Associated with the events, of utmost importance is the **CLM0** specification of the states of the environment that are safe or unsafe. The system should indeed be able to deal with the *postulated initiating events.* The **CLM0** requirement specifications should therefore include the precise description of the *safe or preferred states* into which the system should be driven.

Secondly, the environment – and especially previously installed systems – imposes *constraints*, restrictions and invariant relations that must be maintained in the environment. Examples of such constraints are mass, energy or flow conservation laws; constraints imposed by the behaviour of pre-existing equipment. For example, a typical constraint of nuclear material handling systems (Appendix B) is the need to respect conditions on mass storage and locations determined by considerations on mass nuclear criticality or by limitations on radiation release.

We will see in Chapter 9 that models play an essential and concrete role in the description, documentation and validation of **CLM0** specifications. Particular details will be found in Section 9.3.

5.2.2 Elicitation

"*Principiis obsta*" (Beware beginnings!) wrote Ovid (43BC–17AD); and the artist Piet Mondrian said once that the most difficult brushstroke in a painting is the first one because it is unconstrained. The initial stages of a dependability justification are no exception. The definition and formulation of a complete and coherent set of *initial dependability requirements* has to be carefully done; their specifications need to have received the prior approval of all parties involved in what we referred to as the *dependability case*, possibly including the green light from safety and legal authorities.

This definition and formulation process always takes place in a given *pre-existing* context and environment. The process must take into account the assumptions and constraints dictated by the environment. Models and assumptions therefore play an important role which will be discussed at length in Chapter 8.

The **CLM0** specifications, independent of the actual implementation and those alone, determine and define what the safety of the computer based system is expected to be. Their formulation and truth are taken for granted in this book; so are the assumptions and the models on which these initial specifications are based. Since they originate from prior safety plant and environment analysis work, *validating* this set of initial specifications, *i.e.* asserting the completeness, soundness and coherency of the set, is outside the scope of this work. Plant engineers, regulator and Technical Support Organizations (TSO) specialists must do this critical task in the context of a safety analysis which is a primal and preliminary step of the computer dependability justification.

Of course, these initial requirements do not come from nowhere. The material from which they are elicited and formulated consists of prior safety analyses reports, operational reports, accident reports, regulations, documentation of the system to be replaced or modernized, initial descriptions of the functionality, of the dependability requirements, and of the replacement constraints. A typical sample of this initial documentation is given for the SIP system in Appendix A.

In addition, these requirements are dictated by well-established *fundamental safety concerns or safety goals* specific of the application domain. For example, the design of the protection system of a nuclear reactor has to address the three fundamental concerns:

-   Safe shutdown of the reactor and safe maintenance in the safe shutdown state;
-   Removal of the residual heat;
-   Prescribed upper limits for the exposure to radiation of personnel and environment.

And the first two concerns, for instance, are those that determine dependability requirements on the properties of specific subsystems of the protection system:

-   The reactivity control subsystem;
-   The reactor coolant inventory subsystem;
-   The primary to secondary side heat transfer subsystem;
-   The primary side pressurizing subsystem;

- The primary heat removal subsystem;
- The secondary heat removal subsystem.

5.2.3 Completeness

Hence, advice on how to elicit the initial requirements is basically outside the scope of this book. Computer experts need to understand these requirements but are not competent to define them. This important point is often misunderstood and source of problems. Some computer engineers believe they have to take the responsibility of defining the dependability requirements, confusing them with computer system requirements. Others tend to ignore them in the belief that dependability is nothing else but a computer system property. These two extreme attitudes are not rare but wrong and dangerous.

However, from the viewpoint of a "user", some general criteria can help the elicitation of dependability requirements. To maximize efficiency and pertinence, and to minimize cost, initial requirements must be, as much as possible, limited to functional and non-functional specifications that:

- Address the expected system functionality and dependability,
- Are dictated by the constraints of the environment,
- Are dictated by the use of specific technologies imposed by the environment.

As we shall see in Section 7.6, the attitude coined "enlightened catastrophism" may also provide useful viewpoints to identify initial dependability requirements. In any case, these requirements should be limited to those that are justifiable. This is a delicate issue which will be discussed and will become clearer later.

It is good practice that once defined and approved, initial requirements should be made explicit at the start of the project, and communicated to stakeholders, preferably even as part of the early bidding process.

The completeness of the set of *initial dependability requirements* is of course a major issue. Justifying completeness is always a difficult task; an impossible one, one could even argue. However, the initial requirements alone define the dependability properties that are expected from the system[5]. If there is no agreement on the completeness of this initial set, there is no definition of the system dependability to start with.

Besides, initial requirements delineate the scope of the dependability justification and case. A set of initial requirements – agreed to be appropriate and complete for the purpose – is thus an essential and necessary basis to define what

---

[5] Recently, B. Boehm wrote [10] that when he was working on the NASA High-dependability Computing Program with V. Basili and others, they found that one of their biggest challenges was just defining "dependability". Here, the *initial requirements* are the definition.

the system dependability is and to plan the time and resource needed to complete the case.

One important point to keep in mind is that completeness is a model property; the only way to establish completeness is by reference to a model. In particular to specify the **CLM0** requirements, a model of the system-environment interface is needed to precisely identify the environment states, events and properties of interest. Completeness of the set of **CLM0** specifications can then be established by comparison to this model. Completeness – and the need for precise models in safety justification – is discussed in detail in the descriptive part (Chapters 8 and 9).

Thus, the purpose of our work is to show how to demonstrate that the **CLM0** requirements are satisfied at all lower levels of the computer design and implementation. We shall see later that, in many cases, they can be formulated and justified independently from one another, but that arguments and evidence to support them may have much in common and may then be re-used.

5.3 The Other Foundation: The System Input-Output Preliminary Requirements

The first level of the dependability justification, the level-0, could be compared to the exoskeleton of molluscs such as clams, oysters, snails and many others, which is made up of hard elements totally different from the animal internal structure. Level-0 consists of "hard" application-specific **CLM0** requirements in essence totally different from the computer structures which must implement them. As we said in Section 5.1, the initial requirements do not even assume the use of a computer system.

A huge conceptual gap therefore exists between level-0 and the inner levels of the computer system implementation. This gap is another major difficulty of the justification that the **CLM0** specifications alone do not bridge.

Another level-0 foundation is needed to start the justification, namely the specification of "what the system is expected to do in the environment". That is, a preliminary specification of the relations that the system – viewed as a black box – is expected to maintain in the environment in order to satisfy the requirements, to deal with the effects of the postulated initiating events and comply with the environment constraints. These *system input-output specifications*, as we shall call them, play a pivotal role between the **CLM0** *requirements* and the lower levels of the computer implementation. They are qualified as *preliminary* because at this stage of the justification they are not yet validated; their validation against the *dependability requirements* will be the first step of the justification.

The **CLM0** *requirements* have to be mapped onto functions and properties of a computer system. At this early stage, these functions and properties – for example the functions implementing properties such as reliability, availability, and fail-safeness – cannot yet be precisely specified, nor do they need to be. The role of the computer system that must be assumed and described at this first level is that of creating and maintaining new relations in the environment compatible with the environment constraints.

As an example, the main preliminary *system input-output requirements* which address the replacement constraints and the use of the digital technology for the SIP system replacement described in Appendix A are:

fct.clm0:   The SIP I/O functional relations must conform to the original logic

rel.clm0:   The SIP global reliability and availability must remain at least as good as the original one

cmf.clm0:   The SIP software common mode failure (CMF) level must not be superior to the CMF level of the original SIP hardware being replaced

sgfailure.clm0: No single failure should cause the loss of a protection function

autodetect.clm0: The SIP must detect its own failures (coverages achieved by auto-tests and periodic tests must be complementary)

failsafe.clm0:   The SIP must be fail-safe

Computer system experts, together with application experts should take part in the elaboration of these preliminary specifications. Section 9.4.6 and the following ones in Chapter 9 discuss in detail a model for documenting *system input-output specifications*. The model takes into account the central and interlinking role of these specifications and is adequate for their validation against the **CLM0** specifications. Section 9.4.7, in particular, shows how these specifications can be expressed in terms of:

- A relation between system monitored and controlled variables,
- Properties on the domain and range of this relation,

and how this relation, domain, and range can be validated, *i.e.* shown to *satisfy* the **CLM0** specifications.

5.4 Primary Claims

Thus, the **CLM0** *requirements* and the *system input-output preliminary specifications* constitute the initial material which is given and on which the dependability justification has to be built.

The evidence that this material – as it is given and documented – is correct, complete and implementable is the primary evidence required for the justification. Clearly, not all this evidence is available at this stage. Part of the evidence can be provided at this stage by the documented interface between the environment and the system. The other part is to be provided by the system architecture, design and modes of operation. Nevertheless, all the evidence needed must be either provided or specified and claimed at this stage. This is the role of what we call – at level-1 of the justification – the *primary claims*. In the further stages of the justification, some of these *primary claims* have to be reified into more specific *subclaims* on properties of the system implementation.

More precisely, the primary evidence needed is of two kinds. The first kind is evidence that supports claims on the validity of the **CLM0** documented environment system interface; typically, evidence of:

-   The validity of the documented environment **CLM0** constraints, initiating events and their consequences,
-   The correctness, completeness and feasibility with respect to the environment constraints of the *system input-output preliminary specifications,*
-   The detectability and controllability of the initiating events.

Evidence for these properties must be available at this stage and is typically level-1 evidence. Level-1 *primary claims* requiring this evidence will be referred to as *primary validity claims.* For instance, the first SIP *system input-output requirement* **fct.clm0** mentioned in the previous section, which requires evidence of the validity of the SIP input-output specifications, has to be expanded into level-1 *validity primary claims.*

Evidence of the second kind addresses properties of the implementation of the *system input-output* specifications; typically, the following characteristic properties of the system implementation must be specified as precisely as possible and claimed:

-   Correctness, completeness and accuracy of the implementation,
-   Fail-safeness, diversity, dependability.

This evidence often is not yet available at this stage of the justification and is anyway not level-1 evidence. These *primary claims* of the second kind, referred to as the *implementation primary claims*, have to be subsequently expanded and reified into more specific *subclaims* requiring the proper evidence at the lower levels of the system implementation. For instance, the last five SIP *system input-output requirements* of the previous section have to be expanded into level-1 *implementation primary claims.*

The elicitation and formulation of level-1 *primary claims* are the initial steps which belong to the level-1 of the dependability justification; a more detailed example of these claims will be given in Chapter 9 and can be seen on Figure 9.2, page 147.

These essential founding and constituting tasks of the dependability justification require creativity and expert judgment, but we will show how they can be guided by some principles. Those are some of the most important issues analysed in Chapters 0 and 9.

5.5 Differences Between Dependability Requirements and Claims

At this point, we should clarify the reasons why we find it necessary to make a distinction between "claims" and "requirements", although by some aspects the two notions may appear quite close to each other.

In some respect, indeed, claims are like system and design requirements. They all address properties of the dynamic behaviour of the system implementation and/or properties of the interactions of the implementation with the environment. But a first difference is that, while requirements specify what the system should do and how, claims specify the evidence required (i) for justifying that what must be done is indeed done and (ii) the way it is done.

System architects and designers use the terms "requirement" and "requirement specification" in relation to their design work. They do not normally talk about "claims". One usually starts to make claims on a computer system when an application has to be made for approving or licensing the system for a given usage, or when dealing with regulators, safety authorities or their technical support organizations, or when submitting the system to independent assessment. In those circumstances, another difference is that a requirement is essentially a statement specifying what the computer system is originally designed to do and how, while a claim is a statement addressing the appropriateness of these specifications for a given usage.

To make the relationship between claim and requirement more precise, we can view a claim as being either a *meta-requirement* or an *extra-requirement*.

A claim can be viewed as a *meta-requirement* when it is a statement requesting evidence of a property that a (set of) computer system requirement specification(s) should enjoy: *e.g.* validity, consistency, completeness, correct implementation, or maintainability in operation.

A claim can be viewed as an *extra-requirement* when it is a statement requesting evidence of a property or a function that was not explicitly part of the original computer system requirement specifications; such an *extra-requirement* can be a *functional or a non-functional claim*, depending on whether it claims the existence of an extra functionality or an additional property. For instance, one may have to claim that a monitoring system is fit for operations not only in normal operation but also in post-accidental conditions. It must then be claimed and shown to satisfy some design criteria – *e.g.* a single failure criterion or a maximum reaction time – required for this specific mode of operation, although the system or some of its components may not have been originally specified and designed for this purpose.

Claims and requirements may coincide in contents in the ideal situation when the dependability justification is part of a development project and progresses along with the specifications and the design of the system.

Components off the shelf (COTS) and other pre-existing components used in systems important to safety are a typical case where a distinction between requirements and claims is useful. Claims that need to be made on the use of the component usually do not coincide with the requirement specifications according to which the component was designed.

A third and more subtle difference is that design requirement specifications focus on system functionality while dependability *claims* need to also comprehensively cover failure modes and undesired events like threats or hazards external to the system. For example, a functional specification for a monitoring system may be "to display the pressurizer coolant temperature and level values", while a functional dependability claim may in addition require evidence that silent failure modes are excluded and in this example demand evidence that "pressurizer coolant temperature and level values be displayed *if validated measurements are available and not otherwise*". Thus, a *computer system* may satisfy its functional requirement specifications, and yet not necessarily satisfy a functional dependability claim.

Other examples of claims illustrating these differences can be found in Section 4 of [14].

To sum up, a claim is a statement that needs justification and requests evidence of either:

- An additional functionality or a property that must be enjoyed by the system (*extra-requirements*),

or

- A property that must be enjoyed by (a set of) requirement(s): validity of a specification, completeness, correctness of an implementation, maintainability (*meta-requirements*).

5.6 Evidence and Model Assumptions.

A clear distinction must be made between *"what"* to demonstrate (for instance the satisfaction of the *primary claims* and the *initial dependability requirements*) and *"how"* to demonstrate (in terms of *evidence and arguments)*. So far, we have dealt with *"what"* only. Let us now pay attention to *"how"* which, obviously, is our main concern in this book. What is evidence? A complex question! If "to be believed, a story has to be true", the counterpart also carries truth: "how could a story be true if there is no way to believe it ?"

Evidence and assumptions are the two pillars on which rest our beliefs.

Evidence is what makes a claim assertion true in the model in which the claim is formulated. Formally, in mathematical logic, evidence consists of valuations which make predicates true when they are interpreted in a given model, essentially a mathematical structure and its universe. These concepts, the formal analysis of evidence and its properties are essential in dependability justification and are the major subjects of Chapter 8.

In practice, the evidence material of a dependability justification or a safety case consists of factual data that can be of different types. They may be documented results obtained by real tests or measurements, simulation runs, software static analyses, formal proofs; or feedback data of documented operational experience; or data establishing properties of a hardware component, of an actuator, of a sensing device; or the establishing of the existence of periodic testing or re-calibration procedures, alarms, manual controls and by-pass, *etc*.

In industrial safety cases, evidence is often erroneously confused with assumptions. Assumptions are inherent parts of the models that are used to formulate and to interpret claims and evidence. Chapter 8 helps make this distinction clear and analyses its implications, while Chapter 9 is an illustration of how in practice one can deal with these implications. The confusion obviously can be dangerous since assumptions are taken for granted and not justified. For instance, the assessment of a safety critical software system that would implicitly "assume", without further ado, a perfect hardware, a given range of environment conditions, or a given behaviour of sensing devices can obviously lead to catastrophe.

But assumptions, of course, cannot be dispensed with. A dependability justification has to start somewhere without re-questioning the whole real world every time. Assumptions and the models based on them need thereby to be carefully and explicitly made. This important issue is also dealt with in Chapters 8 and 9.

5.7 How to Organize Evidence? A Four-level Structure

The evidence required for the justification of a single dependability property of a computer system is well-known to be of a quite disparate nature. B. Littlewood and his colleagues [29, 60, 62] must be credited for being amongst the first who recognised and seriously attacked this problem. The difficulties raised by this disparity are exacerbated by the drastic discontinuity between the environmental universe to which the properties to be justified belong and the computer system inner universe where the evidence is to be found. Remember the exoskeleton of molluscs metaphor at the beginning of Section 5.3.

The very first objective of this study was to find an organization for the heterogeneous sources of evidence which would make the justification of a dependability property of the "exoskeleton" easier and more convincing. The most obvious and natural direction to take first was to look at structures which had proved to be the most efficient to specify, design and implement the hardware and software of a complex computer system, and to remember that a well-known principle of engineering is that these structures are necessarily hierarchical (*cf. e.g.* [12, 73, 74, 93, 94]).

Practical experience, accumulated in several safety cases in nuclear engineering, later corroborated by a formal approach [14], suggested that the evidence required to justify *an initial dependability requirement* can be organized in a multi-level structure.

This key observation was fundamental. In all cases, the justification of a correct implementation of a requirement for prevention or mitigation of a hazard or a threat at the plant – computer system interface always appeared to require convincing *evidence of one or more different types*:

- Evidence that the functional and/or non-functional specifications of how the computer system must deal with the hazard/threat are valid, *i.e.* are sound, consistent, complete;
- Evidence that the computer system architecture and design correctly implement these specifications,
- Evidence that the specifications remain valid and correctly implemented in operation and through maintenance and operator control interventions.

In all cases met in practice, the required evidence always consisted of facts and data of one or more of these types. More precisely, any *dependability initial requirement* had to be supported by evidence material at one or more of the four levels:

- *The environment-system interface (system functions and properties expected by the environment)*: evidence of the adequacy of the set of functional and non-functional system specifications to satisfy the dependability requirement and to deal with the environment-system constraints and undesired events;
- *The system architecture:* evidence that the computer system architecture satisfies the functional and non-functional system specifications;
- *The system design*: evidence that the system is designed and implemented so as to perform according to the functional and non-functional system specifications;
- *The control of operation*: evidence that the dependability requirement remains satisfied during the system lifetime, given the environmental constraints and the way the system is integrated and operated in its environment. This includes evidence that the system does not display behaviours unforeseen by the specifications, *i.e.* that behaviours outside the design basis will be detected, controlled, and their consequences mitigated.

These levels are *not arbitrarily* chosen. We will see in Sections 5.8 and 6.3 – and in greater depth in later chapters – how these four levels are necessary and sufficient to construct arguments that relate disparate evidence material from the four levels to justify an initial dependability requirement. In particular, Section 8.12 addresses the more fundamental question of how many levels a formal justification structure should have in general, and how refined these levels should be.

But, from a practical viewpoint, the following considerations are sufficient at this stage. The first and fourth levels are usually easily understood and accepted by experts as being indispensable sources of evidence, although claims for evidence of the fourth type are often not explicit in industrial safety cases.

On the other hand, the necessity of a distinction between an architecture level and a design level may appear less obvious to computer and software engineers for whom the term architecture may take on different meanings; in particular, they often consider the processor architecture (*e.g.* its set of instructions) and the software architecture (*e.g.* its modular organization) as integral parts of the design.

For the justification of dependability, however, it is essential to distinguish between the architecture of the system which is intended to respond to certain failures, and the system internal design. Failures caused by undesired events occurring in the environment may require redundancies and diversities of functions, components and communication lines that dictate specific aspects of the system architecture. Besides, as Peter Neumann pointed out in his ACM "Inside Risks" column [71], specific modes of failure may result from inadequate attention to what he also refers to as being the architecture of the system. Distributed control over networks of redundant or diverse unreliable subsystems and components, with distributed sensors and distributed actuators is particularly vulnerable to widespread outages and perverse failure modes such as deadlocks, race conditions, starvation and other unwanted interdependencies. It is at the level of the system architecture that specific defences must be conceived to detect and tolerate these failures and not only in the design of isolated components. While design correctness is a property of components, reliability, survivability, fault - tolerance, security are properties of component interactions for which architecture is an indispensable source of evidence.

Hence, for our purposes and in the context of dependability, we adopt the following definition:

Definition 5.2 We regard the *system architecture* at level-2 as being the organization in space and in time of the dependable hardware and software components of the system, and of their control and communication interfaces.

To make things more concrete, let us give some examples of evidence material for each level:

- At level 1: a regulation or a regulatory position asserting or justifying the validity of a given system requirement specification; a safety analysis report.
- At level 2: a statement from a certification body guaranteeing certain properties of a hardware/software platform; a (set of) test result(s); the proof of the existence of an architectural hardware or software property such as redundancy or diversity.
- At level 3: the coverage of a (set of) module or integrated test result(s); a conclusion of a code static analysis or failure mode analysis; existence and proof of certain applied defensive programming measures.
- At level 4: existence proof of an invariant property or a system state guaranteed by periodic tests, by operator control interventions or by operational procedures; a (set of) pertinent operational feedback data.

At any of these levels, a same element of evidence can of course be used to support more than one initial dependability requirement.

Other examples of different types of evidence are given in Table B.2 of Appendix B for the justification of a claim on the safe use of a nuclear material handling system.

Evidence material at the four levels must consist of facts and data, taken for granted and agreed upon by all parties involved in the dependability case. There must be confidence beyond any reasonable doubt in the truth of these facts and data, without further evaluation, quantification or demonstration. In order to consider the evidence as being unquestionable, this confidence must be supported by a *consensus* amongst all protagonists. Besides, to be validated, we will see that evidence must be amenable to *plausible* models. *Consensus* and *plausibility* are essential concepts discussed in greater detail Section 7.2.

5.8 How Are the Four Levels Related? Levels of Causality

Hence, four different levels of evidence, each level consisting of possibly various evidence elements, support an *initial dependability requirement.* Clearly, the difference between these levels is not degree, value or priority, but the intrinsic nature of evidence. How, then, can these levels be mutually related so as to build a coherent argument justifying an *initial dependability requirement*?

The four levels can be viewed as four *levels of causality.* To each level correspond specific system properties and functions and also undesired events likely to cause these functions to fail or these properties to be invalidated. A function may rely on properties or functions implemented at lower levels and may thus also fail as a consequence of undesired events occurring at these lower implementation levels. Similarly, a property may rest upon and imply the existence of lower level properties. The consequences of undesired events are therefore dealt with at each level on the basis – more precisely on the "implication" – that consequences of lower level undesired events are (or will be) properly dealt with at these lower levels. Such an implication is similar to claiming at each level that lower levels are somehow free of unsafe failures. Similar implications are also inherent to hierarchical design processes.

Because of these implications a claim on the dependability of a function or a property at a given level usually implies claims on evidence of lower levels. And because evidence material is provided at four different levels, any *initial dependability requirement* or *primary claim* may need to be expanded into claims for evidence at any of the lower levels.

In Section 5.4, we explained that, at the first level of the environment-system interface an *initial dependability requirement* needs to be expanded into a set of *primary claims* of two kinds.

Primary validity claims address properties of the documented environment-system interface; they claim the validity of this documented interface. These claims are entirely supported by level-1 evidence. But the other *primary claims* require evidence of a correct and fail-safe implementation of the system. These claims must be supported by evidence at the lower levels. At level-1 these *primary implementation claims* actually infer properties of the architecture, design or control of operations. They are assumed to be satisfied at level-1 but need to be justified by evidence at lower levels.

At level-2, the architecture level, each *primary implementation claim* not supported by evidence at level-1 and assuming properties of the architecture must be expanded into level-2 claims. As an example, a *primary implementation claim* may imply the correct supervision of the system by an independent watchdog. This *primary claim* would imply at level-2 a claim asserting that:

(5.1)   **watchdog.clm2:**

> (system is in initstate and watchdog is disarmed)
> **or** (system is not in initstate **and** watchdog is armed)

Table 5.1. Dependability Justification Level Organization

Justification and Evidence Level:	Requirements/ Claims:	Inferred from evidence:	Expanded into and inferred from delegation subclaims at lower levels
0	Initial Dependability Requirement	(Out of justification scope)	Plant–System Interface Primary Claims
1	Plant-system Interface Primary Claim	Plant-system interface Evidence	Architecture Subclaims
2	Architecture Subclaim	Architecture Evidence	Design Subclaims
3	Design Subclaim	Design Evidence	Operation Subclaims
4	Operation Subclaim	Operation Evidence	(Out of justification scope)

In turn such a level-2 claim must be supported by level-2 evidence and/or expanded into lower level claim; and so on, down to the lowest level of operation control. For instance, it may be implied at the architecture level that output channels unfailingly detect data that are invalid before delivering them to voting mechanisms or actuators; this implication requires at level-3 a claim asserting that:

(5.2)   **output_channel_register.clm3**:

$$\forall t: (\text{register value}(t) \text{ is in range } \underline{\textbf{and}} \text{ validbit}(t) = 1)$$
$$\underline{\textbf{or}} \ (\text{valid bit}(t) = 0)$$

Every *initial dependability requirement* is therefore the root node of a tree structure of nested subclaims at one or more of four levels: the environment-system interface and the system architecture, design and control of operation.

The four levels are *hierarchical levels of causality* because claims, properties and functions at each level i, $i = 1...4$, depend upon claims, properties and functions at the lower levels $j > i$, and not upon those of the upper levels. The reasons for this one-way and top-down dependency are to be found in the hierarchical design organizations of complex computer-based systems; these reasons are re-examined at the light of the formal developments of Chapter 8 in Sections 8.12 and 8.13. Let us just recall here that the top-down stepwise refinement of design abstractions, the robustness, adaptability and efficiency of the behaviour of hierarchical organizations and the absence of cycles of dependencies are some of the main factors which naturally lead to hierarchical structures, both in engineering and in the natural world; key references on this subject are [1, 12, 24, 74, 94].

This hierarchical dependency is clearly also beneficial for the justification of claims on system properties since it precludes the existence of loops in arguments supporting these claims. This is our first observation in this book – and not the last one – of a system property being essential to both design and to validation, leading to the presumption that only well-designed systems can be validated and conversely.

It is useful at this point to summarize and capture by the following definitions the essential notions of the four-level structure introduced so far:

Definition 5.3 A *claim* stated at some level i, , $i = 1...4$, is a statement that requires *evidence* from some lower level j, $j \geq i$. Claims are referenced by a unique identifier that specifies its name and its level: **<name>.clm** i; a *claim* can also be qualified as being either a *primary claim* if $i = 1$ or a *subclaim* if $i = 2...4$.

A more precise definition of a subclaim will be given in Section 6.3.

Definition 5.4 *Evidence* at some level i, $i = 1...4$ is factual data that make true a *claim* statement **<name>.clm** i. An element of *evidence* at level i is referenced by an identifier that specifies its name and its level: **<name>.evd** i, $i = 1...4$.

While the notion of *claim* is syntactic, *evidence* is part of the semantics of the justification. How evidence makes a *claim* statement "true" requires a model to formulate and valuate the statement, which will be the subject of the Description Part of this book.

We can now also define what we mean by *dependability justification*:

Definition 5.5 A *dependability justification* is the set of *claims* and *evidence* components which justify the implementation and operation of a set of *initial dependability requirements* on a computer system, given the environment constraints and the preliminary input-output specifications of this system.

A *dependability justification* is an essential part of what is often called in industrial practice a *safety case* or more generally a *dependability case*; in line with nuclear regulators' guidelines [28] and other authors [66], we can define :

Definition 5.6 A *safety* or *dependability case* is the set of activities to be undertaken in order to carry out and document a *dependability justification*, including the verification and validation activities, the organizational arrangements to ensure the competence and independence of those undertaking the justification activities, to ensure the legal negotiations and compliance with the applicable standards, guidelines and regulations, and the setting up of a feasible programme for completion of these activities.

If we were to venture a comparison with linguistics, *dependability requirements* and *claims* would be the syntactic elements of the *justification*, *evidence* would be semantics elements, and the *safety case* would cover the pragmatics and the socio-linguistic aspects; the latter aspects, albeit important, are only alluded to in this work.

Table 5.1 gives an overview of the four-level-organization, which relates dependability requirements, primary claims, subclaims and evidence.

5.9 Examples

These concepts can be illustrated by examples drawn from the safety case of the nuclear material automated storage, retrieval and handling vehicle described in Appendix B.

The *initial dependability requirement* considered in Appendix B specifies that the vehicle represented on Figure B.2 and its vertical lift must be deactivated while a pallet loaded with material is being inserted or removed horizontally from its cabinet.

The corresponding level-0 *dependability requirement*, already mentioned in section 5.2 is (requirement (B1), page 297):

(5.3) **no_displacmt.clm0**:

    horizontal vehicle and vertical lift displacements are prevented when pallet is being inserted or extracted from a cabinet

And a *system input-output preliminary requirement* specifies what the logic of the vehicle control microprocessor is expected to do in terms of variables describing the state of different components of the vehicle:

(5.4) **<no_ displacmt_system specification>**:

    if
    (vehicle docked) **or** (vehicle-cabinet clamped) **or** (fork/shuttle engaged) **and** ((drive mode on) **or** (lift mode on))
    then
    <disable vehicle drive and lifting device motors within 0.1 sec>

At level-1, a *primary claim* must claim evidence for the validity of this system specification:

(5.5) **no_displacmt_spec_validity.clm1:**
    specification (5.4) is valid

Part of the level-1 evidence material (plant–system interface), which supports this *primary validity claim* is to be found in the documented specifications of the vehicle (Table B.2, page 305):

    AGV_spec.evd1:
    Documented specification of the automated guided vehicle operation and interlocks

At level-2, the system architecture, a *subclaim* from which is inferred the *primary claim* for evidence of the correct implementation of the system specification (5.4), is (Table B.1, page 302):

(5.6)   **inputch.clm2**:

> set of AGV control unit sensing input channels is complete and adequate; status signals (docked/undocked, sealed/unsealed, fork/shuttle positions, drive mode) are correctly captured

This *claim* requires that states of vehicle drive, lift, fork, shuttle and cabinet clamps are correctly captured by corresponding sensing devices and channels.

One of the level 2 evidence documents supporting this *subclaim* is (Table B.2, page 305):

> **sens_spec.evd2**:
>
> documented specifications, performance and failure characteristics of input sensors

A *subclaim* at level-3 (design) from which is inferred the *primary claim* (5.4) for evidence of the correct implementation of the system specification is (Table B.1, page 303):

(5.7)   **correct_code.clm3**:

> the code implementation of the specification (5.4) is correct

A level-3 design evidence components supporting this *subclaim* is (Table B.2, page 306):

code_utst.evd3:

reports of code unit tests

One of the *subclaims* at level-4 (control of operation) from which are inferred level-2 *claims* for evidence of a fail-safe architecture, is (Table B.1, page 304):

(5.8)   **oper_fma.clm4**:

> adequate anticipation of failure modes of vehicle, fork/shuttle, sensors, actuators, power supplies, motor lock

Claim (5.8) requires a pertinent return of experience from past operation to ensure that all potential failure modes of the implementation have been anticipated. Level-4 operational evidence material supporting this subclaim oper_fma.clm4 is (Table B.2, page 306):

> **opr_rex.evd4**:
>
> data from probation period report, operators reports, operational feedback from other similar AVG's

The detailed expansion of the justification of the dependability requirement **no_displacmt.clm0** is given in Appendix B.

6

Arguments, Syntax and Semantics

Evidence material, like raw data, proves nothing by itself. A machinery of *claims* and *arguments* is needed to organize the material across the different levels and support the *initial dependability requirements*.

6.1 Claim Assertions

From the definitions and examples developed in the previous chapters, a *claim* appears as an *assertion* specifying a property or a function for which evidence is requested; and this assertion is expected to simply return a value *true* or *false,* depending on whether the requested evidence specified by the *claim* is available or not.

Hence, a *claim assertion* can in principle be quite simple. As a matter of fact, there are good reasons to make it as simple as possible, although this recommendation is, in many ways, in opposition with what can be met in actual safety cases where claims are often involved and elaborated descriptions of required properties or functions.

First, whenever it is possible, a *claim assertion* should not be a conjunction of assertions that could be justified separately or that could be used independently in different arguments. The advantage of simplicity is indeed twofold:

- To simplify the argument and the evidence required,
- To allow a *claim* and its argument to be more easily used in different arguments.

Consider, for instance, the level-2 *claim* (5.1); more efficient is a substitution of this compound *assertion* with the two simpler *claim*s:

(6.1)   **watchdog_disarmed.clm2**:   system in initstate $\subset$ watchdog disarmed

             watchdog_armed.clm2:   system not in initstate $\subset$ watchdog armed

When satisfied, each of these *claims* can, if necessary, be re-used to separately justify properties related to individual states in each of the two different system modes: during the initialisation phase, or during cycle execution. Besides, the evidence material to support the two *claims*, as it refers to different operating modes, is likely to be of a quite different nature and better kept separate in the argumentation.

The same comments apply to the compound level-3 *claim* (5.2), which may be easier and more efficient to justify if decomposed into two *claims*:

(6.2)　　　　　**output_channel_register_valid.clm3**:
register value in range $\equiv$ (validbit = 1)

　　　　　output_channel_register_invalid.clm3:
register value not in range $\equiv$ (validbit = 0)

This pair of *claims* (6.2) calls also for another comment in favour of simple *claims*. In a non-safety-critical system, only the second *subclaim* might be necessary to ensure that some procedure is activated upon detection of invalid data, assuming by default that all data not detected invalid are valid. But this implicit assumption often is unacceptable in a dependable system where validity of the data must be guaranteed at any time it can be used, *e.g.* for voting or initiating a safety action; the first *subclaim* explicitly requires evidence of this validity.

A claim formulated as an assertion taking the value true or false is obviously easier to verify than if it is an entangled statement of assertions and implications; in particular, a simple claim is likely to imply a simple expansion into lower level subclaims.

Chapter 8 shows how the predicate calculus is useful to formally express claims and evidence and to analyse some of their logical properties. We shall also see that it is essential to have precise models for unambiguous interpretation of the predicate symbols, the functions, variables, and constants that are used to formulate claims and evidence. A related and particularly important remark is that completeness is a difficult property to claim if it is not a property claimed over a well-defined set of states, events or objects. Therefore, adverbs such as "all", "always", and "never" should be avoided in claim *assertions* except if they unambiguously address properties of well-defined finite sets.

6.2 White, Grey and Black Claims

From Chapter 5, and more specifically because of the dependency between levels discussed at the end of Section 5.8, one can conclude that a *primary claim* or a *subclaim* at a given level is necessarily of one of three different types. Keeping in mind Definition 5.3, page 38, and using colour names for easier reference, one can indeed distinguish three types:

Definition 6.1 A *claim* (Definition 5.3) is either:

- *White*: the *claim* is justified by evidence material from the same level;
- *Black*: the *claim* is not justified by evidence from the same level and is inferred from lower level subclaims;
- *Grey*: the *claim* is justified by evidence at the same level *and* also inferred from lower level subclaims.

A *black* or a *grey* subclaim is therefore a sort of *"logical link"* across two levels of the justification: the level at which the claim is formulated and assumed true, and the level at which it is proven true. In other words, a *black* or *grey* subclaim is a logical construct to relate the level at which evidence is requested and the level at which this evidence belongs and must actually be provided.

Figure 6.1 shows a level-i *claim* being the root of a tree of *subclaims* and evidence components, with instances of *white, grey* and *black subclaims*. The tree is a graphical representation of the argument that supports the root *claim*.

An *initial dependability requirement*, since it is inferred from the *primary claims*, has the status of a *black* claim. We shall come back to expansions of claims and arguments in more detail in the following sections.

6.3 The Inductive Justification Process

Thus, every *claim* at a given level is supported by a set of evidence components from the corresponding level and/or inferred from a set of *subclaims* at the levels below. More precisely, a *claim* (Definition 5.3) at level-i is the consequent of an implication, the antecedents of which are evidence components of the same level i and/or *subclaims* requesting evidence from lower levels $j > i$.

Subclaims or evidence components may be missing in the antecedents. The antecedents of a *white claim* are evidence components from the corresponding level only. The antecedents of a *black claim* are lower level *subclaims* only. The antecedents of a *grey claim* are both evidence components from the corresponding level and lower level *subclaims*.

A more precise definition of a subclaim results from these implications:

Definition 6.2 A *claim* (Definition 5.3) at level j is a *subclaim* of a *grey* or *black claim* at level $i, j \geq i,$ if j is an antecedent of an implication of which the *grey* or *black claim* is the consequent.

By inductively applying these implications, a *claim* (Definition 5.3) appears to be the root of an argument, which is a backward justification tree, of which the intermediary nodes are *subclaims* and the terminal leaves are evidence components. Figure 6.1 is an example of a tree corresponding to an argument supporting a claim at some level i and Figure B.4 at the end of Appendix B is an example of the full expansion of an argument supporting an *initial dependability requirement*.

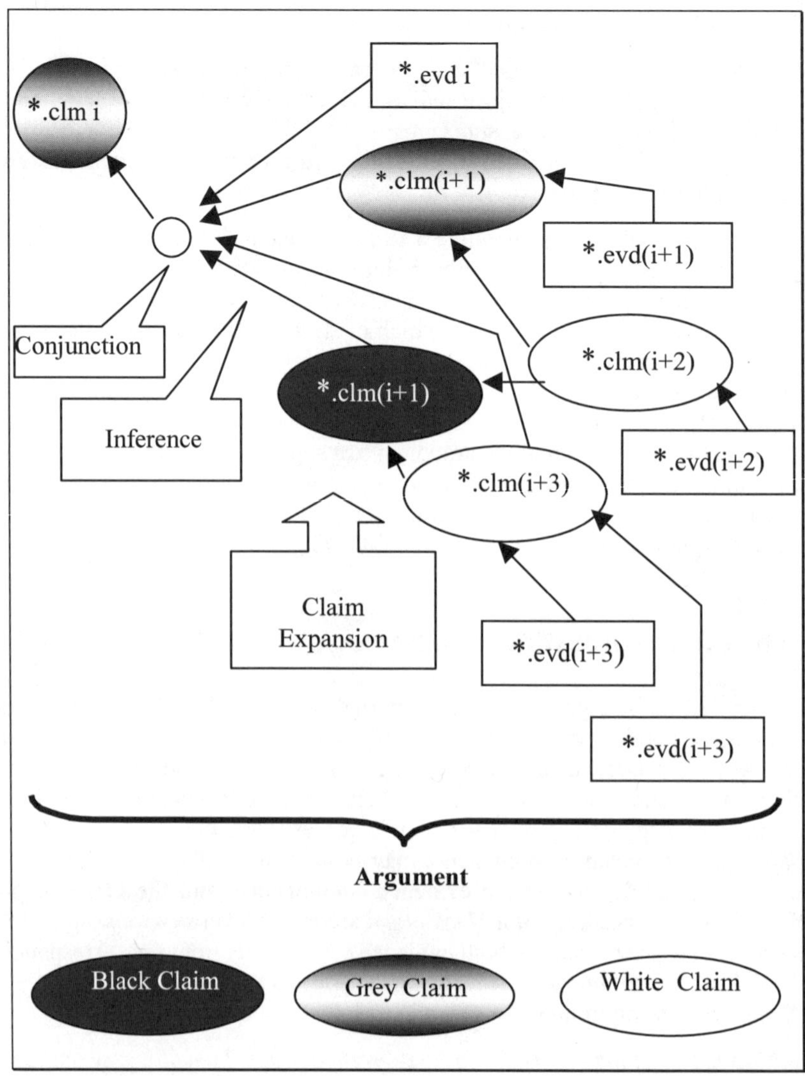

Figure 6.1. Claim expansion

The goal of the justification is to apply these rules of inferences to construct an argument, which is a proof tree whose branches are all terminated by evidence components. This is called a "backward" proof, since we start with a claim that is a conclusion to be proved. By opposition, the natural deduction is a "forward" approach to proof in which we would start with evidence components and work towards a subclaim.

An *argument* is thus, to a certain extent, similar to the backward proof of the sequent calculus (see *e.g.* [32]). The basic idea of sequent calculus is to establish the truth of a proposition φ by searching for a tree, the nodes of which are propositions obtained by inference rules that would lead to falsify φ. The tree is constructed in a logical and orderly way so that the failure to falsify φ is a proof that φ is valid. The difference between the sequent calculus and the claim justification process is that the latter is a search for a tree of inferences which make the proposition true.

Hence, we have the following definition:

Definition 6.3 An *argument* is a backward justification tree. The tree is the complete set of nested *subclaims* and *evidence* components that justifies a *claim*, together with the inferences that relates the *evidence* components and the nested *subclaims* to the *claim* (Figure 6.1).

Although the argument is constructed "backwards", it can, of course, be "read forwards".

A *claim* at level i is justified by evidence from the corresponding level i and/or by "assuming" that certain *subclaims* will be satisfied at lower levels. We thus have the situation where a *subclaim* is invoked, *i.e.* assumed true at one level, and justified by evidence and/or other *subclaims* at a lower level, and possibly at a later stage of the justification process. It must therefore be possible to formulate a same *subclaim* at two different levels. We shall see (Chapter 8) that this constraint has a strong impact on the models and languages used at the different levels to formulate subclaims and valuate evidence.

In other words, the justification of an *initial dependability requirement* appears to be a *backward or reverse inductive* process, rather than a *direct* inductive or a deductive process (see Figure 6.2). Criminal officers, like inspectors Maigret and Poirot, make direct inferences. They start by collecting evidence, if possible in absence of any presumption, until enough is accumulated to argue that someone is guilty. In safety justification, we have to start by stating what the safety properties to justify are, and only then do we construct arguments and identify and request whatever evidence is necessary. It would be unacceptable and dangerous to work the other way round, as it would somehow mean adapting the dependability requirements and the definition of the system safety to the available evidence.

We can sum up the above discussion by the following:

Definition 6.4 The *justification process* of an *initial dependability requirement* or a *claim* is an inductive process by which a backward justification is constructed as a tree of implications, whose root is the *requirement* or the *claim* to be justified, whose intermediary nodes are *primary claims* and/or *subclaims*, and whose end nodes (leaves) are *evidence* components.

Precisely because of its inductive nature, the process of justifying a dependability requirement and/or of building a safety case is easier when it is an integrated part of the design process. Carried out hand in hand with the design, the justification can specify at each stage of the design what evidence is needed, thus guiding the design on how to make it available. This is another process property supporting the presumption that only carefully designed systems can be efficiently validated.

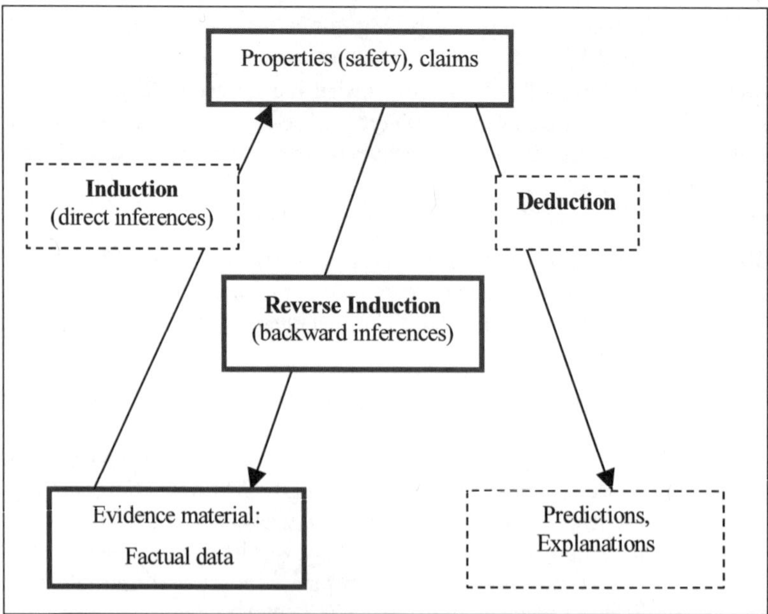

Figure 6.2. Inductive processes

6.4 The Conjunctive Property

Hence, rules similar to the sequent calculus apply to claim expansion and claim proving. A backward justification is a tree of implications, whose root is the *claim* to be justified, whose intermediary nodes are *subclaims*, and whose end nodes (leaves) are evidence components. The justification tree is grown from the root by applying inference rules of the form:

(6.3) $\gamma.\mathbf{clm}\,i \;\Leftarrow\; (\upsilon.\mathbf{evd}\,i \wedge \ldots \omega.\mathbf{evd}\,i) \wedge (\alpha.\mathbf{clm}\,j \wedge \beta.\mathbf{clm}\,k \wedge \ldots)$

with levels $j,\,k,\ldots > i$.

The meaning of that rule is that the *claim* of name γ at level i is inferred (we use the symbol $\Leftarrow$ for the backward implication) from a conjunction of evidence components with names υ, ω,... from the corresponding level i and from *subclaims* with names α, β,... at lower levels. The *claim* γ is called the *consequent* of the rule.   The *subclaims* α, β,.. and the evidence components υ, ω,.. are the *antecedents* of the rule.

It is essential to understand why the right hand side of the inference rule (6.3) is necessarily a *conjunction* of *claims* and of evidence components, and not a *disjunction*. A disjunction would mean that alternative *subclaims* and/or evidence components would exist to justify the *claim*, *i.e.* that alternative arguments are available.

To understand why it cannot be a disjunction, suppose that alternative arguments exist. Then two possibilities would need to be considered. Either one of the alternative arguments is sufficient to justify the claim. In this case the dependability justification would be simpler by just retaining that argument. Or none of these alternative arguments is sufficient by itself. In this case, all arguments are necessary, and we must have a conjunction of antecedents in a single argument.

As a concrete example (*cf.* Figure 6.3), and to grasp the full significance of this *conjunctive property*, consider some claim at architecture level-2 that would assert that a processing unit, say <CPU#1>, is fail-safe in its output behaviour:

(6.4)   **CPU#1_failsafe.clm 2:** processing unit <CPU#1> is failsafe

Suppose that this claim (6.4) can be inferred from either one of the two following *subclaims* at design level 3:

- A *subclaim* for evidence that a failure mode and consequence analysis (FMECA) of the hardware and software running on <CPU#1> has identified potential errors and has rightly concluded that consequent failures are safe,
- A *subclaim* for evidence that a hardware maximum cycle timer (watchdog) together with the self-tests of the processing unit trap all potential hardware and software errors that may occur in this unit, leaving the system in a safe state.

Then, three possibilities may arise:

- Each of these two *subclaims* is shown with convincing evidence to cover the whole set of potential errors. In this case (which, as a matter of fact, is very unlikely in practice), only one *subclaim* (for example, the most convincing one) is necessary. In terms of complexity, resources or costs, one could perhaps even question the appropriateness of having both the FMECA and the watchdog.
- Neither the FMCA analysis nor the hardware timer can be claimed to cover alone the whole set of potential errors, but the two subclaims justifiably (*i.e.* with supporting convincing evidence) complement each other by addressing complementary sets of potential errors. In this case, we clearly have a conjunction of three *subclaims* and the two *subclaims* must conjunct with a third *subclaim* requesting evidence of the complementarities of the two sets of potential errors.
- Each subclaim intends to cover the complete set of potential errors, but with a certain degree of uncertainty (probability). In this case, the two *subclaims* together are intended to re-enforce statistical confidence in the completeness of the coverage of potential errors, on the basis for instance that they use independent and different means of detection and protection.

In this case, the claim on the fail safeness of the unit <CPU#1> must be stated as a probabilistic one, and must be clearly inferred from a conjunction of three *subclaims*: the two *subclaims*, expressed as probabilistic claims, conjunct with a *subclaim* for evidence that properties of independence and complementarities of means are satisfied.

Thus, when two or more *subclaims* – together – or two or more pieces of plausible evidence are combined and intend to re-enforce confidence, on the basis that they are independent and of a different nature, then claims and evidence are required to explicitly assert and justify the re-enforcement, independence and difference.

In other words, if alternative arguments are needed, the necessity and sufficiency of these arguments has to be explicitly claimed and justified.

In all cases of course, including the probabilistic claim here above, evidence must be convincing, *i.e.* must have a sufficient plausibility. Plausibility – not to be confused with the mathematical notions of probability and statistical confidence mentioned in the third case above – was already evoked in Section 5.7, and will be further discussed in Sections 7.2.

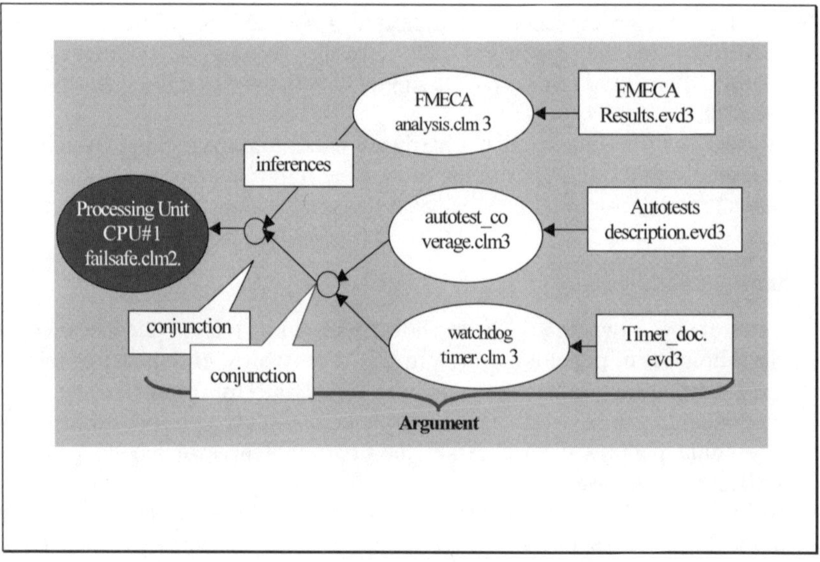

Figure 6.3. The conjunctive property

It is also important to note that the inference mechanism used in (6.3) is not the so-called material implication of first order logic, which is true if the antecedent is false, whatever the value of the consequent may be. In a dependability justification, there need to be a relevant "causal connection" between the conjunction of antecedents and the consequent. Relevance logic and the relevance implication modal operator (see *e.g.* [26, 31, 33]) might, for instance, be more appropriate to formalize the type of implication used here in what could be called "*claim proving*".

In order to keep arguments as simple as possible, it was recommended in Section 6.1 to have *claim* assertions that are – in so far as possible – atomic sentences. For the same reason, the conjunctive property obviously imposes a careful selection of the antecedents of a *claim*, *i.e.* the *subclaims* and evidence components; they must be restricted to those of a conjunction that is necessary and sufficient to make the implication true. In other words, despite the temptation and contrary to some regrettable practices, superfluous and unessential evidence material must be banned; it would be a waste of resources creating confusion only.

6.5 The Syntax of an Argument

At any level j, $j = 1...4$, an argument must justify every *claim* of level j as well as all *claims* made at upper levels $i < j$ that are not – or not completely – justified by evidence from these upper levels and that require evidence of level j. A mechanism is needed to transfer obligations of providing evidence from one level to a lower one.

This mechanism involves three constructs that are used at each level and that we call *delegations, expansions* and *justifications*. These constructs are all backward inferences and make use of the same symbol "$\Leftarrow$". They differ only by the nature of their antecedents.

6.5.1 Delegations and Expansions

We need first to introduce the notion of *delegation*. We define a *delegation claim* or, in short a *delegation* as:

Definition 6.5 A *delegation claim* denoted **δ.clm[i,j]** is a *subclaim* (Definition 6.2) of a level-i *claim*; it is made at level i, $i= 1...3$, to request evidence material of a certain type identified by the name δ from a lower level j, $j > i$.

The evidence of type δ to be claimed at the lower level j, $j > i$, is not necessarily precisely known and identified at the level i where it is found necessary. It is at level j only that this required evidence can be precisely defined, and that the delegation claim **δ.clm[i,j]** can be "replaced" by – and refined into – level- j precise *claims*.

We call this level-j refinement the *expansion* (Figure 6.4) of the *delegation claim*. By virtue of the *conjunctive property*, this expansion is necessarily either a single level-j claim or a *conjunction* of level-j *claims*. Thus, we have:

Definition 6.6 The level-j *expansion* of a *delegation claim* **δ.clm[i,j]**, $i < j$, is a conjunction of level-j *claims* requesting the evidence of the type δ specified at level-i and to be reified at level j.

Expansion is the *syntactical* construction by which claims are generated at each level. It is therefore the first construction that should be made at every level j, j = 1…4 of a justification.

More precisely, the *expansions* of all *delegation claim(s)* **δ.clm[i,j]**, made at the upper levels $i < j$ to a level-j must be performed at level *j*. The elicitation of *primary claims* from an *initial dependability requirement* (Section 5.4) is the level-1 *expansion* of the justification.

This elicitation and the *expansions* at all levels are essential steps of the justification requiring creativity and expert judgment. *Expansions* are not mere "replacements" or "substitutions"; they are "refinements" instead. They must enjoy the *syntactic* property of *consistency*: no two *claims* in any *expansion* can negate each other. Every *expansion* must also be *valid* and *complete*. The validity and the completeness of an expansion are *semantic* properties of the dependability justification. These properties cannot be established and verified without models; these issues are the main subject of Chapters 8 and those following.

6.5.2 Claim Justifications

Once expansions are completed at a level *j*, every *claim* produced by the set of expansions of the *delegation claims* **δ.clm[i,j]**, $i < j$, has to be considered against the available evidence material at the corresponding level.

Claims that can be totally justified by evidence of the current level have been labelled *white* in Section 6.2. These *subclaims* should be considered first and justified by a conjunction of level-*j* evidence components.

Claims that cannot be totally supported by evidence of the current level must then be considered; these subclaims have been labelled *grey* and *black* in Section 6.2.

Grey subclaims are those that are partially inferred from evidence at the same level and partially inferred from *delegation subclaims* that must be expanded at the lower levels. These subclaims have to be justified by a *conjunction* of level-*j* components and delegation claims to appropriate lower levels $j + k, k = 1…4 - j$.

Black subclaims are not supported by evidence material of the same level and are entirely justified by a *conjunction* of *delegation subclaims* expanded at the lower levels.

Hence, in agreement with the conjunctive property, we have the following definition:

Definition 6.7 The *justification* of a level-*j* *claim* is the *conjunction* of level-j evidence components specifications and/or of *delegation subclaims* **δ.clm[j,j+k]**, $k = 1…4$-*j*, from which the claim is *inferred*.

Like *delegation* and *expansion, claim justification* as defined here is a purely syntactic construct: the evidence being specified is not presumed to be complete, coherent or valid. The justification of these semantic properties require the valuation of models, and is the main subject of Chapter 8 and those following.

The syntax of the level *j, j* = 1,..,4

of the argument justifying the requirement CLM0

argument (CLM0, level-j)

<u>**begin**</u>

  expansions : $\forall \delta, \forall\ i, i = 0... j\text{-}1$:

$$\delta.\text{clm[i,j]} \Leftarrow (\text{*.clm j} \wedge \text{*.clm j} \wedge \text{*.clm j} \wedge ...) ;$$

  white justifications: *. clm j $\Leftarrow$ **(*.evd j** $\wedge$ **... *. evd j);**

  grey justifications: *. clm j $\Leftarrow$ **(*.evd j** $\wedge$ **... *. evd j)**

$$\wedge\ (\text{*.clm[j,j+1]} \wedge ...\wedge\text{*.clm[j,j+k]});$$

  black justifications: *. clm j $\Leftarrow$ **(*.clm[j,j+1]** $\wedge$ **...**

$$\wedge\text{*.clm[j,j+k]});$$

$$(k = 4 - j)$$

<u>**end**</u> **argument (CLM0,j)**

Figure 6.4. The syntax of the generic structure at level j of the argument of a requirement **CLM0**. δ**.clm[i,j]** denotes a *delegation claim* with predicate δ made at level i and specifying evidence of type δ to be claimed at a lower level $j, j > i$. The character "*" stands for the wild card symbol

6.5.3 Argument Unicity and Termination

A syntactical consequence of Definition 6.3 and of the Conjunctive Property introduced in Section 6.4 is:

Lemma 6.1 Every *claim* is supported by a unique *argument,* every *claim* being the consequent of a unique *justification* and every *delegation subclaim* having a unique *expansion.*

Proof. The Conjunctive property not only implies that the right hand side of (6.3) is a *conjunction* but also that, for each *claim expansion* or *justification*, the inference rule (6.3) is unique; indeed, two distinct inference rules with the same consequent would be equivalent to a single implication with a disjunction of antecedents. Thus every *claim* is the consequent of a unique *justification* (Definition 6.7); and every *delegation subclaim* has a unique *expansion* (Definition 6.6) and at one single lower level only. ∎

In Chapter 10, this unique argument property will prove to be a condition for guaranteeing the consistency of the dependability justification, and for allowing *subclaims* to be re-used with their arguments in different safety cases.

Obviously, for an argument to be *complete*, all terminal nodes of the justification tree of a *dependability requirement* must be evidence components. In particular, all *subclaims* in the *expansions* of the lowest level (level 4) must be of the *white* type, inferred from evidence components only.

But at the semantic level, a trickier condition for an argument to *terminate* is that all evidence components must be *plausible*. If the plausibility of an evidence component remains questionable, additional claims and arguments are needed to assert this plausibility; plausibility is discussed in Section 7.2.

6.6 Claim and Argument Semantic Aspects

Without waiting for the formal and detailed treatment of valuation models given in Chapters 0 and 9, we can already anticipate some of the pragmatic aspects of the justification framework that are related to its semantics.

6.6.1 Delegation and Expansion as Tools for Refinement of Evidence and Argument Incremental Construction

Claim delegation, *expansion* and *justification* are basic reification mechanisms for the elicitation of evidence and the incremental construction of multi-level arguments.

Section 5.8 explained that the four levels are levels of causality. To each level are associated specific system properties, functions and undesired events susceptible to cause these functions to fail. A property or a function may rely on lower levels; it may also fail as a consequence of undesired events occurring at these levels.

This is why a *claim* on the dependability of a function at a given level may imply *claims* requesting evidence from lower levels. During the construction of an argument at some given level-i, claims on lower level evidence usually cannot yet be precisely stated. Precise descriptions of evidence and of undesired events that may occur at these lower levels are not yet necessarily available. And without these descriptions, it is difficult or even impossible to predict undesired events and to make valid and complete *claims* from above on lower levels. This problem is inherent to the construction of dependability cases and is tackled with by the *model*s and *interpretation structures* developed in Chapters 8 and 9. The construction of the dependability case in pace with the design is obviously a great advantage to cope with this problem.

This is precisely where *delegation claims* are useful. *Grey* or *black claims justifications* use delegation claims to specify types of evidence that must be claimed at lower levels and for which no accurate model of validation need to be yet available at the stage where the claim is formulated. *Delegation subclaims* α.clm[i,j] made at level-i specify that evidence of a certain type specified by a name α is required and must be modelled and provided at some lower level j. Only the lower level(s) of the evidence that is (are) required and their type need be identified while elaborating a given upper level-i.

What is required from the argument is to ensure that this delegated evidence will eventually be precisely defined and provided at the specified lower level. These are the refinement and reification roles of *expansions*. Formally, the link between the *delegation claim* and its *expansion* is neither an identity nor a mere substitution. At the semantic level, we will see in Chapter 8 that the *expansion* of a *delegation claim* [i,j] j > i, must be a *logical consequence* of the *delegation claim* in a model of level j. Moreover, as the expansion is required to be unique, it has the language status of a *validity form*, a status which, provided that the underlying models satisfy certain conditions, justifies the use of expansions in the different languages in which claims are formulated at the different levels (Section 8.10).

As stated earlier, constructing a valid and complete *claim expansion* for a *delegation subclaim* is one of the delicate tasks of the justification which requires much engineering discernment. An *expansion,* however, is not arbitrary and not created from the void. *Expansions* at a given level are the links between on the one hand the evidence required and specified by an upper level, and on the other hand the properties of validity, robustness, and dependability that need to be demonstrated at that level.

Thereby, a multi-level justification scheme provides some guidance for the elaboration of an *expansion*. We already indicated in Section 5.4 that *primary claims* are basically of two kinds: validity and implementation. More precisely, experience shows that any single *claim* in a valid and complete *expansion* is essentially of one of three main kinds:

- *Validity claims*: claims on the validity of the model of the level at which the *expansion* takes place; all these claims are by definition *white*; or
- *Dependability claims*: claims for the existence of system dependability properties (robustness, redundancy, independency, diversity, safety, reliability, ...) at the level at which the *expansion* takes place; or
- *Implementation acceptability claims*: claims on the acceptability of the implementation of these dependability properties at the lower levels.

Validity claims are *white*: they address evidence that is required by the upper levels and must be provided at the current level. *Dependability claims* also address evidence required from upper levels; but this evidence may be available at the current level and/or at levels below; these claims may be *white, grey* or *black*. *Implementation acceptability* claims require evidence from lower levels; these claims are necessarily *black*. Examples of the three kinds will be given by the *expansions* constructed in the generic models of Chapter 9.

Validity claims are determined by the model structures of the level at which the *expansion* takes place. These models are the source of the evidence that validates *white* claims and therefore determines the nature of the *validity claims*. To the extent that they are already defined when the expansion is elaborated, the models of the lower levels also determine the selection of the *implementation acceptability claims*. Reciprocally, these implementation claims somehow determine what the lower levels should be. As for the *dependability claims,* they address properties (robustness, redundancy, independency, diversity, safety, reliability, availability, security) that are to a large extent determined by the model at which the *expansion* takes place.

An *expansion* is therefore not built without pre-existing information: it is in part constrained by the three kinds of claims mentioned above and by the underlying model structures. The validity and the completeness of an *expansion* are properties of the model structure at the level of the *expansion*, and only as such can be evaluated and justified. Thus, *expansions* and models must be conceived hand in hand; a necessity that will be illustrated and become clear in Chapter 9.

Once made complete and valid, the right hand side of the *expansion* of a *delegation claim – i.e.* its antecedents (*cf.* Figure 6.4) – can replace its occurrence in all *justifications*. These replacements can be started at level 4 where all *subclaims* are white. When all *delegation subclaims* are replaced, starting from the bottom level and proceeding bottom-up, *primary claims* at level-1 "become" white and the *justification* of the *dependability requirement* is complete. A concrete illustration of this two-phase downwards-upwards process is provided by the example of Appendix B.

Arguments, therefore, although unique and monolithic, can be incrementally constructed, one level at a time. *Delegation* and *Expansion* appear as two basic mechanisms of a top down *"claim reification"* process. *Claims*, once reified, can then be *justified* with the specified evidence, level by level, in a bottom up process, eventually reducing all *primary claims* to *white claims*.

This dual *expansion – primary justification* reduction process is at the core of the method presented in this study, and motivated the choice of G. Severini's painting, the "centrifuga y centripeta" expansion of light, to illustrate the first page of this book.

6.6.2 Universal and Existential Claim Assertions

Let us come back to *claim* assertions and some aspects of their semantics. In Section 6.1 we argued that assertions should be as "simple" as possible. Chapter 8 will show that, in the predicate calculus (also called quantification theory or first-order logic), a *claim* or a *delegation claim* is expressed in its simplest form as a *well-formed formula (wf)* using predicates, logical connectives, and the quantifiers $\forall$ (the universal quantifier "for all") and $\exists$ (the existential quantifier "there exists").

More precisely, the two simplest forms that a *claim* assertion can take are:

$$\exists\ (y_1,\ y_2...y_n)\colon \varphi$$
$$\text{or}$$
$$\forall(y_1,\ y_2,...y_n)\colon \varphi,$$

where $y_1,\ y_2,\ ...y_n$ are values of variables of φ, the symbol ":" stands for "such that" and φ is the well-formed formula of a simple assertion containing no quantifiers and using the variables y, the values of which may refer to environment or system states, events, or data.

By definition, the domain of the values of the y variables that follow a quantifier symbol is the scope of this quantifier. In a multi-level justification the domain of a claim assertion may belong to the universe of a level different from the level at which the claim is made.

A claim assertion of the first kind (*existential claim assertion*) at some level i claims the existence of evidence for a property φ established at that same level i and for values of the variables y belonging to a domain of the universe of the same level. This is typical of a *white claim,* the evidence to support this existence belonging to the same level i. It is important to note that this claim, true at level i, must of course remain true at the lower levels $j > i$. Functional diversity is an example: justified by evidence at the environment-system interface, its existence must be preserved in the lower levels of architecture and design where separation and independence properties of the implementation are to be claimed.

Black claims are *universal assertions* of the second kind; *grey* claims necessarily include *universal assertions.* These claims are *justified* by *delegation claims* that at some level i require that a property is true at some lower level j, for *all* possible states, data, events or conditions at that lower level j and below (*universal claim assertion*). Here, the scope of the variables y belongs to those lower levels. For example, *delegation claims* generated at the environment-system interface for the justification of *black claims* requiring evidence on a correct implementation of system requirements are of that kind; they require evidence that architecture, design, or operation control satisfies some property for *all* possible behaviours.

More precisely, the following lemma states – with respect to these universal and existential assertions – the obligations of a multi-level justification:

Lemma 6.2 *White claim* assertions at any level are formulated by predicates using *existential quantifiers* over specific values of a domain of the same level universe. Justified by evidence at that level, they have to remain satisfied at the lower levels. *Grey* and *black claims* assertions are formulated at levels 1, 2 and 3 by predicates with *universal quantifiers* over all values of a domain of a lower level universe. If they are justified by evidence at this lower level, they have to be satisfied also at the upper level where they are formulated.

Chapter 8, in particular Section 8.9, will determine the semantic inter-relations that must exist between models at the different levels so as to ensure the obligations expressed by the lemma. These model inter-relations must guarantee that *universal claim assertions justified* at some level remain true at higher levels and that *existential claim assertions justified* at some level remain true at lower levels.

6.6.3 Claim Refutation

In the twenty fifth stratagem of his book "Die Kunst, Recht zu behalten" [92], Arthur Schopenhauer notes that induction (επαγωγη) needs a great number of cases to establish a proposition while an *exemplum in contrarium* (απαγωγη) – *i.e.* a proof *a contrario* – needs one instance only to which the proposition does not apply. He gives the example of the sentence: "All ruminating animals have horns", refuted by the single instance of the camel.

This stratagem does not apply directly to claims. Claims are existential or universal sentences that need to be shown true and not refuted. Induction is the only way to do it and a difficult task indeed since all applicable instances need to be considered. Only false, inappropriate or unnecessary claims need to be refuted.

But this is precisely where Schopenhauer's stratagem can be of some use. It gives the possibility, before embarking on a perhaps lengthy argument and search for evidence, to check at reduced cost that no counter-example refutes the existence or the universality of a claim assertion for which evidence is claimed. Of course, the validity of the counter-example must be ensured. In fact, three conditions must be satisfied by the counter-example: the counter-example must be true, must belong to the same universe, and must be inconsistent with the proposition; otherwise, using model-checking terminology, the counter-example is spurious.

A typical application is the refutation of properties claimed over subsets of system states; a single state instance may be enough to prove the property wrong.

6.6.4 Absence of Level-1 Grey Primary Claims

Section 5.8 explained how an *initial dependability requirement* (level-0) is expanded at level-1 into *primary claims* which are basically of two kinds. Those of the first kind address properties of the requirement specifications of the system (validity, completeness, consistency); those of the second kind address properties of the implementation of the requirements. Two disjoint types of evidence generally support these two types of primary claims. Primary claims of the first kind are white and entirely supported by evidence that belongs to the environment-system interface; primary claims of the second kind are black and supported by evidence that belongs to the lower implementation and operation control layers. Thus, in general, all *primary claims* at level-1 are either *white* or *black*.

This is a semantic property of the models used at level-1 which will be discussed further and more formally established in Sections 8.8.3 and 9.4.13. It is also a useful guiding principle for the elicitation of *primary claims*.

6.6.5 Adjacent Delegation Sub-claims

By definition, a *delegation subclaim* made at level-i specifies evidence that must be provided by some lower level-j, j > i. In principle, any lower level may be

addressed; for instance, in Figure 6.1, the subclaim **.clm**(i+3) is specified by –
and part of the expansion of a delegation claim at level-i. Which lower level is
addressed by a *delegation subclaim* depends on the models used at the different
levels and on the relations between these models. In principle, these models can
be arbitrarily different from each other. However, they must be such that the
delegation claim can be formulated and interpreted and valuated at the two levels i
and j.

We will see in Chapter 8 that, to comply with this constraint, the construction
of an argument across the different levels requires that the models be related in
certain ways. To make a long story short, the models of every level must be
"direct extensions" of the models of the level just above, this "direct extension"
relation between levels being transitive.

As a consequence, a delegation claim at level-i requesting evidence from a
lower level-k need not but can always be formulated as a claim **α.clm**[i,j], j = i+1,
addressing the next adjacent lower level, and from there expanded again to the
next lower level until the level-k providing evidence is reached.

Despite an increase in length of expansions and justification trees, this practice
has some advantages for the design of dependability cases: it may be easier and
safer for the designer to systematically delegate requests for evidence to the
immediately lower level. As explained in Section 6.6.1, when expanding a claim
at a given level, one is not necessarily aware of the exact evidence needed from the
lower levels. Systematic delegation to the next lower level forces the designer to
consider all delegation claims remaining to be expanded at every level. Some of
the delegations may then just be "short circuits", but the risk of missing potential
useful sources of evidence is reduced. Besides, these "short circuit" delegations
may be removed after all levels have been expanded. This is illustrated by the
level-1 and level-2 models developed in Chapter 9.

7

Axiomatic Principles and Limits

One could expect a chapter with this title to be right at the beginning of the study pursued in this book. We found it easier, however, to postpone the discussion of axiomatic principles until some of the most important syntactical and semantics concepts have been introduced.

The justification of the dependability of a real system is based, like other disciplines, on general accepted principles and assumptions that are most of the time considered as self-evident and kept implicit. But dependability is at stake; therefore, it is important that these principles and assumptions get questioned and clarified so as to ensure the robustness of the approach and delineate the limits of its applicability. This chapter examines some of the most basic principles and limits of the method presented here, namely those of claim justifiability, evidence plausibility, consensus and epistemic uncertainty.

7.1 Claim Justifiability

One might object that not all claims can be dealt with by the dependability evaluation framework; and that sound claims for which the framework would be inadequate might exist. Alternatively, one might also argue that the framework is so general that unsound or unrealistic claims might get justified.

The answers to these objections call for a definition of what a 'justifiable claim' is in practice. Let us first not forget of course that Kurt Gödel showed in 1931 that there will always exist propositions that cannot be demonstrated in mathematics. A justification, however, as it is meant here, is something else than a mathematical proof. It is a weaker and less formal concept: a claim assertion may not be mathematically provable and still be justifiable.

The *justification* of an *initial dependability requirement* or a *claim* was introduced by Definition 6.7, page 50, as a purely syntactical notion: an *expansion*, *i.e.* a conjunction of evidence and/or *delegation claims* without regard to their actual truth. In practice, however, we must also be concerned by the semantic counterpart which includes the concept of evidence *validity* and also, as we shall see, the less formal aspect of model *plausibility*.

Strictly speaking, a claim is *justified* by the framework if the supporting *argument* is logically correct and if the layered evidence defined by the *expansions* and *delegations* of the argument is *valid*. Thus, claims that are not justifiable are *unsound* claims, *i.e.* claims for which no correct and valid argument exists.

Definition 5.4, page 38 states that evidence is factual data that make claims true. How evidence makes a *claim* statement "true" requires a model to formulate and to interpret the claim statement and the evidence. *Valid* evidence is the set of values which make the claim true in the model. This definition of formal semantics will be a subject of the Description Part of this book.

Evidence may therefore be *valid* regardless of whether the interpretation model is a true representation of reality or not. If we limit ourselves to these formal definition, an *unrealistic* claim could therefore be supported by formal valid evidence and be justifiable. The meaning of justification generally accepted in practice for certification, licensing and system validation is of course different. The true representativeness of the interpretation models is indispensable. We shall say that the models must be *plausible*, *i.e.* worthy of belief. *Plausibility* is discussed in the next section; but, adopting provisionally its usual intuitive meaning, we come to the conclusion that, to be *justifiable*, a claim must satisfy the following conditions:

Definition 7.1 A claim is *justifiable* by the framework if its *argument* is logically correct, the layered evidence *valid* and the interpretation models *plausible*.

Following this definition, typical claims that are *not justifiable* are claims that are unsound or unrealistic. Unsound claims have no logically correct and validated argument; unrealistic claims have no *plausible* model to valuate evidence.

A problem, however, remains.

Claims which are formulated in a *plausible* interpretation model but for which no evidence can be found to valuate clearly are not justifiable according to the definition and have to be refuted and excluded from the *dependability case*. Alternatively, claims for which no *plausible* model can be found to formulate the claim and to valuate evidence are an issue; these claims are not justifiable according to the definition, but may nevertheless hold true in reality.

Completeness, for instance, in practice is seldom easily *justifiable* in this sense. A claim on completeness may be realistic but hard or impossible justify. Consider for example a property of a digital system that cannot be established by analysis, and would need to be established by testing. A typical situation met in practice is that the available test results are positive but evidence that the set of test runs is exhaustive (complete) is not obtainable. The claim for the existence of the property and the argument based on testing are *a priori* sound; but the models to valuate evidence of test coverage exhaustiveness are missing or lack *plausibility*.

Such claims also should not be part of a dependability justification. This is the price to be paid for dependability. These claims need to be substituted with alternative justifiable claims, by perhaps lowering pretence to a more realistic and justifiable level. For example, the following pair of *justifiable* claims might be an alternative to establish the property: to claim (i) that a set of failure modes are covered by a given test plan, and (ii) that these failure modes are the set of failure modes anticipated by a FMECA (failure mode and effect consequences) analysis and identified as violating the property. If uncertainty remains as to whether the identification of these failure modes by the FMECA analysis is complete, then there should also be a (probabilistic) claim supported by *valid* evidence and a *plausible* interpretation model that this uncertainty is acceptable.

7.2 Evidence Plausibility and Weight

Claim justifiability is therefore inseparable from the notion of model *plausibility*. What is exactly model plausibility? The answer is not simple.

The end of Section 5.7 on the organization of evidence into levels already alluded to the notion. Evidence material at the four levels consists of facts or data that must be taken for granted and agreed upon by all parties involved in a dependability case. There must be confidence beyond any reasonable doubt in *the axiomatic truth* of these facts and data, without further evaluation, quantification or demonstration. This confidence must be supported by a *consensus* of all parties involved to consider the evidence as being unquestionable; *consensus* is discussed in the next section. Confidence and *consensus* are necessarily obtained by means of representations of the real world, *i.e.* by models, that are believed true.

Plausible models are therefore instrumental not only for validating evidence but also for obtaining confidence and achieving consensus.

How is the *plausibility* of models evaluated? The formal developments of Chapter 8 will facilitate a more precise apprehension of the concept. In a nutshell, let us say that the evidence that justifies a claim can be defined as the set of values that valuate to true a claim assertion within the model in which the claim and the evidence are formulated and interpreted. If the model is precisely defined, the formulation, the interpretation and the valuation of evidence become well-defined transformations. The plausibility of evidence is therefore a quality of the interpretation *model*. It is the model with its underlying assumptions, which must be unquestionable, as result of expert judgment and a *consensus* between all parties involved. The authors of [51] note that an argument taking the form of a chain of reasoning cannot be stronger than its weakest link. An argument being a chain of claims and evidence components of different levels, thus also a chain of interpretation models, cannot be more valid than its less plausible model.

Plausibility being an attribute of the underlying models should not be confused, as it is sometimes the case, with the mathematical statistical notions of confidence when *claims* are probabilistic assertions. Whether a deterministic or probabilistic assertion, *plausibility* is required for the model that valuates the evidence that supports the claim. For instance, a probabilistic claim that takes the form: "pfd must be less than X with confidence interval less than Y", can be supported by an adequate set of statistical test results. But in addition to the statistical validity of the samples, *plausibility* also is required for the underlying model: the representativeness of the operational behaviour and the correctness of the test harness are essential factors of this *plausibility*.

Plausibility is therefore essentially a measure of the validity of the underlying models in which *claim* assertions and evidence are formulated and valuated. This validity is in part the result of expert judgment and *consensus*; hence, *plausibility* is also a measure of the trust one has in expert judgment and consensus. The next two sections examine these issues.

Plausibility is of course also a necessary condition for the termination of an argument. Only evidence validated by *plausible* models can be a terminal node of an argument (Section 6.5.3). If the *plausibility* of a model remains questionable, it means that additional arguments are needed to assert this plausibility: *primary claims* or *subclaims* are missing and must be added and justified to assert the representativeness, the completeness or other attributes of the model(s). If the uncertainty cannot be removed by justifiable claims, the data and facts cannot be taken as evidence.

Evidence has another attribute. An evidence component can support more than one *primary claim* or *subclaim* in an *argument*, and can also be used in more than one *argument*. More precisely, since a backward justification tree is acyclic, we can define the *weight* of an evidence component in an *argument* (Definition 6.3):

Definition 7.2 The *weight* of an *evidence component* within a *claim argument* can be defined as the number of "distinct" paths between the *evidence component* and the *claim* in the argument backward justification tree.

By "distinct" paths we mean paths which differ by at least one node. For instance, in the *.clm i *argument* of Figure 6.1, the evidence components of level-i and level-(i+1) have *weight* 1, and those of level-(i+2) and level-(i+3) have *weight* 2.

This *weight* metric is an estimation of the number of distinct lines of reasoning that rely on a given evidence component in the justification of a *claim* or a *dependability requirement*. We will see in Section 7.4, however, that one must be careful to use this metric for prioritizing evidence material, giving for instance relatively more resources to evidence of greater *weight*, with the objective of balancing risk with resources. Remark 7.1 explains why it can make sense for an evidence component used in different arguments of a dependability case, but makes no sense within a same argument.

7.3 The Ineluctability of Consensus

Claim *justifiability* and model *plausibility* appear therefore to be inseparable from the notion of *consensus*. In previous sections, and already in Chapter 3, we observed that, ultimately, some form of underlying *consensus* amongst experts remains inescapable, whatever degree of rationalization is achieved in a dependability justification. There is an irreducible part of the justification which must be posited as axioms and assumptions, and requires a consensus between those involved in a safety case: the licensee and regulator experts, the system architect, the supplier.

In the very first place there must be a common agreement to work within the framework presented in this book; this implies the acceptance of the notion of dependability requirement, of claim, of a four level of evidence structure, of the axioms of logic. In second place there are several aspects of the dependability justification that we already identified and involve expert judgment and ultimately require consensus:

- The elicitation of the *initial dependability requirements* and the acceptance of their completeness and *expansions*;
- The acceptance of the models and their assumptions to formulate these requirements; for instance: the acceptance of the choice of parameters and their ranges, of conservative hypotheses, of the selection of accident scenarii;
- The acceptance of the logical correctness of *arguments*;
- The *plausibility* of models (Section 7.2): the acceptance of the interpretation models used at every level of the justification to formulate and interpret *claims*, and to valuate *evidence*.

Should any of these *acceptances* prove questionable in the course of the construction of the justification, then – as stated earlier – additional claims or more appropriate models, assumptions, arguments or evidence may be needed to resolve the remaining doubts. Ultimately, however, after all precautions have been taken, some *consensus* will always be needed. In a way, this should not be a surprise. Even in mathematics, it is known that all proofs rest on some sort of social consensus (see *e.g.* [19, 75, 86]).

Consensus must be based on experience and expert judgment. Consensus can be assisted by normative approaches such as established standards and benchmarks. One should not forget, however, that it is the usage, experience and experts that make the norms and not the opposite. Clearly, in normal practices, the sincerity of experts has not to be questioned. If we follow some of the principles of social science communicative theory [3, 35], we have to posit that under normal circumstances there is in every expert claim assertion or evidence statement an inherent intention and expectation to be understood, credible, in agreement with the current value system, true (in relation to what the expert thinks he knows) and sincere. There would be no point, and it would certainly be outside the scope of this work, to assume that experts would be malevolent and have different attitudes, expectations or intentions.

Nevertheless, consensus remains subjective, and subjectivity can lead to substantial incoherencies among experts of different origins.

So, whatever level of consensus, whatever degree of realism, detail and certainty has been achieved, some *uncertainty* that some assumptions, models, arguments or evidence might be invalid, incorrect or incomplete will always remain. Understandably, but somewhat paradoxically, all those involved in a consensus *de facto* consider this residual uncertainty, which they can't eliminate, as being *physically and logically unreasonable*; otherwise they would strive to further improve models or arguments.

The objective of the justification framework is of course to reduce the uncertainty that requires consensus to a minimum and to make consensus easier and more cost effective to achieve, by maximizing the part of the justification that can be rationalized by logical arguments. Identifying and making explicit, at each of the four levels of evidence, those aspects of the justification process that require expert judgment and consensus is certainly not the most negligible advantage of the approach.

The necessity of a consensus emphasizes the need for interactions between stakeholders in the construction of the safety justification and in the different phases of the dependability case. The above considerations also indicate how the choice of models, assumptions, justifiable *subclaims*, correct arguments and plausible evidence are mutually interdependent.

Let us finally remark that the achievement of consensus through the exposition of criticisms and conflicting views also has its virtues. As noted by the social scientists, authors of [3], "*dissensus*" also can be beneficial. They quote the arguments developed in [21, 64] that a lack of consensus is healthy as it provides the possibility of communicating and confronting different perceptions of the world, thus revealing in the present case a multiplicity of aspects of the environment constraints and the system design. Following these arguments, the generation of *dissensus* and conflicting views, as opposed to the achievement of consensus, or at least as a preliminary structured phase to this achievement, should be a non necessarily negative ingredient in the construction of a dependability case.

In this context, it is worth mentioning Habermas' work [36], a major contribution to social science critical and communication theory. Habermas distinguishes and analyses two ways of achieving consensus:

- A normative achievement, which results from common perceptions and the uncritical acceptance of cultural traditions – as norms and standards – or from the dictate of experts and the manipulation by an elite;
- A communicative achievement supported by critical inquiries and investigations, dialogue, confrontation of conflicting views and mediation.

The second type of consensus is considered by social scientists as being one of the processes at the very heart of the "rationalization" of the real world [3]. More detailed treatment in this direction is out of our scope; but it would be worth investigating whether and how results of research in social communication science may in practice improve the efficiency and the safety of the empirical consensus achievement processes inherent to the construction of a dependability case.

7.4 Epistemic Versus Stochastic Uncertainty

The residual uncertainty mentioned in the previous section, which may appear to experts as being *physically and logically unreasonable* after a consensus has been achieved, is sometimes coined *epistemic uncertainty*[6] [99]. Epistemic uncertainty is attached to expert judgment; it is difficult – if not impossible – to quantify, by opposition to the well-posed uncertainty – so called stochastic or aleatory uncertainty in [99] – that can be well understood by laws of physics and logics,

---

[6] επιστημων meaning expert in Ancient Greek

and can be quantified by sound probabilistic models, statistical inferences, statistical testing or experiments.

The two types, epistemic and stochastic uncertainties, often co-exist in a dependability justification, and can even affect the same evidence material. As an example, consider again the probabilistic claim: "the probability pfd is less than X with confidence interval less than Y". Such a claim must be supported by evidence (*e.g.* test results), for which, as well as statistical validity, *plausibility* is required: *plausibility* of the operational behaviour representativeness, of the test harness correctness, of the calculation correctness, *etc.* The assurance of these attributes, to the extent that it involves expert judgment and consensus, can be affected by *epistemic uncertainty*.

Epistemic uncertainty affects and decreases the *plausibility* of models asserted on the basis of expert judgment, and this impact is hard to *quantify*[7]. By opposition, a confidence interval specified in a probabilistic claim assertion is a rational measure of the stochastic uncertainty that may be calculated from the coverage and the outcome of the tests. For the above probabilistic claim, one might think of finding a relation between the claimed confidence interval and the *plausibility* of the stochastic model supporting its calculation. Even such relations, however, would be difficult or impossible to establish.

Nevertheless, the dependability evaluation framework clearly separates stochastic uncertainty, which can be precisely expressed by probabilistic claims, from the epistemic uncertainty affecting the *plausibility* of interpretation models. The separation should allow a better assessment of the impact of the *epistemic uncertainty* on the arguments.

The impact can, if necessary, be estimated and controlled in the following *qualitative* way. Suppose that we can define different sets of criteria characterizing the ways evidence and consensus have been obtained, and derive from these criteria various "*degrees of plausibility*" for the evidence interpretation models. Criteria might specify the type of evidence (formal proof or tests), the type of consensus achieved (normative or communicative), the value of the expertise, the level of independence of verifiers, or any other criterion drawn from social sciences studies on consensus achievement.

Then, the *degree of justifiability* of a *claim* could be related via its *argument* to the *degree of plausibility* of the evidence interpretation models that support this claim. Remember that all arguments are conjunctions (Section 6.4); also that when two or more *subclaims* – or two or more components of evidence – are combined and intended to re-enforce confidence, because they are considered as being independent and diverse, then a claim is required to explicitly assert and justify this re-enforcement.

---

[7] A very interesting attempt, however, is made in [91] at a quantitative assessment of software reliability based on the engineering judgment of experts in the absence of evidence from the operational feedback of previous usages.

Thus, because of this conjunctive property (Section 6.4), a coarse relation between claim *justifiability* and *plausibility* can be obtained by the brute force application of the following definition:

Definition 7.3 The degree of *justifiability* of a (*delegation*) *claim* is equal to the degree of the antecedent(s) of least *justifiability/plausibility*, either in the *expansion* or in the *white, grey* or *black justification conjunction*, of which the (*delegation*) *claim* is the consequent.

A rough estimation of the *degree of justifiability* of a *dependability requirement*, a *primary claim* or a *subclaim* is obtained by recursively applying this definition backwards in the corresponding justification tree, starting from the evidence terminal nodes.

Definition 7.3 is quite severe, conditioning the strength of the justification of a *claim* on the evidence models of least *plausibility*. This severity is the direct consequence of (i) the conjunctive property which constrains an *argument* to be as simple as possible, necessary and sufficient, without superfluous or redundant *claims* or evidence; and (ii) of the rigid preciseness required, any complementarities between diverse pieces of evidence being explicitly claimed and justified.

It is worth noting that the brute force application of Definition 7.3 takes on a special meaning when diverse types of evidence are supposed to re-enforce each other and one of these types becomes less plausible or even gets refuted. Take the example of a software module, the dependability of which is claimed to be supported by the combined evidence provided by operational testing on the one hand and by a formal correctness proof on the other. Suppose that testing reveals a defect. What is then the value of the *plausibility* of the formal proof evidence models when this defect is discovered? Different aspects of this question are discussed in [62]. Definition 7.3 remains however applicable in such cases. It may even lead to considering the need to introduce levels of negative *plausibility*. At any rate, such cases illustrate the necessity of explicitly claiming the support of combined evidence, or diversity, and not leave this support implicit in the justification.

Remark 7.1   One might erroneously think that in order to maximise the *justifiability* of a *dependability requirement* or a *claim*, a simple rule would consist in ensuring that the evidence components of greater *weight* in the *argument* – as defined in Section 7.2 – should also be those validated in models with a higher *degree of* plausibility; this would suggest that evidence material of greater *weight* should receive a relatively higher level of assurance, the resources dedicated to the elicitation of evidence and the criteria applied to achieve consensus being allocated in proportion with the *weight*. This strategy, however, would not comply with Definition 7.3. Consider, as a counter-example, the *expansion* of Figure 6.1. The evidence components of level-(i+3) have weight 2. Augmenting their associated model *plausibility* beyond that of level-i and level-(i+1) *evidence* which are of *weight* 1 would not increase the *justifiability* of the level-i *claim*. The strategy conflicts with the conjunction principle which makes all claims and evidence components necessary in an argument and all distinct paths of an argument equally important.

In some ways it is paradoxical that these very same exigencies for making an argument explicit and simple also make it fragile and sensitive to the evidence material of least plausibility.

Finally, an evidence component may be used in the justifications of different *initial dependability requirements*. Hence, the *total weight* of this component can be defined for the arguments of the different *dependability requirements* of a *dependability case* (Definition 5.6):

Definition 7.4 The *total weight* of an evidence component in a dependability justification or a *dependability case* can be defined as the number of *initial dependability requirement arguments* which use the component in the *dependability case*.

This definition allocates a unit weight per *initial dependability requirement argument* which uses the evidence component because, contrary to Definition 7.2, the conjunctive property does not apply here. In a *dependability case*, there may be some advantage – in view of balancing resources and risk – in prioritising and increasing the *plausibility* of evidence models on the basis of their use in the different *requirement* arguments.

7.5 Claims on Product Versus Evidence from Process

In order to come to a mutual agreement, licensing organizations and software suppliers often define standard requirements which focus on the software development process, defining phases, criteria and attributes that this process should meet. However, a development process does not on its own guarantee a good product; besides, a process defined on paper can be faked in reality [77]. Recently [65], members of McMaster University' Software Quality Research Laboratory mentioned that the US Food and Drug Administration guidelines on the validation of medical software are not precise or explicit on measurable attributes of the software subject to assessment, nor on criteria on which their staff can base its decision to approve or reject medical software. Instead they focus on attributes of a development process likely to produce satisfactory software. The capability Maturity Model (CMM) is another example of a certification approach based on maturity and focused on characteristics of the development process. Although these characteristics are important, an assessment strictly based on the process misses the objective of ensuring the adequacy of a system for a given usage.

Conversely, it has sometimes been argued that a goal- or claim-oriented approach – because it aims at the justification of specific dependability properties – is biased towards the product and neglects evidence that can be obtained from the qualities of the development process.

Level-0 *initial dependability requirements,* indeed, address functionalities and properties of the end product system implementation and do not specify how they should be achieved and implemented. But the qualities of the development process are essential sources of evidence to support the claims and the arguments into which these requirements are expanded at architecture and design levels. This distinction is often not made by standards and guidelines when they address both the process and the product. They may confusingly mix requirements on process characteristics with requirements on the evidence that the process has to provide. The consequence is often a wrong, inefficient or unconvincing use in *dependability cases* of the evidence that can be drawn from the development methods and processes.

The clear separation between requirements/claims and evidence, and the structuring into arguments provide a method to use more efficiently the evidence material provided by adequate development processes, and by requiring from these processes specific evidence to support well-defined dependability properties.

7.6 Logics of Prevention, Precaution and Enlightened Catastrophism

At least three distinct logics (sometimes pedantically called "theories") can be invoked to minimize risks and to protect ourselves from accidents when designing or using systems: Prevention, Precaution and Enlightened Catastrophism. It is interesting to briefly discuss these different approaches, and to see how their principles translate into the *dependability justification* framework.

The most classical approach, *prevention*, is based on the anticipation of a set of accident scenarios and failures (or *Postulated Initiating Events,* as defined in Section 5.2.1), and on the design of features or procedures, which make their occurrence impossible or acceptable. Once prevention is implemented and judged sufficient, these accidents and failures are in some way no longer considered as being possible or even probable. Uncertainty is not considered, or at best is relegated to the occurrence of accidents and failures, which have not been anticipated and are considered as being outside the "design basis". In relation to the framework, the logic of *Prevention* should play a role to guide the capture of level-0 *initial dependability requirement*s and, at each lower level, to help the elicitation of *expansions* into the *claims* that need to cover the *postulated initiating events* and anticipated failures that may occur at those levels.

The *precautionary principle* takes into account the presence of uncertainty and risk, including the accidents outside the design basis, and recommends minimizing both (see for instance [59]). When projected on the *dependability justification* conceptual framework, one could argue that uncertainty is unlikely to affect the logic of the deterministic arguments, but may result from incomplete or incorrect elicitations of initial *dependability requirements* or *claim expansions,* or from evidence validation models lacking *plausibility*. In the context of the framework, the residual risk would be determined by the consequences of these deficiencies; and the application of the precautionary principle would require focusing attention on uncertainties affecting the *initial dependability requirements* capture, the *claim expansion* correctness and evidence valuation models' *plausibility*[8].

More recently, some people have argued that prevention and precaution are principles that have been shown inadequate and incommensurable with the scale of accidents and catastrophes made possible by modern technology. These principles do not give sufficient protection from a catastrophe in the sense that, once they have been applied, the catastrophe is somehow erased from the field of the possible events.

They advocate another way of dealing with risk (see [27] for instance). Contrary to the precaution principle, the *enlightened catastrophism* considers the accident as being certain, ineluctable. This certitude of the accidental event is an intellectual simulation, or better stimulation, which is supposed to generate the method to avoid it. The approach is based on the paradoxical principle that accidents will not happen if we act with the absolute certitude that they will. This attitude should in some sense be the right one to identify the *initial dependability requirements* and to construct their *expansions* into *claims* at lower levels.

7.7 Concluding Remarks on Claims, Arguments, Evidence

One is tempted to draw an analogy between the mechanisms of claim expansion and delegation in a safety justification on the one hand, and those of stepwise specification refinement or structured decomposition of functions in structured programming on the other hand (*cf.* for instance [1, 12, 23, 24, 69, 74]). Sections 8.12 and 8.13 in the following Part III discuss this analogy in detail with the help of some notions of model theory. A few comments are in order, however, before concluding the first two parts of the study.

Claim *expansion* and *delegation* are different from the stepwise refinement of the specifications of a computer-based system in at least two essential ways:

- Unlike a hierarchy of levels of abstraction or abstract machines, the real universe in which a *subclaim* is specified and interpreted is disjoint from the real universe in which is specified and interpreted the upper level parent claim from which it is expanded, and from the real universes of the *subclaims* into which it is expanded.;
- The four levels of an expansion are fixed, unlike the arbitrary number of refinement steps.

Besides, unlike the nested functions of a compound function in structured programming (see *e.g.* [69]), the subclaims of a given expansion can seldom be

---

[8] A more exacting formulation of the precautionary principle is sometimes: "Claims other than those proven safe are unsafe". The extreme and pessimistic character of this formulation, postulating the entire system environment as being *a priori* adverse, and thus requiring an impossible justification of the contrary, makes it useless in practice.

identified and specified as independent entities with well-defined interfaces. In particular, they are not independent from the evidence components at the level above that are antecedents of the same inference rule.

We conclude, however, the first two parts of this study by observing that the construction of an argument is *hierarchic* (by its level structure), *backward* (by its proof), *and hologramatic* (each *claim* "encompassing" all lower level *subclaims* of its supporting argument). This construction will later appear to be also *recursive*: The design and implementation of safety prevention and hazard mitigation mechanisms at one level may introduce new undesired events and new modes of failure at lower levels, which may require new claims to be made and new evidence to be found.

Claims, evidence and argument are of course quite general concepts that have been in use for a long time under various forms and in many different contexts of science and human activity. They have been formally and mathematically studied in various models of causality and counterfactual inference, in both deterministic and probabilistic contexts, see *e.g.* [100]. They were more recently applied under different forms to the safety justification of computer-based systems, for example in [7, 14, 17, 22, 85]. When they are not algebraically studied as in [86] these concepts often vehicle general and imprecise notions related to the rationale and modalities for the choice of claims, the validation of the arguments, the backing of the evidence, and the assumptions behind the models.

The present framework proposes mechanisms, which should facilitate the use by regulator and licensee experts of a claim based axiomatic approach in the industrial context and for engineering purposes.

These new concepts are mainly a 4-level evidence structure, the priority given to the declination of dependability properties into functional claims at these levels, inference relations between these levels, the mechanisms of claim conjunction, expansion and delegation and the recognition of the essential role of consensus between licensor and licensee experts on argument correctness and evidence plausibility.

A rather paradoxical finding is that those very same exigencies which force an argument to be explicit and simple, at the same time make the argument sensitive to evidence material of lowest interpretation model plausibility. As we shall see in Part III, well-defined level models, and the relations between these models play an important role in the justification of dependability and the specification of the necessary documentation.

PART III

Descriptions

8

Structures and Interpretations

Le formel pur est stérile et sans intérêt.
Telles sont les mathématiques, jongleries de symboles abstraits.
La valeur de la forme commence
lorsqu'un peu de matière la leste et la gauchit.

Michel Tournier, "L'image Abîmée" in *petites proses*, 1986.
© Éditions GALLIMARD

A critical gap between human perception and reality must be crossed to understand and assess the safety of an engineered system. The gap is is the territory of models and interpretations and is vast and full of ambushes. In many places, the previous chapters indicated that, in addition to relying on multi-level and disparate evidence, the justification of the dependable behaviour of a computer system critically depend on truthful models of the real world.

Models are indispensable. They enable us to analyse, to reason, to evaluate and to communicate about the real world. In physics and in engineering, models are simplified representations of reality that highlight some aspects and by necessity ignore others. In logical systems, models are mathematical objects that assign truth-values to sentences in a given language, where each sentence represents some aspect of reality.

The two functions, analysis and valuation, are essential in the justification of the dependability of an engineered system. Surprisingly, they have been almost completely neglected by engineers and researchers working on computer system safety cases. And in those cases where models are evoked, they are often confused with exact copies of reality.

This chapter explores models apt at fulfilling these two functions. We are seeking models "pour leur vérité et non leur vraisemblance": for their ability to express truth, not for their verism and similitude to reality.

8.1 The Roles of Models in Dependability Assessment

In several places (in particular in the previous Section 7 on axiomatics), we came across situations where models are essential to the demonstration of dependability. In practice, models play at least three different roles.

First, safety, reliability, availability and other attributes of a system can only be apprehended by means of observations and by models, the latter being necessary to give meaning and purpose to the former.

Models are then needed because they alone – despite the fact that some engineers or parishioners may not like the idea – can apprehend and convey what we mean by dependability. Consider, for instance, the property of completeness, which often turns out to be an essential ingredient in the definition of safety. Many claims include assertions with "all", "none", "always", "never" *etc*… Such assertions which imply some form of completeness are very hard, and in practice impossible to justify in the absence of well-defined models. Without a proper definition of a set, it is impossible to define or to discuss any form of completeness.

Besides, misconceived, ambiguous, or incomplete models, or the sheer absence of an agreed model may lead to misinterpretations, design defects, and even project or system failures.

When we formulate a *claim* (Definition 5.3) or describe a piece of evidence, we necessarily have in mind a particular interpretation. A *claim* is an assertion taking the value true or false. Depending on the interpretation that one has in mind, truth may have quite different meanings. Particular attention must therefore be paid to validating the premises of these interpretations. A precise and rigorous definition of what we mean by true and false in the context of a dependable system and its environment is needed. Because of its need to formulate explicit claims, the claim-oriented approach makes the importance of precise models even more obvious. Standard, norm or rule based approaches have not the same need and tend to mask this importance dangerously.

In [79], D.L. Parnas shows how natural language alone is inadequate for the task of precise requirement specification. He gives the example of the statement: "the pumps must be shut off if the water level remains above x meters for more than t seconds". Although this sentence – a rather usual type of claim in fact – may seem clear and precise, at least 4 interpretations can be found depending on whether the water level is taken as a mean, median, *rms* or a minimum value. Some of these interpretations may lead to implementations that would not trigger a control system in case of rapid sizeable oscillations of the water level. A computer program can implement only one of the interpretations; and which one is implemented may go unnoticed if no precise model supports the requirement specification.

The second practical role of models is to provide a logical basis for the construction of arguments. Arguments relate claims to evidence components. As we have seen, evidence must be obtained at several levels: from the system environment, the system architecture, design and operation. Arguments must be carried over these different levels, and each level needs different model representations. These models cannot be independent of each other, otherwise the thread of the argument cannot be consistently carried over from one level to the other. The justification of dependability is a logical analysis based on the inter-relations between these models.

We believe that certain basic elements of model theory [1, 6] which is a branch of mathematic logic dealing with the connections between a formal language and its interpretation can help in several ways. The intention, of course, is not to claim that the entire semantics of the justification of the dependability of a computer embedded system can be formalized nor that dependability properties could be rigorously proved in the logic sense. *A contrario*, by contrast the simplicity and elegance of model theory throws light on the complexity of these problems when real systems must be addressed.

We will see, however, that model theory helps understand what a model is, what a model can do and also what a model cannot do. It reveals basic concepts and properties that tell us how to combine models with different universes, how to expand and how to simplify them while keeping the necessary inter-relations to prove properties expressed in any one of them. It clarifies what the interpretation of a language is, how it can be formalized, and also what escapes the realm of formalization. Above all, familiarity – even elementary – with this discipline nurtures confidence; it provides guidance for designing models that are sound, consistent, simple and general enough and which can be used together although they cover different universes.

This guidance is all the more essential when the evidence required for the justification of the dependability of a computer based system spans four distinct universes (*cf.* Section 5.7), which, although quite different by their technical nature, must nevertheless be related to each other to support claims and arguments.

Last but not least, models of the system and of the dependability justification provide – as we shall see – accurate means to produce the documentation required for the safety case. Miscommunication between engineers, programmers and non-programmers is frequent. Models not only give structure to documentation; they also establish terminology and vocabulary and constrain people to share the same meaning.

All these reasons explain why we need truthful representations of the structures on which – at each level – we formulate claims and evidence, and on which arguments are built. These representations must be carefully selected for their usefulness in interpreting, explaining, predicting and proving. Their effectiveness, of course, *i.e.* the cost of designing and using them, must also be taken into account.

8.2 Basic Model Notions

We first recall a few notions from model theory and mathematical logic, useful for the more detailed analysis that will follow.

8.2.1 Model Structures and Relations

In model theory, the most elementary and general form of a model is often called a structure [6]. We shall deal with structures in more detail in Section 8.4. Let us just say here that a structure essentially consists of a well-defined set of elements (which correspond to the entities of the universe of the structure) and of a set of relations between these elements.

For instance a *structure* can be the model of a (not necessarily finite) *state machine*. The universe of the structure is the set (not necessarily finite) of state and time values, the behaviour of the machine being described by a set of *relations* over these state and time values. *Relations* on the elements of a *structure* are also instrumental for representing model assumptions, constraints imposed by the environment on system entities, claims and supporting evidence.

More precisely, a *binary relation* between two sets A and B is any subset R (possibly empty) of the *Cartesian product* A $\times$ B, *i.e.* of the set of all ordered pairs (see Figure 8.1):

$$(8.1) \qquad \{< a,b > \mid a \in A, b \in B\}.$$

Given a *binary relation* R between A and B, the set

$$\{x \in A \mid \exists\, y \in B, <x,y> \in R\}$$

is the "domain" *of* R and denoted *domain* (R). The set

$$\{y \in B \mid \exists\, x \in A, <x,y> \in R\}$$

is the "range" of R and is denoted *range* (R).

We shall indifferently use the notations R, R(A,B) or A(R)B; and if the context leaves no possibility of confusion, R(A) will be used in certain expressions to denote the *range* B of R(A,B); similarly, if $<x,y> \in R$, it is sometimes convenient to use R(x) to denote y. As most relations used in this book are binary, we shall simply use the term *relation*.

The union of the *domains* and *ranges* of the *relations* of a *structure* is the *universe* of the *structure*. Note that to avoid confusion we use "universe" to refer to this union, and "domain" for the input set of *relations*, although some authors indistinctly use "domain" in both cases.

A *functional relation*, or a *function*, is a *relation* such that every element of the *domain* has at most one *image* in the *range*. A *function* R(A,B) is *one-one* (an *injection*), sometimes noted A $\rightarrow$ B, if R(x) = R(y) implies x = y for all x,y $\in$ A.

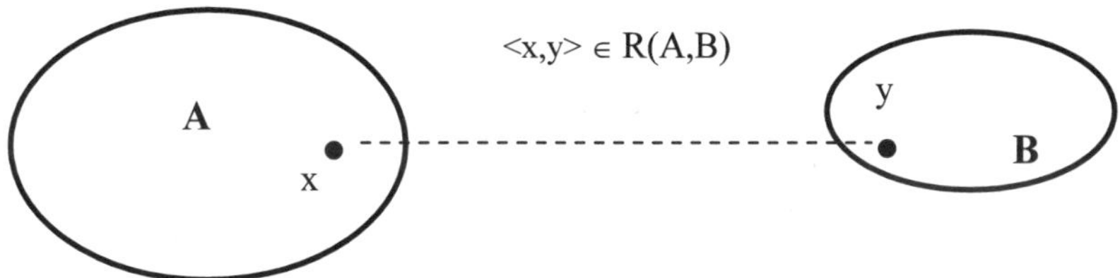

Figure 8.1. Functional relation R(A,B)

Given two relations R between A and B, and S between B and C, their composition (Figure 8.2) denoted by R ° S is a relation between A and C defined by the following set of ordered pairs:

$$\{<a,c> \mid \exists\, b \in B,\ <a,b> \in R \ and \ <b,c> \in S\}.$$

Given a *relation* R between A and B, its *converse* is the *relation* between B and A denoted R^{-1} defined by the set

$$\{<b,a> \in B \times A \mid <a,b> \in R\}.$$

Given two *functions* f(y) and g(x), according to our notation, their *composition* denoted by **f °g** is equal to g(f(x)), that is, f is applied first.

More generally, we will also need *n*-ary *relations*; for $n \geq 1$, an *n-ary relation* on a set A is a collection of *n*-tuples of elements of A, *i.e.* a subset of the *Cartesian product* A^n. If R is an *n*-ary relation on A and $B \subseteq A$, the *restriction* of R to B is defined to be the *n*-ary relation $R \cap B^n$ on B.

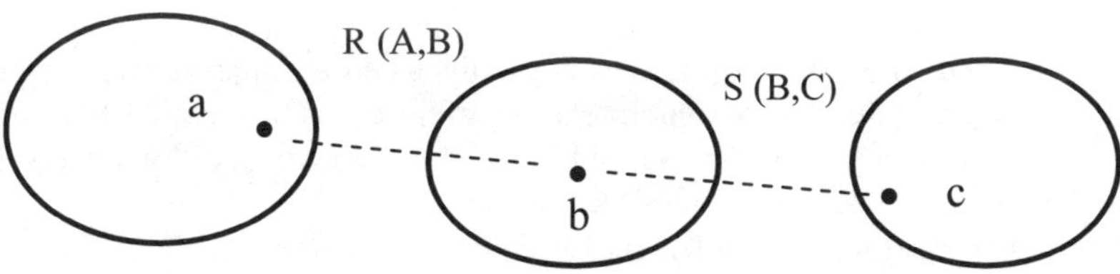

Figure 8.2. Functional composition R ° S

8.2.2 Predicates

The predicate calculus is useful and powerful enough to formally express claims and evidence and to analyse their logical properties. We recall hereafter the few basic notions we need; more details can be found in [32] or [67] for instance.

The predicate calculus is strictly logical. Its logical *symbols* include the propositional connectives (negation, conjunction, disjunction, implication, and bicondition), the universal quantifier $\forall$, the existential quantifier $\exists$, and the following groups of *symbols*: constants, variables, predicates, and functions.

Function symbols apply to individual variables and constants and generate *terms*: a *term* is a constant, a variable or a function applied to terms.

Predicate symbols applied to terms yield *atomic formulae* (*i.e.* formulae without internal structure). For instance, if **P** is a predicate symbol, the *atomic formula* **P**(x) asserts that x has the property P. In particular, if **R** is a predicate symbol for the relation defined by (8.1) between two sets A and B, then **R**(a,b) is an atomic formula asserting that the pair <a,b> satisfies R or that

$$\{a \in A, b \in B, <a,b> \in R\}.$$

Note that we use bold characters to designate symbols and use regular characters to designate *structure* elements; the purpose is to maintain a clear distinction between structure elements and their symbolic representation.

In the predicate calculus, a claim, a subclaim, or a statement asserting some evidence can be expressed as a *well-formed formula* (in short: *wf*).

A *well-formed formula* is made of *terms* and *atomic formulae*. A *wf* is an *atomic formula* or several *atomic formulae* combined by propositional connectives, or an expression of the form $(\forall x : \varphi)$ or $(\exists x : \varphi)$, where x is a variable, φ a *wf*, $\forall$ the universal quantifier and $\exists$ the existential quantifier. We shall freely use the generic name *formula* when no distinction between *wf* and *atomic* is needed.

In $(\forall x : \varphi)$ or $(\exists x : \varphi)$, φ is called the *scope* of the quantifier. Occurrences of a variable which follow or fall under the scope of the universal operator $\forall$ or the existential operator $\exists$ are said to be *bound* occurrences of the variable; all other occurrences in a *wf* are said to be *free*. Note that the same variable may have both free and bound occurrences in the same *wf*.

A *wf* is *open* if it contains *free* occurrences of a variable; otherwise it is *closed;* closed *wf's* are sometimes called *sentences* in the predicate calculus.

Well-formed formulae acquire meaning only when an interpretation is given for their symbols. For a given interpretation, a closed *wf* is a proposition that is true or false, while a *wf* with free variables may be true (satisfied) for some values in the domain and false (not satisfied) for others.

The *claim* assertions α.**clm[i]**, the *delegation subclaim* assertions δ.**clm[i,j]**, and the evidence component υ.**evd[i]** assertions that were introduced in Chapter 6 are *wf's* of the predicate calculus; formulae such as the *expansion* (Definition 6.6, page 49) of a *claim* into evidence components and *delegation subclaims* are also *wf's*. Since we do not need in this chapter to describe and refer to the specific technical content of a *claim* or an evidence component, we shall use here the generic and simpler indexed notation α_i, β_i, γ_i, υ_i, to denote any white, black, gery claim and evidence component at a level i respectively, and δ_{ij} for any delegation claim from a level i to a level j.

Operation and function symbols are in principle not strictly necessary in *claim* and *subclaim* formulae; what can be expressed as a *n*-ary operation or function can be expressed as $(n+1)$-ary relation.

8.2.3 Languages and Proofs

The formal proof in some formal language **L** of a *formula* φ from a set of *formulae* Γ is a finite sequence $\varphi_1...\varphi_n$ of formulae such that $\varphi_n = \varphi$, and where each formula is either an axiom, a formula from Γ, or else follows from earlier formulae by rules of inference. The formula φ is said to be *provable* from the formulae Γ (written: $\Gamma \vdash \varphi$).

A *theory* in **L** is a set of **L** *closed wf's* (*sentences*) Γ which is closed under deducibility, *i.e.* for each *sentence* ψ, if $\Gamma \vdash \psi$, then $\psi \in \Gamma$. A system is *consistent* if it contains no formula ψ such that both ψ and $\neg \psi$ are provable.

8.3 On Descriptions and Interpretations

In Chapter 6, we studied properties of claims, subclaims, and arguments most of which hold independently of their actual technical content and of the evidence components that support them. No understanding of the real domain is required to verify these properties, which were coined "syntactical" for that reason. We "simply" considered claims as assertions (Section 6.1) formulated in some more or less formalised language and returning a value true or false. In practice, of course, these assertions address and are related to the real implementation of the system and to its environment through descriptions and interpretations of the real world.

Mental representations and interpretations of the real system and of its interactions with the environment play indeed an essential role in dependability cases. "The best evidence in the world does no good if it is misinterpreted and misapplied" wrote Max Houck in reply[9] to a reader's comment on his paper [41]. While languages are needed to manipulate symbols for predicates, operators and functions, variables, and constants, underlying descriptions are needed to allow a precise interpretation of these symbols in terms of entities and properties of the relevant domain of the real world. These descriptions, however, remain abstractions of the real world, consisting of a well-defined sets of basic entities, and of assumptions – often simplifications – on the behaviour of these entities.

The purpose of the following sections is to explore and clarify, as rigorously as possible, the properties that these descriptions or models must have, and in

---

[9] Letters. Scientific American, p.7, November 2006.

particular to identify some of the necessary relations that must exist between the different models supporting the four levels of subclaims and evidence. One is tempted to use the term "semantics" to refer to these properties and relations. With D.L. Parnas [80], we believe, however, that the term is misleading in the context of computer systems and programming languages; for the reasons he advocates and also because it may leave the false impression that the real system and its environment could be totally and perfectly apprehended. We prefer instead to refer to descriptions, models and interpretations of the real system and its environment and restrict the use of "semantics" to notions discussed in Section 6.6 and Chapter 7

As stated earlier, the elicitation of the set of initial *dependability requirements* is application-dependent and outside the scope of this work. However, we have to deal with the means to ensure that:

- The expansion of a dependability claim into a tree of subclaims and evidence components across the four levels is "right", *i.e.* corresponds to what is expected from and executed by the real system;
- The evidence components are "true", *i.e.* they reflect what the real system and its environment actually are;
- The evidence components actually support the subclaims.

These issues cannot be *proven* in a formal system or *verified* against a higher level of specifications. They must be answered through a process of *validation* against the system and the real world.

However, as stated above, the difficulty and the frustration – especially when risk and dependability are at stake – are that we cannot exhibit and directly deal with real entities. We are limited to using their descriptions. This fundamental limitation makes a "strictly true" validation of *real* dependability claims impossible. Carefully defined interpretations of the real entity behaviours must be defined. It is in terms of these interpretations only that the decomposition of a dependability claim into distinct inter-related levels of subclaims and evidence components can be shown to be "right" and "true".

A formal concept of *interpretation* is defined and carefully studied in model theory. An *interpretation* establishes a correspondence between a description of a real world domain and a language by associating a true or a false valuation of the real world description with each formula of a language. The purpose is to make the notions of theorem and proof correspond to the notions of truth and model. More precisely, if the axioms and assumptions of the description correspond to true informal statements about the real system and its environment, then the theorems – *e.g. claims* and *subclaims* – that could be proven in a formal system would also correspond to true informal statements about the real domain.

Thus, an *interpretation* is a bridge between the concepts *valid* and *provable*. However, we must be careful again: *interpretations* are not a panacea. An understanding of the real universe remains necessary to select the descriptions, the axioms and the assumptions. Besides, no formal or mathematical system can encompass the whole complexity of a real system. Thus, formal interpretations obviously could at best formally validate only those parts of a system that can be formally described, a subset of what we usually need in practice to establish as being "true".

Nevertheless, we shall see in the next sections that the concepts of *structure* and *interpretation* are useful to identify basic properties that the descriptions underlying a multi-level dependability justification are required to have.

To sum up, the notions introduced so far indicate that the multi-level justification of the dependability of a computer-based system must be developed at each of the four levels along three axes:

- A *real* dimension spanning the four real entities: (i) the physical environment, (ii) the hardware and software architecture, (iii) the design and implementation, (iv) the mode of use, operation and maintenance.
- A *formal* and *logical* dimension: Four languages and a logical system – *e.g.* the predicate calculus – to formulate *claims*, statements of evidence, and rules of inference.
- *Interpretations* of the real entities, *i.e.* mappings between structures associated with the four real entities and the four languages respectively.

8.4 Mathematical Structures

In model theory, see *e.g.* [5, 6], a *structure* merely consists of (1) a non-empty set we shall call the *universe* of the *structure*, the members of which are the *individuals* of the *structure*, (2) possibly various *basic operations* and (3) various *basic relations* on the *universe*. More formally:

Definition 8.1 A *structure* U is defined as an ordered triple:

$$(8.2) \qquad U = <A \; ; \; \{R_i\}_{i \in I} \; ; \; \{c_j\}_{j \in J}>$$

where A is the *universe* of U, $\{R_i\}_{i \in I}$ is the set of all basic $\lambda(i)$-ary relations on A, $\{c_j\}_{j \in J}$ is the set of designated members of A that are constants of U, and I and J are sets indexing the sets R_i of relations and the constants. The domains and the ranges of the relations $\{R_i\}_{i \in I}$ are in A.

As an example of a formal *structure*, elementary arithmetic can be defined as the study of a particular structure, that of natural numbers, basic operations being the addition and the multiplication, and the only relation being the identity relation.

Set theory can also be regarded as a *structure* whose individuals are all sets, identity and membership being the only relations. Elementary Euclidean geometry can be viewed as the study of a *structure*: the elementary Euclidean plane whose individuals are points and straight lines, with the properties (unary relations) of being a point, a straight line, and the ternary relation of collinearity between three points.

Table 8.1. Structures for the material handling machine introduced in Section 4.3 and described in Appendix B

Dependability Initial Requirement:	Vehicle and lift displacements are prevented when pallet is being inserted or extracted from a cabinet	
Environment-System Interface Structure	*Universe:*	- States of handling machine components, of transported material, data communication links and power supplies;
	Relations:	- Mechanical relations and constraints; real-time constraints; effects of failures and environment undesired events on states of monitored variables;
Architecture Structure	*Universe:*	- States of sensing devices, actuators, A/D converters and PLC input and output registers;
	Relations:	- Devices, actuators and A/D converters input/output relations;
Design Structure	*Universe:*	- States of PLC input/output registers
	Relations:	- Input data conversion and validation checks;
		- PLC input-output register relations;
		- Software functional relations;
Operation and Control Structure	*Universe:*	- States of handling machine components, of operator, of control console;
	Relations:	- Operator control actions;
		- Effects and consequences of component failures, of operator failures.

8.5 System Structures

The mathematical concept of structure is of the most elementary kind and of the most general applicability. It is able to capture most aspects of a real system with a minimal set of restrictive assumptions. This is why, for the justification of the dependability of a computer system, we chose to base the description of real world entities on this concept, extending it to non purely mathematical and more tangible uses; hence, the term *system structure*.

This structural viewpoint can be opposed to a constructive or intuitionist viewpoint in which descriptions and proofs would be based on feasible constructions rather than on structures. This choice is not arbitrary. It reflects the fact that, in the case of a computer system analysis, many structures like those defined by the environment and the implementation hardware and software equipments, are given as pre-existing "out there", fixed and completed before we use them in our analysis.

Table 8.2. Examples of System Structures for the monitoring system introduced in Section 4.3

Dependability Initial Requirement:	Abnormal behaviour of the pressurizer must be detected and signaled to operators within 6 minutes	
Environment-System Interface Structure	*Universe:*	- Set of Postulated Initiating Events (PIE set); - Pressurizer, data communication links;
	Relations:	- Pressure (P), Temperature (T) and Level (L) flow and thermodynamic equations; - Postulated Initiated Events effects on T, L values.
Architecture Structure	*Universe:*	- States of sensors, I/O devices, data links, processors, A/D converters, power supplies;
	Relations:	- Redundancy, diversity, segregation, and other dependencies between components;
Design Structure	*Universe:*	- States of input-output registers;
	Relations:	- Input data conversion and validation; - Input-ouput register relations; - Software functional relations.
Operation and Control Structure	*Universe:*	- States of sensors, of calibration and test equipment, of power supplies, of other interfacing plant instrumentation, of operator; - Set of postulated undesired events.
	Relations:	- Operator actions, scripts, plant procedures; - Effects of postulated undesired events.

For a dependability justification, we need descriptions of four real universes: (i) the system environment, (ii) the system hardware and software architecture, (iii) the system design and implementation, and (iii) its mode of use, operations and maintenance. As an example, *structures, universes and relations* that would be relevant to the material handling machine introduced in Section 4.3 are given in Table 8.1. Those that support the initial dependability requirement of the monitoring system introduced in the same section are given in Table 8.2.

The mathematical concept of structure is of the most elementary kind and of the most general applicability. It is able to capture most aspects of a real system with a minimal set of restrictive assumptions. This is why, for the justification of the dependability of a computer system, we chose to base the description of real world entities on this concept, extending it to non purely mathematical and more tangible uses; hence, the term *system structure.*

This structural viewpoint can be opposed to a constructive or intuitionist viewpoint in which descriptions and proofs would be based on feasible constructions rather than on existing structures. The choice, not arbitrary, reflects the fact that, in the case of the justification of system properties, many structures like those defined by the environment and the implementation hardware and software equipments, are given as pre-existing "out there", fixed and completed before being used in the analysis.

The universes and relations of the architecture and design structures can be regarded as time-independent descriptions of the computer system functionality, while the universes and relations of the environment and operations structures capture the time-dependent interactions of the system with the environment and the users.

Note that, since *initial dependability requirements* can be dealt with individually, the *structures* to be used need not be the same for all initial requirements of a same dependability case.

We will study in more detail in Chapter 9 how generic structures and functional relations can be used to model the specifications and the behaviour of an embedded computer system.

8.6 L-Interpretations

We can now introduce the concept of *interpretation* as a formal connection between a *structure* U corresponding to a perception of a real entity and a language **L** used to formulate properties and statements about this real entity.

We again use some of the basic notions presented in J. Bell and M. Machover's course [6]. An **L**-*interpretation* or an **L**-*structure* can be defined as a *structure* U with a *universe* A, a set of constants associated with the domain, and a mapping which assigns the basic elements of the *structure* to the symbols of the language **L**. This mapping allows *values, basic operations and basic relations* of U to be assigned to, respectively, the *variable, function and predicate symbols of* **L**. Terms of **L**, which do not contain variables, denote the constants of U. Remember that we use bold type letters to denote language symbols and plain letters to denote *structure* elements. The denominations **L**-*interpretation* or **L**-*structure* will be used indifferently, depending on the context and whether attention is focused on the language or the *structure* aspects.

In terms of these concepts, we shall see that every single initial *dependability requirement* is described and interpreted by means of:

- Four sets of *structures* $\{U_i\}$, i=1...4, to describe the environment, architecture, design, and operation levels;
- Four sets of languages $\{L_i\}$, i=1...4, to formulate claim, subclaim and evidence *wf* formulae and predicates on the system environment, architecture, design and operation; and a logical system of rules of inference to make arguments in these languages.
- **L**-*interpretations, i.e.* mappings <U - **L**> which allow values and relations of *structure*s to be assigned to the variable and predicate symbols of the languages.

8.7 The Formal Definition of a Model

In order to evaluate a *wf* in some language **L**, we need an **L**-interpretation together with an assignment of values to some variables in U.

Definition 8.2 An **L**-*valuation* is defined as an **L**-*interpretation* (or an **L**-*structure*) together with a particular assignment to each predicate symbol of the language **L** of a relation on the elements of the *universe* A of the *structure*, to each function symbol of **L** of an operation on elements of A and to each constant of **L** of some fixed element of A. Given such an **L**-*valuation,* variables are considered as ranging over the set A.

Recall the distinction made in Section 8.2.2 between the two types of occurrence of a variable in a *wf.* Occurrences of a variable which follow or fall under the scope of the universal operator $\forall$ or the existential operator $\exists$ are bound occurrences; all other occurrences in a *wf* are free. Note that the same variable may have both free and bound occurrences in the same predicate.

For a given **L**-*valuation*, a *wf* without free variables (called a *closed wf* or a *sentence*) is a proposition that is true or false, while a *wf* with free variables may be true for some values in the *universe* and false for others.

Definition 8.3 If an **L**-*valuation* σ valuates to true a formula φ when all φ free variables $v_0, v_1 \ldots v_n$ are assigned values $a_0 \ldots a_n$ from the *universe* A of U, or from U set of designated constants $\{c_j\}$, we say that this particular valuation *satisfies* φ in U, and we write:

$$U \models \varphi \, [a_0 \ldots a_n].$$

The evidence requested by a claim formulated as a *wf* φ $[a_0 \ldots a_n]$ is nothing else but a valuation σ which makes this *wf* true.

Satisfiability with respect to a *structure* is a fundamental notion on which the formal notions of truth and validity are based; we have:

Definition 8.4 If *every* valuation *satisfying* a set Γ of formulae also *satisfies* a formula φ, we say that φ is a *logical consequence* of Γ, or that Γ *logically implies* or *entails* φ, and we write:

$$(8.3) \qquad\qquad \Gamma \models_U \varphi.$$

Thus, the two symbols $\models$ and $\vdash$ are used in this chapter; $\models$ is used for semantic notions and $\vdash$ for syntactical notions.

As stated above, the value true or false of a *sentence* ψ in **L**, *i.e.* a formula without free occurrences of variable, depends only on the *structure* U. If the sentence takes the value true for a given **L**-*valuation,* the sentence is *valid* or *holds* for the **L**-*interpretation* U, or the *structure* U is a *model* for ψ; we write:

$$(8.4) \qquad\qquad U \models \psi.$$

We shall also write (8.4) when ψ denotes a *wf* that can be either closed or with free variables all of which are assigned values from U.

An **L**-*structure* is said to be a *model* for a set of *sentences* Γ if and only if it is a model for every *sentence* in Γ.

A formal system is *sound* if $\Gamma \vDash \varphi$ whenever φ is provable from the formulae Γ (written: $\Gamma \vdash \varphi$). It is *complete* if $\Gamma \vdash \varphi$ whenever $\Gamma \vDash \varphi$.

In plain words, soundness guarantees that every provable formula or sentence is semantically true. Semantic truth is of course essential in the demonstration of safety. An *inconsistent* system cannot be sound. Mathematics and formal systems are sound systems. Thus, whenever we require a proof in those formal systems, the proof implies soundness. Completeness on the other hand ensures that every true fact is provable. Most formal systems of practical interest are not complete. This is why we cannot expect provability whenever we require semantic truth. In other words, not every sound and necessary claim of a dependability case can be expected to be provable.

8.8 Validation and Satisfiability Obligations

The purpose of the following sections is to show that, in terms of the predicate calculus, the justification of dependability requires at each level-i, $i = 1, 2, 3, 4$ that every claim/subclaim *wf* expressed in some language $\mathbf{L}_i$ be satisfied by some $\mathbf{L}_i$-*structure* and/or be the logical consequence of *wfs* that correspond to *delegation subclaims* which in turn are satisfied by *structures* at lower levels $j > i$.

Evidence at each level-i is provided by $\mathbf{L}_i$-*interpretations* that valuate the claim *wf's* to true by assigning relations, operations and *universe* elements of the *structure* to predicates, functions and constant symbols of the language $\mathbf{L}_i$ used to formulate the *subclaims*.

If a *subclaim*, at some level-i, is a closed *wf* (sentence), the $\mathbf{L}_i$-*interpretation* and those at the levels below provide a *model* for this *subclaim*. If the *subclaim* is an open *wf* with free variables, values must be assigned to the free variables by *interpretations* possibly at different upper levels j≠i.

Thus, as we shall see, establishing satisfiability requires consistency and other properties between the *interpretations/models* of the different levels.

8.8.1 Language Expansion/Reduction

In order to understand how the *structures* U_i and the languages $\mathbf{L}_i$ for their *interpretations* must be inter-related and the properties that they should satisfy with respect to each other, we need first to introduce a few additional basic notions of model theory, related to the concept of *substructure*; our terminology is classical and similar to [1, 6, 11, 67].

Let **L'** be a language which is an *expansion* of a language **L**; that is, every symbol of L is also a symbol of **L'**; and, in addition to the predicate symbols and constant symbols of **L**, **L'** contains a disjoint set $\{\mathbf{R_i}: i \in I'\}$ of predicate symbols and a disjoint set $\{\mathbf{c_j}: j \in J'\}$ of constant symbols. Then:

Definition 8.5 *(reduction and expansion)* Given a **L'**- *structure*

$$U' = <A \; ; \; \{R_i\}_{i \in I \cup I'} \; ; \; \{c_j\}_{j \in J \cup J'} >,$$

with disjoint sets I, I' of relations and disjoint sets J, J' of constants, the *structure*

$$U = <A \; ; \; \{R_i\}_{i \in I} \; ; \; \{c_j\}_{j \in J} >$$

is called the *reduction* of U' to **L** or the **L**-*reduction* of U', and U' the **L'**- *expansion* of U; that is, U has the same *universe* as U' but a subset only of the constants and relations of U'.

The **L**-*reduction/expansion* is the most general relation that can exist between two *structures*. An **L**-*structure* can have many distinct **L'**-*expansions*. On the other hand, an **L'**-*structure* has only one **L**-*reduction* to a given language **L**.

Clearly, if U' is an **L'**- *expansion* of U, then for any valuation (*cf.* Definition 8.2) of a *well formed formula* φ of **L**:

(8.5) $\qquad\qquad U \vDash \varphi[a_0,\dots a_n]$ if and only if U' $\vDash \varphi[a_0,\dots a_n]$,

for any $a_0,\dots a_n$ in A, since the same relations and constants are assigned to the same symbols in both *structures*.

8.8.2 Substructures and Extensions

The processes of *expansion* and *reduction* do not change the universe of the *structure*. Property (8.5) implies *structures* that have the same universe; it is therefore not a property that exists between *structures* and *interpretations* belonging to *different* levels of the justification. However, we will see in this chapter and in Chapter 9, that it is useful to deal with **L**- *reductions* and **L**-*expansion* over a same universe at a same level of the justification. In those cases we shall sometimes use the notation U^- and U^+ for an **L**-*reduction* and **L**-*expansion* of U, and more simply refer to *reductions* or *expansions* of U.

Let us now consider the two *structures*:

$$U = <A \; ; \; \{R_i\}_{i \in I} \; ; \; \{c_j\}_{j \in J} >$$

(8.6) $\qquad\qquad U' = <A'; \; \{R'_i\}_{i \in I} \; ; \; \{c_j'\}_{j \in J} >.$

Definition 8.6 *(isomorphism)* The U *structure* would be *isomorphic* with U' if and only if there is a one - one mapping g (the isomorphism) of the universe A of U with the universe A' of U' such that:

$$g(c_j) = c_j' \text{ for all } j \in J,$$

$$(<a_1,...,a_{\lambda(i)}> \in R_i \,) \Leftrightarrow (<g(a_1),...,g(a_{\lambda(i)}) > \in R_i'),$$

for any relation R_i $i \in I$, and all $a_1,...,a_{\lambda(i)} \in A$. The symbol $\Leftrightarrow$ is the dual implication (or bi-implication). To be isomorphic, the universes of the two *structures* must have the same cardinality. Isomorphic *structures* differ in no essential way.

But more useful relations between *structures* exist in model theory when dissimilar universes are considered and not only the *structures* but also their **L**-interpretations are taken into consideration. A *substructure* is a structure whose components (universe, relations, constants) are contained within the corresponding components of another *structure*. More precisely:

Definition 8.7 *(substructure and extension)* Consider again the two *structures* U and U' given in (8.6) with universes A and A' of arbitrary shape and cardinality. For a given language **L**, U is said to be a **L**-*substructure* of U' and U' a **L**-*extension* of U, and we write:

$$U \subseteq U',$$

if:

- $A \subseteq A'$,
- for each $j \in J$: $c_j = c_j'$,
- for each $i \in I$, the relation R_i is the restriction of R_i' to A; *i.e.* if R_i is a $\lambda(i)$-ary relation, $R_i = R_i' \cap A^{\lambda(i)}$, where $A^{\lambda(i)}$ is the set of $\lambda(i)$-tuples of A.

In other words, $U \subseteq U'$ is a partial order relation. If U and U' are *interpretations* of a language **L** and $U \subseteq U'$, then the universe of U is a subset of the universe of U', the interpretation of each relation symbol in U is the restriction of the corresponding symbol interpretation in U' and each constant of U is the corresponding constant of U'.

If $U \subseteq U'$, it results immediately from Definition 8.7 that for any *operation, function or predicate symbol* **P** of the language **L**, any $a_0...a_n$, in A, and any *atomic formula* $\mathbf{P}(a_0,...,a_n)$:

$$(8.7) \qquad (U \vDash \mathbf{P}(a_0,...,a_n)) \Leftrightarrow (U' \vDash \mathbf{P}(a_0,...,a_n)).$$

For example, the set of integers is a *substructure* of the set of rational numbers, if **L** has only the relation symbol $=$, the function symbols $+$ and $\times$, and the constant symbols 0 and 1.

The following proposition follows from (8.7):

Lemma 8.1 If $U \subseteq U'$ and $U' \subseteq U''$, then $U \subseteq U''$.

Between *substructures* and *reductions*, one also has:

Lemma 8.2 If, for a language **L**, $U \subseteq U'^-$ and U'^- is the *reduction* of U' to **L**, then, for **L**, $U \subseteq U'$ with the relations and constants of U.

Proof. Direct consequence of (8.5) and of the fact that sets of relations and constants being in U' but not in U'^- are disjoint from the sets of U'^- (Definition 8.5). ∎

Let us now consider a stronger and more restrictive relation between *structures*:

Definition 8.8 *(elementary substructure and extension)* U is said to be an **L**-*elementary substructure* of U', and U' *an* **L**-*elementary extension* of U if $U \subseteq U'$ and for any *well formed formula* φ expressed in **L**, all free variables of which are assigned values from the *universe* A of U, we have:

$$(8.8) \qquad (U \vDash \varphi[a_0,\ldots a_n]) \quad \Leftrightarrow \quad (U' \vDash \varphi[a_0,\ldots a_n])$$

for all possible value assignments $[a_0,\ldots,a_n] \in A$ of those free variables; in this case, we write:

$$U \prec U'.$$

In other words, if $U \prec U'$, any *well-formed formula* (or *sentence*) φ is satisfied (holds) in U, if and only if it is also satisfied (holds) in U' for assignments of values in A, and reciprocally.

It is obvious that:

$$(8.9) \qquad \text{if } U \prec U', \text{ then } U \subseteq U',$$

but the converse is not true, as these two examples show:

Example 8.1 Let U_Q be the **L**-*structure* of the rationals with **L** having the predicate symbols +, −, ×, < and the constant symbols 0 and 1; and let U_R be the **L**-*structure* of the reals with the same predicate and constant symbols. Clearly, we have $U_Q \subseteq U_R$, but not $U_Q \prec U_R$ since, for example, $U_R \vDash (\exists\, x: x \times x = 1+1)$, while U_Q does not.

Example 8.2 Mendelson [67] gives another simple example. If **L** has the predicate symbol =, the function symbol + and the constant symbol 0, then the *structure* U_{2I} of even integers is an **L**-*substructure* of the *structure* U_I of integers ($U_{2I} \subseteq U_I$), but is not an **L**-*elementary substructure*. Indeed, consider the *wf* $\varphi[y] = (\exists x: x+x=y)$; we have that $U_I \vDash \varphi[2]$, but not $U_{2I} \vDash \varphi[2]$. Nevertheless, note that U_{2I} is a *substructure* also *isomorphic* with U_I with the *isomorphism* g such that g(x)=2x for all x in I. Thus, a *substructure* or an *isomorphic substructure* is not necessarily an *elementary substructure*.

The following proposition between *substructures* and *elementary substructures* will be useful:

Lemma 8.3   If $U \prec U'$ and $U' \subseteq U''$ then $U \subseteq U''$.

Proof. Direct consequence of (8.9) and Lemma 8.1. ∎

Substructures are characterized by properties of their respective *universes* and *relations* (R_i) while *elementary substructures (models)* are also determined by the formulae *(sentences)* they satisfy.

The following theorem due to Tarski and Vaught (1957) [97], see also [6] or [67], provides a useful criterion to determine whether $U \prec U'$:

Theorem 8.1 Assume that $U \subseteq U'$. Then, $U \prec U'$ if the following condition holds: for any formula φ whose free variables are all among $v_0, v_1 \ldots v_n$ and any values $a_0 \ldots a_{n-1} \in A$ of U, if there is $a' \in A'$ for which $U' \vDash \varphi[a_0, \ldots a_{n-1}, a']$, then there is also a $\in A$ for which $U' \vDash \varphi[a_0, \ldots a_{n-1}, a]$.

Note that the condition requires satisfactions in the *extension* U' only. In the example ($U_{2I} \subseteq U_I$), the condition is not satisfied since the only possible value for satisfying $\varphi[2]$ is the uneven value a' = 1 which does not belong to the universe of U_{2I}.

The following two properties will also be useful:

Lemma 8.4 If $U \prec U'$ and $U' \prec U''$ then $U \prec U''$.

Lemma 8.5 If $U' \prec U$ and $U'' \prec U$ and $U' \subseteq U''$, then $U' \prec U''$.

The equivalent of Lemma 8.2 for *elementary structures* is:

Lemma 8.6 If for some language **L**, $U \prec U'$ and U' is a *reduction* of U'' to **L**, then $U \prec U''$.

Proof. By (8.9), we have $U \subseteq U'$, and by Lemma 8.2 $U \subseteq U''$; hence, because of (8.5), the Tarski–Vaught condition of Theorem 8.1 applies to U'' as it does to U'. ∎

8.8.3 L₁-*Interpretation* (Environment-System Interface)

For reasons that will become clear later, the *wf* of a *dependability requirement* is denoted ξ_0 in this chapter. ξ_0 has the predicate calculus status of a *closed* well-formed formula or of a formula all free variables of which would be assigned values from the level-1 *structure*. By definition, there is indeed within the scope of the justification no outer level from which free variables could be assigned values.

The *dependability requirement wf* ξ_0, expressed in some language, say $L_1(\xi_0)$, must be *expanded* (see Section 6.5.1) into a set of *primary* level-1 claims *wf*'s.

Call Γ_1 this set of primary claims *wf*'s; we have at the syntactic level:

$$(8.10) \qquad\qquad \xi_0 \Leftarrow \Gamma_1,$$

where the symbol $\Leftarrow$ stands for the backward inference of the syntactical expansion introduced in Section 6.5.1.

Let $U_1(\xi_0)$ be the $\mathbf{L_1}(\xi_0)$-*interpretation* which provides the valuation of these *wf*'s. Since there is no outer level, ξ_0 and the Γ_1 *wf* variables, predicates and constants are all assigned values from the *universe* and the relations of the *structure* $U_1(\xi_0)$.

For the *expansion* (8.10) to be semantically true, every valuation of the *structure* $U_1(\xi_0)$ *satisfying* the set Γ_1 of *primary* claims must also *satisfy* the dependability requirement $wf \xi_0$; in other words, at the semantic level, ξ_0 must be a *logical consequence* of Γ_1, or Γ_1 must *logically implies* or *entails* ξ_0, or:

$$(8.11) \qquad\qquad \Gamma_1 \vDash_{U_1(\xi_0)} \xi_0,$$

which defines in terms of a $\mathbf{L_1}(\xi_0)$-*interpretation* $U_1(\xi_0)$ the semantic obligations of the syntactical *expansion* (8.10) at level-1. Providing evidence for establishing the *logical consequence* (8.11) is the essence of the justification of the *initial dependability requirement* ξ_0. The single argument property (Section 6.5.3, page 51) requires that the $\mathbf{L_1}(\xi_0)$-*structure* $U_1(\xi_0)$ be unique.

Definition 8.9 It will be convenient to refer to this *structure* $U_1(\xi_0)$ as being the *expansion structure* of the *dependability requirement wf* ξ_0.

Now, remember from Section 6.6.4 that *primary* claims at level-1 are *white* or *black* and not *grey*; a *primary* claim is thus justified by level-1 evidence components or (exclusive) by *delegation* subclaims upon lower levels. Therefore, because the expansion Γ_1 is a conjunction (*cf.* Sections 6.3 and 6.4), we must have if α_1 and β_1 designate an arbitrary level-1 *white* and *black* primary claim well-formed formula respectively:

$$(8.12) \qquad (\forall \alpha_1 \in \Gamma_1: \quad U_1(\alpha) \vDash \alpha_1 [a_0 \ldots a_n]) \quad \text{(white claim justifications),}$$

$$(8.13) \qquad (\forall \beta_1 \in \Gamma_1: \quad \delta_{12}(\xi_2) \wedge \delta_{13}(\xi_3) \wedge \delta_{14}(\xi_4) \vDash_{U_1(\beta)} \beta_1 [b_0 \ldots b_n])$$

$$\text{(black claim justifications),}$$

where $\delta_{1j}(\xi_j)$, $j = 2, 3, 4$, is the predicate of a delegation subclaim from level-1 to level-j where it must be expanded. At least one of these delegation predicates must be present amongst the terms of the left hand side of (8.13), which is not trivially equal to true, a black claim being not a tautology.

8.8.3.1 Primary White Claims Justifications

The *structure* $U_1(\alpha)$ in (8.12) is an $\mathbf{L_1}(\alpha)$ *-interpretation* which *satisfies, i.e.* valuate to true the level-1 white claim *wf* α_1, when all n free variables of α_1 are assigned the same values $a_0\ldots a_n$ as the values required in the logical implication (8.11) by which ξ_0 is the *logical consequence* of Γ_1. The universe of the *structure* $U_1(\alpha)$ need not be identical to $U_1(\xi_0)$ universe, but must be included in this universe. The evidence provided by the valuation of $U_1(\alpha)$ is valid, *i.e.* correct, consistent and complete if and only if, for any values $a_0\ldots a_n$ within the universe of $U_1(\alpha)$, the *wf* α holds true in both *structures* $U_1(\alpha)$ and $U_1(\xi_0)$. More precisely:

Lemma 8.7 The evidence provided by the valuation of the $\mathbf{L_1}(\alpha)$-*interpretation* $U_1(\alpha)$, $\alpha_1 \in \Gamma_1$, is valid (correct, complete and consistent) if and only if $U_1(\alpha)$ is a $\mathbf{L_1}(\alpha)$ - *elementary substructure* of the *structure* $U_1(\xi_0)$.

Proof. The proof is in five steps.

(i) For *correctness*, we first prove that $U_1(\alpha)$ must be a *substructure* (Definition 8.7) of $U_1(\xi_0)$. For the evidence to be *correct*, the $\mathbf{L_1}$-*interpretation* of the *wf* α_1 must hold true in $U_1(\alpha)$ if and only if it holds true in $U_1(\xi_0)$; this is possible iff the universe of $U_1(\alpha)$ is included in $U_1(\xi_0)$ universe and for any operation, function or *predicate symbol* $\mathbf{P}$ of $\mathbf{L_1}(\alpha)$, any $a_0\ldots a_n,$ within the universe of $U_1(\alpha)$, and any *atomic formula* $\mathbf{P}(a_0,\ldots,a_n)$ of $\mathbf{L_1}(\alpha)$, (8.7) is satisfied:

$$(8.14) \quad U_1(\alpha) \vDash \mathbf{P}(a_0,\ldots,a_n) \text{ if and only if } U_1(\xi_0) \vDash \mathbf{P}(a_0,\ldots,a_n);$$

these conditions are satisfied if $U_1(\alpha)$ is a of $\mathbf{L_1}(\alpha)$-*substructure* (Definition 8.7) of $U_1(\xi_0)$:

$$(8.15) \qquad\qquad U_1(\alpha) \subseteq U_1(\xi_0).$$

(ii) For the evidence to be *complete*, any set of values $a_0\ldots a_n$ that satisfies α_1 in $U_1(\xi_0)$ must also satisfy α_1 in $U_1(\alpha)$. This is precisely the application to the *wf* α_1 of the criterion stated in Theorem 8.1 for identifying *elementary substructures*:

$$(\forall\, a_0\ldots a_n: \quad \text{if } U_1(\xi_0) \vDash \alpha_1\,[a_0\ldots a_n] \text{ then } U_1(\alpha) \vDash \alpha_1\,[a_0\ldots a_n]).$$

(iii) For the evidence supporting ξ_0 to be *consistent*, the following proposition must hold:

$$(U_1(\alpha) \vDash \alpha_1 \text{ and } U_1(\neg\alpha) \vDash \neg\alpha_1) \text{ iff } (U_1(\xi_0) \vDash \alpha_1 \text{ and } U_1(\xi_0) \vDash \neg\alpha_1);$$

i.e. a claim *wf* α_1 ($\alpha_1 \in \Gamma_1$) can hold true in one valuation of a $\mathbf{L_1}$-*interpretation* $U_1(\alpha)$ and be false in another if and only if ξ_0 is not consistent.

(iv) *Sufficiency.* For the conditions (i), (ii), and (iii) to be satisfied, it is sufficient that for *any wf* $\alpha_1, \alpha_1 \in \Gamma_1$, and any values of the free variables in the universe of $U_1(\alpha)$, the dual implication (8.8) holds:

$$U_1(\xi_0) \vDash \alpha_1 \, [a_0...a_n] \quad \Leftrightarrow \quad U_1(\alpha) \vDash \alpha_1 \, [a_0...a_n]),$$

i.e. that the $\mathbf{L_1}(\alpha)$-*structure* $U_1(\alpha)$, $\alpha_1 \in \Gamma_1$, be an *elementary structure* of $U_1(\xi_0)$, (Definition 8.8).

(v) *Necessity.* This part of the proof is based on Tarski and Vaught Theorem 8.1. Suppose that $U_1(\alpha)$, $\alpha_1 \in \Gamma_1$, is not a $\mathbf{L_1}(\alpha)$-*elementary structure* of $U_1(\xi_0)$. Then, for any formula α_1 whose free variables are all among $\mathbf{v_0, v_1}... \mathbf{v_n}$ and some values $a_0...a_{n-1}$ in $U_1(\alpha)$ universe, there may exist a value a' from $U_1(\xi_0)$ universe for which $U_1(\xi_0) \vDash \alpha_1[a_0,...a_{n-1}, a']$, but no value a_n in $U_1(\alpha)$ universe for which $U_1(\alpha) \vDash \alpha_1 \, [a_0,...a_{n-1,} a_n]$. In this case, completeness of the evidence provided by the $\mathbf{L_1}(\alpha)$-*interpretation* $U_1(\alpha)$ would not be guaranteed. ∎

Remark 8.1 In step (v) of the above proof, the value a' may very well not exist for some formula α_1. The necessity of the *elementary substructure* property results from the need to *ensure* that the evidence provided by the valuations of the $\mathbf{L_1}(\alpha)$-*interpretations* $U_1(\alpha)$ is complete; *i.e.* it is necessary to have an interpretation which excludes the possibility of the existence of a value a'.

Example 8.3 To illustrate steps (i) and (ii), suppose that integers designate system states and that $\varphi[x]$ is a *wf* expressing some property of state *x*, *e.g.* the property of being a fail-safe state. Suppose also that U is the $\mathbf{L}$-*interpretation* of the system and $U(\varphi)$ the $\mathbf{L}(\varphi)$-*interpretation* with the "smaller" state space (universe) in which $\varphi[x]$ valuates to true if *x* is a fail-safe state. The evidence provided by the valuations of $U(\varphi)$ to identify all failsafe states of the system is correct if every value *x* satisfying $\varphi[x]$ in $U(\varphi)$ also satisfies $\varphi[x]$ in U, and is complete if every *x* satisfying $\varphi[x]$ in U also satisfies $\varphi[x]$ in $U(\varphi)$, *i.e.* if:

$$(\forall x: U \vDash \varphi[x] \quad \Leftrightarrow \quad U(\varphi) \vDash \varphi[x]).$$

As opposed to Example 8.2, page 89, this dual implication holds for $\varphi[x]$ if the $\mathbf{L}(\varphi)$-*interpretation* satisfies: $U(\varphi) \prec U$.

Example 8.4 To illustrate step (iii), consider two *primary white* claims which address the detectability and the controllability of undesired events that may occur in the system environment (claims of this type and their evidence will be detailed in Section 9.4.10, page 130). Detectability requires that undesired event values be univocally related with monitored variables. Controllability requires that there exist within the system range controlled values which are allowed by the environment constraints and lead to safe states of the environment. The *structure* to valuate detectability is a "smaller" *structure* than the controllability *structure* which includes as complements the additional universe and relations of the environment constraints. The evidence for the two *primary claims* would not be consistent should the detectability claim not hold true in the two *structures*.

If α_1 is a closed *wf*, the condition of Lemma 8.7 is that $U_1(\alpha)$ be a *model* for the *sentence* α_1 which is then *valid* or *hold* in the *elementary submodel* $U_1(\alpha)$ of $U_1(\xi_0)$. In more intuitive terms, the interpretation and valuations provided by the **$L_1(\alpha)$**-*elementary substructures* $U_1(\alpha)$ constitute *the complete, correct and consistent evidence that must validate* the *white* claims of level-1.

8.8.3.2 Primary Black Claim Justifications and Delegations

The *structure* $U_1(\beta)$ in (8.13) is an **$L_1(\beta)$**-*interpretation* which must valuate to true the black claim *wf* β, $(\beta_1 \in \Gamma_1)$, if all *delegation* claims are *satisfied* and the free variables of β are assigned in $U_1(\beta)$ the same values $b_0...b_n$ as those required in (8.11) for ξ_0 to be the *logical consequence* of Γ_1. The universe of the *structure* $U_1(\beta)$ need not be the same as $U_1(\xi_0)$ universe, but must be included in this universe.

Besides, conditions similar to those of Lemma 6.2 apply also on the *structures* $U_1(\beta)$. The evidence which is predicated by the delegation claims $\delta_{1j}(\xi_j)$, $j = 2,3,4$, and valuates the *structures* $U_1(\beta)$ is correct, consistent and complete if and only if, for any set of values $b_0...b_n$ in the universe of $U_1(\beta)$, the *wf* β is satisfied in $U_1(\xi_0)$ if it is satisfied in $U_1(\beta)$. And reciprocally, evidence is complete if any set of values $b_0...b_n$ that satisfies β in $U_1(\xi_0)$ also satisfies β in $U_1(\beta)$. Moreover, the evidence supporting ξ_0 would not be consistent if a same *wf* β, $\beta_1 \in \Gamma_1$ could hold true in one **$L_1(\beta)$**-*interpretation* $U_1(\beta)$ and be false in another, with $\beta_1 \in \Gamma_1$. Therefore, Lemma 6.2 applies and the condition for the evidence for β to be valid is that the *structure* $U_1(\beta)$ be an *elementary substructure* of $U_1(\xi_0)$ (Definition 8.8). In the sequel:

Definition 8.10 It will be convenient to refer to the *structures* $U_1(\alpha)$, $U_1(\beta)$ as being the *justification structures* of the primary claims α, β.

A delegation subclaim $\delta_{1j}(\xi_j)$, $j = 2, 3, 4$, merely specifies that a black claim $(\beta_1 \in \Gamma_1)$ implies properties that must be satisfied at some lower level with a required assignment of values $b_0,...b_1$ from the *universe* of $U_1(\beta)$ (and thus of $U_1(\xi_0)$ universe) to the free variables of ξ_j. These properties are to be refined and expanded into claim *wf's* at the lower level-j. Without loss of generality, a delegation subclaim can therefore be formulated in terms of:

- the predicate symbol ξ_j,
- the parameter indicating the level-j where this predicate must be expanded and *satisfied*,
- and the $U_1(\xi_0)$ value assignment $b_0...b_1$ to the free variables of ξ_j , values that must also prevail and be used in its expansion.

Wherever necessary we shall therefore use the explicit notation δ_{1j} ($\xi_j[b_0...b_n]$), $j = 2, 3, 4$.

The role of the delegation subclaims is to specify the predicates that must be *expanded* and *satisfied* by $\mathbf{L}_j$-*structures* and $\mathbf{L}_j$-*valuations* at the lower level-j, $j = 2, 3, 4$, together with the values for the free variables of these predicates that are determined at the upper level. This specification of predicates and free variables values across levels is essential in a multilevel justification, and requires from the underlying structures special properties that we shall investigate in Section 8.9.

To conclude, we have the following two lemmas:

Lemma 8.8 At level-1, *wf's* are closed or have free variables that are assigned values from level-1. There must be a *single* $\mathbf{L}_1(\xi_0)$-*structure* $U_1(\xi_0)$ in which the *wf* ξ_0 of the initial *dependability requirement* is a *logical consequence* of the *wf's* of an expansion Γ_1 into *primary white* and *black* claims. The free variables of the *dependability requirement* and the *primary* claims *wf's* are assigned values from the universe of the *structure* $U_1(\xi_0)$. The valuations of $\mathbf{L}_1(\alpha)$-*elementary substructures (submodels)* $U_1(\alpha)$ of $U_1(\xi_0)$, $(U_1(\alpha) \prec U_1(\xi_0))$, provide *correct, complete and consistent evidence that satisfies* or *validates* the *primary white* claims *wf's* α_1 ($\alpha_1 \in \Gamma_1$). The *black primary* claims *wf's* β_1, ($\beta_1 \in \Gamma_1$), are *logical consequences* of delegation subclaims δ_{1j} [ξ_j ($b_0...b_n$)], $j = 2...4$, that must hold true in $\mathbf{L}_1(\beta)$–*elementary substructures (submodels)* $U_1(\beta)$ of $U_1(\xi_0)$, $(U_1(\beta) \prec U_1(\xi_0))$.

Lemma 8.9 A delegation subclaim δ_{1j} (ξ_j [$b_0...b_n$]) specifies at level-1 for a lower level-j, $j = 2...4$ a predicate ξ_j with a level-1 value assignment $b_0...b_n$ from the *universe* of $U_1(\beta)$ for its free variables; ξ_j must be expanded at level-j into *wf's* that must be *satisfied* by *valuations* of level-j *structures*.

Remark 8.2 Remember that the set of lower level-2, 3 and 4 is generally viewed from level-1 as one single aggregate, *i.e.* as a black box, so that, in practice, only adjacent delegation subclaims $\delta_{12}[\xi_2]$ are formulated, *i.e.* from level-1 to level-2 (*cf.* Section 6.6.5).

A straightforward consequence of Lemma 8.8 is:

Theorem 8.2 *(elicitation)* The $\mathbf{L}_1(\xi_0)$-*interpretation* $U_1(\xi_0)$ is an *extension* (Definition 8.7) of all the $\mathbf{L}_1(\alpha)$–*elementary substructures (submodels)* $U_1(\alpha)$ and the $\mathbf{L}_1(\beta)$- *elementary substructures (submodels)* $U_1(\beta)$ of the white and black claims of the expansion Γ_1 ($\alpha_1, \beta_1 \in \Gamma_1$).

This theorem has an important practical impact on the elicitation and formulation of the initial dependability requirements and of the white and black primary claims of their expansions. The language $L_1(\xi_0)$ must be an *expansion* (Definition 8.5) of the languages $L_1(\alpha)$ and $L_1(\beta)$. Besides, the formulation of an initial dependability requirement and the determination of its *interpretation structure* $U_1(\xi_0)$ cannot be dissociated from the elicitation and formulation of the *primary* claims of its *expansion*. These initial steps of the design of the justification cannot be conceived separately and have somehow to be carried out hand in hand. We will come back to the practical implications of this observation in the following chapters.

8.8.4 L_i-*Interpretations* (i = 2, 3, 4)

Mutatis mutandis, the semantics at the three lower levels is similar to level-1 but differs in four ways:

- Values for free variables may be assigned from any upper level;
- Grey claims can exist at level-2 and 3;
- White claims only exist at level-4;
- Contrary to level-1 where the *expansion structure* $U_1(\xi_0)$ is unique, different structures $U_i(\xi_i)$ may be involved within every level-2, 3 and 4.

Let us extend to every level-i the definitions already introduced for level-1; let:

- Γ_i be the set of white, black or grey subclaims well-formed formulae ξ_i at level-i;
- $\alpha_i, \beta_i, \gamma_i$, refer to an arbitrary white, black and grey subclaim at level-i, respectively;
- B_k, be the set of values, possibly empty, assigned by a L_k-*valuation,* $k = 1...i\text{-}1$, to free variables of a *wf* from the *universe* of a *structure* at level-k; thus, the values $a_0...a_n$ $b_0...b_n$ introduced in the previous section belong to B_1;
- $\delta_{ki}(\xi_i[B_1...B_k])$ be a *delegation* subclaim from level-k, $k = 1...i\text{-}1$ to level-i; the predicate ξ_i must be *expanded* and *satisfied* at level-i with assignments of the values of the sets $B_1,...B_k$ to its free variables; any of these sets may be empty; if all of them are empty, ξ_i is a closed predicate;
- $\Gamma_i(\delta\xi_i[B_1...B_k])$ be the subset of *wf's* in Γ_i which is the *expansion* of a *delegation* subclaim $\delta_{ki}(\xi_i[B_1...B_k])$ from any level-k, $k = 1...i\text{-}1$; the notation $\delta\xi_i$ is used to designate the well-formed formula of a delegation claim to be expanded at level-i.

If, in agreement with the definitions of Section 6.5.1, we generalize the level-1 *expansion* (8.10), *justification* (8.12) and *delegation* (8.13) to the levels-i, i = 2, 3, 4, we obtain for each level-i the following obligations in terms of *logical consequences* and validations.

First, we have the *expansions* of the *delegation* subclaims:

$$(8.16) \quad \forall\, \delta_{ki}(\xi_i[B_1...B_k]),\ k = 1...i-1: \quad \Gamma_i(\delta\xi_i[B_1...B_k]) \vdash_{U_i(\delta\xi_i)} \delta_{ki}(\xi_i[B_1...B_k]),$$

which state that every *delegation* subclaim $\delta_{ki}(\xi_i[B_1...B_k])$ from level-k to i, $i = 2,3,4$, must be a *logical consequence* of its *expansion* $\Gamma_i(\delta\xi_i[B_1...B_k])$; this logical consequence must hold true in a $L_i(\delta\xi_i)$-*interpretation* $U_i(\delta\xi_i)$ of level-i for the value assignment $[B_1...B_k]$. The single argument property (Section 6.5.3, page 51) requires that every *structure* $U_i(\delta\xi_i)$ be unique.

Definition 8.11 It will be convenient to refer to the *structure* $U_i(\delta\xi_i)$ as being the *expansion structure* of the delegation claim with *wf* $\delta\xi_i$.

Then, we have the *justifications* of the *white* subclaims:

$$(8.17) \qquad \forall\, \alpha_i \in \Gamma_i(\delta\xi_i[B_1...B_k]),\ k = 1...i-1:\ U_i(\alpha) \vdash \alpha_i[B_1...B_k, B_i],$$

where the *structure* $U_i(\alpha)$ is a $L_i(\alpha)$-*interpretation* which must *satisfy* – *i.e.* provide evidence for – and valuate to true the *white* subclaim *wf* α, when all free variables are assigned the values required by the upper levels in addition to those from the universe of $U_i(\alpha)$. If all sets $B_1...B_k$, B_i are empty, $U_i(\alpha)$ is a *model* for α.

In principle, the *structures* $U_i(\delta\xi_i)$ which valuate the *expansions* (8.16) need not be identical. But, for the values of the sets $B_1...B_k$, B_i, a *white claim wf* α, $\alpha_i \in \Gamma_i(\delta\xi_i[B_1...B_k])$, must hold true in the level-i *structure* $U_i(\alpha)$ which provides evidence and valuates this claim to true as well as in the level-i expansion *structure* $U_i(\delta\xi_i)$ defined in (8.16).

The universe of $U_i(\alpha)$ need not be the same as the $U_i(\delta\xi_i)$ universe but must be included in it. Moreover, the evidence provided by the valuation of $U_i(\alpha)$ is valid if and only if, for any set of values from the sets $B_1...B_k\ B_i$, α holds true in $U_i(\alpha)$ if it holds true in $U_i(\delta\xi_i)$ and conversely. Besides, the evidence supporting ξ_i would not be consistent if a same well-formed formula α could hold true in one $L_i(\alpha)$-*interpretation* $U_i(\alpha)$ and be false in another with $\alpha_i \in \Gamma_i(\delta\xi_i[B_1...B_k])$ Therefore, conditions similar to Lemma 8.7 apply and the evidence for $\delta\xi_i$ $[B_1...B_k]$ is correct, complete and consistent iff for α, $\alpha_i \in \Gamma_i(\delta\xi_i[B_1...B_k])$, $U_i(\alpha)$ is an *elementary substructure* (Definition 8.8) of $U_i(\delta\xi_i)$:

$$(8.18) \qquad \forall\, \alpha_i \in \Gamma_i(\delta\xi_i[B_1...B_k]),\ k = 1...i-1: \qquad U_i(\alpha) \prec U_i(\delta\xi_i).$$

Finally, for the level-i, $i = 2$ and 3, the *justifications* of the *black* and *grey* subclaims of Γ_i are the *logical consequences*:

(8.19) $\forall \, \beta_i \in \Gamma_i \, (\delta\xi_i \, [B_1...B_k]), \, k = 1...i\text{-}1, \, j = (4 - i):$

$$\delta_{i,i+1} \, (\xi_{i+1}[B_1...B_k, \, B_i]) \wedge ... \wedge \delta_{i,i+j} \, (\xi_{i+j}[B_1...B_k, \, B_i])$$
$$\vdash_{U_i(\beta)} \beta_i[B_1...B_k, B_i],$$

(8.20) $\forall \, \gamma_i \in \Gamma_i \, (\delta\xi_i \, [B_1...B_k]), \, k = 1... \, i\text{-}1, \, j = (4\text{--}i):$

$$U_i^e(\gamma) \wedge \delta_{i,i+1} \, (\xi_{i+1}[B_1...B_k, \, B_i]) \wedge ... \wedge \delta_{i,i+j} \, (\xi_{i+j}[B_1...B_k, \, B_i])$$
$$\vdash_{U_i(\gamma)} \gamma_i[B_1...B_k, B_i],$$

where a *delegation* claim $\delta_{i,i+j}(\xi_{i+j}[B_1...B_k, \, B_i])$ specifies at level-i the predicate ξ_{i+j} and the value assignments $B_1...B_k, \, B_i$ that must be expanded and satisfied at the lower level-(i+j).

Again, and for reasons similar to those of the previous cases, the *structures* $U_i(\beta)$, $U_i(\gamma)$ which valuate the justifications (8.19) and (8.20) need not be all identical, but must be *elementary substructures* (Definition 8.8) of the *expansion structure* $U_i(\delta\xi_i)$ with ξ_i such that $\beta, \gamma \in \Gamma_i(\delta\xi_i \, [B_1...B_k])$, respectively:

(8.21) $(\forall \, \beta_i, \gamma_i \in \Gamma_i(\delta\xi_i \, [B_1...B_k], \, k = 1... \, i\text{-}1):$

$$U_i(\beta) \prec U_i(\delta\xi_i), \, U_i(\gamma) \prec U_i(\delta\xi_i).$$

Moreover, one observes in (8.20) that, for any set of values from the set $B_1...B_k, \, B_i,$ the *grey* claim γ_i must hold true in the *structure* $(U_i(\gamma))$ which valuates the logical consequence as well as in the *structure* $(U_i^e(\gamma))$ which provides level-i evidence for it. Clearly, these two *structures* are coupled. Any *wf* γ valuated to true by evidence in the *structure* $U_i^e(\gamma)$ must also be satisfied in $U_i(\gamma)$; and reciprocally any *wf* γ which holds true in $U_i(\gamma)$ for an assignment of values from the $U_i^e(\gamma)$ universe must also hold true in $U_i^e(\gamma)$. For any *grey claim* γ, this condition is satisfied if:

(8.22) $U_i^e(\gamma) \prec U_i(\gamma).$

Hence, because of Lemma 8.4, (8.21) and (8.22), we also have for ξ_i such that $\gamma_i \in \Gamma_i(\delta\xi_i \, [B_1...B_k], \, k = 1... \, i\text{-}1:$

(8.23) $U_i^e(\gamma) \prec U_i(\delta\xi_i).$

Definition 8.12 It will be convenient to refer to the *structures* $U_i(\beta)$, $U_i(\gamma)$ and $U_i^e(\gamma)$ as being the *justification structures* of level-i claims β, γ.

To conclude, Lemma 8.8, Lemma 8.9 and the Propositions (8.16) to (8.23) prove the following three central theorems:

Theorem 8.3 (*expansion*) Let Γ_i be the set of *subclaims wf's* at level-i, i=1...4. These *wf's* are closed or have free variables that are assigned values from levels 1...i. For the *dependability requirement* ξ_0 and for every *delegation claim* $\delta_{ki}(\xi_i[B_1...B_k])$ from level-k, k < i, to level-i, there is a unique *expansion structure (model)* $U_i(\delta\xi_i)$ in which this *requirement/delegation claim* holds true and its *wf* is a *logical consequence* of *wf's* in Γ_i .

Theorem 8.4 (*justification*) At level-i, i = 1...4:

(i)   Every *white* claims $\alpha_i (\alpha_i \in \Gamma_i)$ holds true in a *justification structure (model)* $U_i(\alpha)$, which provides valid, *i.e.* correct, complete and consistent evidence if and only if $U_i(\alpha)$ is a $L_i(\alpha)$-*elementary substructure* of the $L_i(\delta\xi_i)$-*expansion structure* $U_i(\delta\xi_i)$ with ξ_i such that $\alpha_i \in \Gamma_i(\delta\xi_i [B_1...B_k])$, $k = 0...i-1$.

(ii) *Every black* and *grey* claim (β_i, $\gamma_i \in \Gamma_i$) is a *logical consequence* which involves *delegation* subclaims $\delta_{i,i+j}(\xi_{i+j}[B_1... B_i])$, i = 1...3, to lower levels j = 1...(4-i); the *logical consequence* holds true in a *justification structure* $U_i(\beta)$, $U_i(\gamma)$ respectively, which provides valid evidence if and only if this *justification structure* is a $L_i(\beta/\gamma)$-*elementary substructure* of the $L_i(\delta\xi_i)$-*expansion structure* $U_i(\delta\xi_i)$ with ξ_i such that $\beta, \gamma \in \Gamma_i(\delta\xi_i [B_1...B_k])$.

(iii) Level-i evidence for a grey claim $\gamma_i \in \Gamma_i$ is provided by the valuation of a *justification structure* $U_i^e(\gamma)$; this evidence is valid iff $U_i^e(\gamma)$ is an *elementary substructure* of $U_i(\gamma)$ and thus also of $U_i(\delta\xi_i)$.

Theorem 8.5 (*delegation*) At level-i, i = 1...3, a *delegation* subclaim $\delta_{i,i+j}(\xi_{i+j}[B_1...B_k, B_i])$, j=1...(4-i) specifies at level-i a predicate ξ_{i+j} to be expanded at level-(i+j) together with an assignment of values from the sets $B_1... B_i$ for the free variables of ξ_{i+j}; B_k, k=1...i, is the set of values assigned to the free variables of ξ_{i+j} from the universe of a *justification structure* $U_k(\beta)$ or $U_k(\gamma)$. The predicate ξ_{i+j} is *expanded* at a lower level-(i+j) into *wf's* which must be *satisfied* by valuations of level-(i+j) *structures* (Theorem 8.3). All sets $B_1...B_i$ are empty when ξ_{i+j} is a closed predicate.

Remark 8.3 These theorems allow the *dependability requirement* ξ_0 to be treated as a *delegation claim* $\delta_{01} (\xi_0)$ from an outer level-0 to level-1.

Thus, the main results of this section are that at each level-i, i = 1...4, every *delegation claim* from lower levels must be the logical consequence of level-i *claims*; the logical consequence must hold true in an *expansion structure*; level-i claims must hold true in *justification structures* that must be *elementary substructures* of these *expansion structures*.

A direct consequence of Theorem 8.4 is the generalisation of Theorem 8.2 to all levels-i, i = 1...4:

Theorem 8.6 *(elicitation)* The $L_i(\delta\xi_i)$-*interpretation* $U_i(\delta\xi_i)$ of a *delegation claim* $\delta_{ki}(\xi_i[B_1...B_k])$ from level-k, $k < i$, to level-i, is an *extension* (Definition 8.7) of all $L_i(\alpha/\beta/\gamma)$ *–elementary substructures (submodels)* $U_i(\alpha/\beta/\gamma)$ of the white, black and grey claims of the *expansion* $\Gamma_i(\delta\xi_i[B_1...B_k])$, $(\alpha_i, \beta_i, \gamma_i \in \Gamma_i)$, $i = 1...4$.

Hence, the language $L_i(\delta\xi_i)$ must be an *expansion* (Definition 8.5) of the languages $L_i(\alpha)$, $L_i(\beta)$ and $L_i(\gamma)$. The formulation of a delegation claim $\delta_{ki}(\xi_i[B_1...B_k])$ and its *interpretation* $U_i(\delta\xi_i)$ cannot therefore be dissociated from the elicitation and formulation of the claims of its *expansion*. We will come back in the following chapters to the practical implications this theorem bears on the implementation of safety cases.

8.9 Interdependencies Between Structures of Different Levels

The theorems of the previous section show the nature and roles of *expansion* and *justification structures (models)* at each level, and how these *structures* must be related to each other *within* each level. The theorems also imply that *expansion* and *justification structures* of different levels have to share common properties. In particular, predicates and values for the free variables must be transferred from upper to lower levels for expansion and justification. A claim or a subclaim is formulated at a given level as a predicate that takes the value true or false, while the direct evidence that supports this predicate is provided by the valuation of well-formed formulae in *structures* at this level and/or below.

Therefore, although covering totally different *universes,* the *structures* and their *interpretations* at the four levels are not independent of each other. In this section, we study how they must inter-relate so that their valuations can validate the upper level *dependability requirement.*

8.9.1 Properties of Justification Structures

Let us first identify a set of properties that need to be satisfied by the *structures* of the different levels for allowing their *valuations/interpretations* to validate a given initial *dependability requirement* ξ_0. We will be able in a second stage to refine these properties and obtain necessary and sufficient conditions.

A *delegation claim* (Theorem 8.5) is formulated at a given level as a predicate that must hold true for a given assignment of values in a *justification structure* at that level and, at a lower level, is expanded and inferred as a logical consequence of *wf's* that must hold true in an *expansion structure* for the same assignment. This *justification* and this *expansion structure* must therefore be related in some ways; to see how, we start by stating sufficient conditions for validating an *initial dependability requirement* well-formed formula; that is, sufficient conditions for the Proposition (8.11) to hold true are given by the following:

Lemma 8.10 For $(\Gamma_1 \vDash_{U_1(\xi_0)} \xi_0)$ to hold true in a multi-level justification framework, it is sufficient that:

(i) Evidence for validating every *white* claim *wf* α_i, i $=1\ldots4$, and every *grey* claim *wf* γ_i, i $= 2,3$ is provided by the valuation of a *justification structure,* $U_i(\alpha)$ and $U_i^e(\gamma)$ respectively, which is an *elementary substructure* of the $\mathbf{L_i(\delta\xi_i)}$-*expansion structure* $U_i(\delta\xi_i)$ with ξ_i being such that
$\alpha_i, \gamma_i \in \Gamma_i(\delta\xi_i\,[B_1...B_k])$, k $= 1\ldots$i-1;

(ii) *Every black* and *grey* claim *wf* $(\beta_i, \gamma_i \in \Gamma_i)$ is a *logical consequence* which involves *delegation* subclaims $\delta_{i,i+j}(\xi_{i+j}[B_1...B_i])$, i $= 1\ldots3$, to lower levels j, j $= 1\ldots(4\text{-}i)$; the *logical consequence* holds true in *justification structures* $U_i(\beta)$, $U_i(\gamma)$ respectively, which are *elementary substructures* of the $\mathbf{L_i(\delta\xi_i)}$-*expansion structure* $U_i(\delta\xi_i)$ with $\delta\xi_i$ such that
$\beta_i, \gamma_i \in \Gamma_i(\delta\xi_i\,[B_1...B_k])$, k $= 1\ldots$i-1;

(iii) The $\mathbf{L_i}$-*justification structure* $U_i(\xi_i)$, i $= 1\ldots3$, $\xi_i = \beta_i, \gamma_i$, in which a *delegation* subclaim $\delta_{i,i+j}(\xi_{i+j}[B_1...B_i])$ is predicated and inferred,
is an *elementary substructure* of the lower level
$\mathbf{L_{i+j}(\delta\xi_{i+j})}$-*expansion structure* $U_{i+j}(\delta\xi_{i+j})$, j$=1\ldots(4\text{-}i)$, in which the predicate $\delta\xi_{i+j}$ is expanded.

Proof. Condition (i) ensures that there is valid evidence for justifying all white and grey claims well formed formulas (Theorem 8.4).

Condition (ii) ensures that there is valid evidence for justifying the logical consequences of grey and black claims expansions (Theorem 8.4).

Condition (iii) ensures that any well-formed formula that holds true in a level-i, i $= 1\ldots3$, *justification structure* in which a delegation claim predicate is assumed true also holds true in the lower level-(i+j) *expansion structure* of this delegation claim. And conversely (by definition of an *elementary structure*), any *wf* holding true in the lower level-(i+j) *expansion structure* also holds true in the level-i *justification structure*. Furthermore, by virtue of the transitivity of the *elementary substructure* (Lemma 8.4) and the reciprocal implication (8.8), any *wf* being satisfied by the lower level-(i+j) expansion *structure* is also satisfied by its *elementary substructures, i.e.* its (i+j) *justification structures,* and thus conversely, any *wf* satisfied in these (i+j) *justification structures* is also satisfied in the level-i justification *structure.* ∎

These conditions are clearly sufficient for the validation of the well-formed formula ξ_0. However, the lemma third condition requiring that *justification structures* be *elementary substructures* of *expansion structures* at lower levels is unnecessarily too restrictive. According to this condition, because of the dual implication (8.8), *any* well-formed formula satisfied at any given level would be satisfied at all lower levels, and *conversely.* But this is not necessary.

All *wf's* satisfied at level-i for a given assignment of values from a universe of this level need to remain satisfied at all lower levels for this assignment but need not be satisfied at the higher levels $1...(i-1)$. The reason is a semantic property already discussed in Section 6.6, and more specifically stated by Lemma 6.2, page 55. *Delegation claims* predicates require properties that must hold true at lower levels but not at the level where they are predicated. Formally, this implies that the reciprocal implication (8.8) between the *wf's* of an *elementary structure* and its *elementary extension* need not be required to relate the *justification structure* where a *delegation claim* is predicated and the lower level *expansion structure* where it is expanded. To illustrate this point, once more consider the example of the *structure* of the integer numbers in Section 8.8.1:

Example 8.5 Suppose that integer numbers identify the states of a system. In the multi-level justification of properties of this system, the *structure* U_{2I} of even integers with its restricted (less detailed) universe would be a higher level *structure* compared to the *structure* U_I of integers. A system property as $\varphi[2]$ holding true for a state $x = 1$ in the lower level *structure* U_I system need not be satisfied in the restricted higher level universe. All properties true in the restricted universe need to remain true in the larger universe but not reciprocally.

Two classical propositions on *substructures* provide the tighter conditions we need (see for instance exercise 2.99 in [67] or 1.3.5 in [11]); they can be formulated as follows:

Lemma 8.11 If $U \subseteq U'$, then:

(i) Let $\varphi[v_0,...v_n]$ be a *wf* of the form $(\forall y_0...\forall y_m: \psi[v_0,...v_n, y_0,...y_m])$ where ψ has no quantifiers. Then, for any $a_0...a_n$ in the *universe* of U, if $U' \vDash \varphi[a_0,...a_n]$, then $U \vDash \varphi[a_0,...a_n]$. In particular, U is a *model* for any *sentence* $(\forall y_0...\forall y_m: \psi[y_0,...y_m])$ where ψ contains no quantifier, if U' is a model for this sentence.

(ii) Let $\varphi[v_0,...v_n]$ be a formula of the form $(\exists y_0...\exists y_m: \psi[v_0,...v_n, y_0,...y_m])$ where ψ has no quantifiers. Then, for any $a_0,...,a_n$ in the *universe* of U, if $U \vDash \varphi[a_0,...a_n]$, then $U' \vDash \varphi[a_0,...a_n]$. In particular, U' is a *model* for any *sentence* $(\exists y_0...\exists y_m: \psi[y_0,...y_m])$ where ψ contains no quantifier, if U is a model for this *sentence*.

Proof. Proposition (i): $\varphi[a_0,...a_n]$ has universal quantifiers only, and since $U' \vDash \varphi[a_0,...a_n]$ the scope of the universal variables y_i for which $\varphi[a_0,...a_n]$ holds is in the universe of U', which includes the universe of U; thus, for $a_0,...,a_n$ also in the universe of U, we have $U \vDash \varphi[a_0,...a_n]$ since, U being a *substructure* of U', $\varphi[a_0,...a_n]$ involves restricted relations of U' only.

Proposition (ii): $\varphi[a_0,...a_n]$ has existential quantifiers only, and since $U \vDash \varphi[a_0,...a_n]$, the scope of the existential variables y_i for which $\varphi[a_0,...a_n]$ holds is in the *universe* of U, which is included in the *universe* of U'. Thus, for $a_0,...a_n$ in the *universe* of U we also have $U' \vDash \varphi[a_0,...a_n]$, since (8.7). ∎

Hence, universal formulae are preserved under *substructures* and existential formulae are preserved under *extensions* in larger *structures*. Lemma 8.11 extends to universal and existential *well-formed formulae (sentences)* the partial relation property that *substructures (models)* by definition enjoy for operations, functions, predicate symbols and *atomic formulae* (see Definition 8.7 and condition (8.7)). Universal (existential) formulae true in a *structure* (in a *substructure*) are also true in its *substructure* (in its *structure*) if they have no free variables, or if all their free variables receive values from the universe of the *substructure*.

It is quite remarkable that these truth conditions of Lemma 8.11 are precisely the conditions that universal assertions of *delegation* claims and existential assertions of *white* claims must satisfy at their respective levels (*cf.* Lemma 6.2, Section 6.6.2). A *white* claim well-formed formula at some level is indeed of the form $(\exists y_0...\exists y_m: \psi[v_0,...v_n, y_0,...y_m])$, where the *wf* ψ has no quantifiers and the values of $y_0...y_m$ are in the scope of the universe of a **L**- *structure* of the same the level. Supported by evidence and satisfied at this level-i, the *white* claim must also remain satisfied in lower level *structures* with extended universes.

On the other hand, *grey* and *black* claim *wf's* at some level-i are inferred from *delegation* claims which are *universal assertions* of the form $(\forall y_0...\forall y_m: \psi[v_0,...v_n, y_0,...y_m])$, where ψ is a *wf* expressed with level-i language predicates and has no quantifiers, and the values of $y_0...y_m$ are taken as belonging to the extended universe of some lower level-j *structure*, $j > i$. Satisfied by evidence at this lower level-j, the *delegation* claim must also be satisfied in the level-i *structure.*

With regard to the way values are assigned to free variables $v_0,...v_n$, Theorem 8.5 guarantees that the *universes* from which these values originate comply with the assumptions of Lemma 8.11. Hence, for the lemma to be applicable, it remains to ensure that it is sufficient that the *substructure* condition $U \subseteq U'$ is satisfied between the *structures* involved at the different levels.

Finally, from Theorem 8.3, Theorem 8.4, Theorem 8.5, and Lemma 8.11, we obtain the central result on *justification* and *expansion structures*:

Theorem 8.7 *(structures):* $\Gamma_1 \vdash_{U_1(\xi_0)} \xi_0$ in a multi-level justification framework if and only if:

(i) Evidence for every *white* claim well-formed formula $\alpha_i,$ $i = 1...4$, and every *grey* claim well-formed formula γ_i, $i = 2,3$ is provided by the valuation of a *justification structure* $U_i(\alpha)$ and $U_i^e(\gamma)$ respectively, which is an *elementary substructure* of the $\mathbf{L}_i(\delta\xi_i)$- *expansion structure* $U_i(\delta\xi_i)$ such that $\alpha_i, \gamma_i \in \Gamma_i(\delta\xi_i [B_1...B_k])$, $k = 1...i$-1;

(ii) *Every black* and *grey* claim $(\beta_i, \gamma_i \in \Gamma_i)$ *wf*, $i = 1...3$, is a *logical consequence* which involves *delegation* subclaims $\delta_{i,i+j}(\xi_{i+j}[B_1...B_i])$ to lower levels $j = 1...(4-i)$; these logical consequences hold true in *justification structures* $U_i(\beta)$, $U_i(\gamma)$ that are *elementary substructures* of the $\mathbf{L}_i(\delta\xi_i)$ - *expansion structure* $U_i(\delta\xi_i)$, $k = 0...i$-1, with $\delta\xi_i$ such that $\beta_i,\gamma_i \in \Gamma_i(\delta\xi_i[B_1...B_k])$;

(iii)   Every *delegation* subclaim $\delta_{i,i+j}(\xi_{i+j}[B_1...B_i])$, $i = 0...3$, predicated by and interpreted in a $\mathbf{L}_i(\xi_i)$-*justification structure* $U_i(\xi_i)$, holds true in a $\mathbf{L}_{i+j}(\delta\xi_{i+j})$ - *expansion structure* $U_{i+j}(\delta\xi_{i+j})$, at some lower level-(i+j), $j = 1...(4 - i)$, and $U_i(\xi_i)$ is a *substructure* of $U_{i+j}(\delta\xi_{i+j})$.

Proof. Sufficiency: The proof for sufficiency is a direct consequence of Lemma 8.11 and
Lemma 8.10 with the condition that delegation subclaim *justification structures* be *elementary substructures* of lower level *expansions structures* being merely replaced by a *substructure* relationship condition between these *structures*.

Necessity: The necessity of conditions (i) and (ii) was established by the justification Theorem 8.4. As for condition (iii), suppose that some $U_i(\xi_i)$ is not a *substructure* of some lower level expansion *structure* $U_{i+j}(\xi_{i+j})$, $j = 1...(4-i)$. Then, either the $U_i(\xi_i)$ domain is not included in any $U_{i+j}(\xi_{i+j})$ domain and/or the $U_i(\xi_i)$ relations are not restricted relations of any $U_{i+j}(\xi_{i+j})$ relations; in either case, the *wf*

ξ_i, whether existential or universal, cannot hold true at two levels i and i+j. ∎

The condition (iii) of Theorem 8.7 is in agreement with the so-called Łoś–Tarski *preservation* theorem. A *theory* T – that is, a set of sentences – is said to be *preserved* under *substructures* if whenever U is a *model* of T, every *substructure* of U is a *model* of T. Łoś–Tarski theorem states that a *theory* – *i.e.* a set of sentences in a language $\mathbf{L}$ – is *preserved* under *substructures* if and only if the *theory* is universal, *i.e.* all sentences of the set are universal sentences (see *e.g.* Appendix A2 in [1]).

8.9.2 Properties of Expansion Structures

In short, Theorem 8.7 provides necessary and sufficient conditions for the multi-level justification of an *initial dependability requirement* well-formed formula, *i.e.* conditions which guarantee (i) that *white* and *grey subclaim* existential assertions are valuated at every level by valid evidence, (ii) that *grey* and *black* claims are validated as logical consequences of *delegation* claims, and (iii) that *delegation* claim universal assertions hold true at the level where they are valuated by valid evidence and at higher levels where they are inferred and predicated.

Theorem 8.7 also defines the conditions which must prevail between the different *structures* of a multi-level justification framework. At every level i, i = 1...4, the *expansion structures* of the *delegation* claims to this level must be such that the upper level *black* and *grey* claims *justification structures* inferring these delegation claims are their *substructures,* and the *justification structures* of the level-i claims generated by their expansions are their *elementary structures*.

Some properties can be derived from this theorem. For reasons of simplicity and by extension of our notation, let:

Definition 8.13 $\Gamma_i(\xi_i)$, $\xi_i = \beta_i$, γ_i, designates the set of the *delegation* claims from level-i to lower levels (i+j), j=1 ...(4-i), which belong to the *justification* of ξ_i .

With this notation, we can formulate the following corollary, as an immediate application of Theorem 8.7:

Corollary 8.1 The *justification structure* $U_i(\xi_i)$, $\xi_i = \beta_i$, γ_i, $i = 0...3$, of a level-i black or grey claim is a $L_i(\xi_i)$-*elementary structure* of the level-i *expansion structure* $U_i(\delta\xi_i)$ with $\delta\xi_i$ such that β_i, $\gamma_i \in \Gamma_i(\delta\xi_i\ [B_1...B_k])$, $k = 0...i-1$, and is a $L_i(\xi_i)$ - *substructure* of the *expansion structures* $U_{i+j}(\delta\xi_{i+j})$, j=1 ...(4-i), such that $\exists\ \delta_{i,i+j}(\xi_{i+j}[B_1...B_i])\in\Gamma_i(\xi_i[B_1...B_k])$, $\xi_i = \beta_i$, γ_i.

We also have the property:

Corollary 8.2 If $U_i(\xi_i)$ and $U_i(\xi'_i)$ are the *justification structures* of two level-i claims of a same expansion $\Gamma_i(\delta\xi_i[B_1...B_k])$, $k = 0...i-1$, and $U_i(\xi_i)$ is a *substructure* of $U_i(\xi'_i)$, then $U_i(\xi_i)$ is an *elementary structure* of $U_i(\xi'_i)$.
Proof.   Direct consequence of Lemma 8.5 and the property of *justification structures* to be *elementary structures* of *expansion structures* at the same level. ∎

Structures of different levels are bound by the following conditions that are also immediate consequences of Theorem 8.7:

Corollary 8.3 *Justification structures* in which a *delegation* claim is inferred and predicated must be *substructures* of the lower level *structure* in which the *expansion* of the *delegation* claim is valuated.

And an *expansion structure* must be an *extension* (Definition 8.7, page 88) with the following property:

Corollary 8.4 The *expansion structure* of a delegation claim is an *extension* (i) of the *justification elementary substructures* of the claims of the delegation claim *expansion* and (ii) of the *justification structures* of the upper level *grey* and *black* claims which infer the delegation claim.
Proof. Direct consequence of Theorem 8.7 and Proposition (8.9).∎

This last corollary tells us how to construct an *expansion structure*. Let us construct at a level-i, i =1...4, the *structure* $U_i(\delta(\xi_i))$ – introduced by Definition 8.11 – in the following way. Let $K_k(\xi_k)$ be the universe of the *justification structure* $U_k(\xi_k)$ of a *grey* or *black* claim $\xi_k \in \{\beta_k, \gamma_k\}$, at level-k, $k = 0...(i-1)$. Then, consider the *justification structures* $U_k(\xi_k)$ of all those level-k *grey* or *black* claims which entail (Definition 8.4, *i.e.* the logical consequence of which predicates) a given delegation claim $\delta_{k,i}(\xi_i[B_1...B_k])$ to level-i. Let $K_i(\xi_i)$ be the union without repetition of these universes $K_k(\xi_k)$:

$$(8.24) \qquad K_i(\xi_i) = \bigcup_k K_k(\xi_k),$$

$$\text{for } k \in \{0...(i-1)\}, \text{ and for all } \xi_k \in \{\xi_k: \Gamma(\xi_k) \subset \delta_{k,i}(\xi_i[B_1...B_k])\}\}.$$

Now, at level-i, consider also the union without repetition of the universes $D_n(\xi_n)$ of the *justification structures* $U_i(\xi_n)$ of the level-i claims ξ_n that belong to $\Gamma_i(\delta\xi_i[B_1...B_k])$, which is the level-i *expansion* of a *delegation* subclaim $\delta_{ki}(\xi_i[B_1...B_k])$; then let $D_i(\xi_i)$ be the union:

$$(8.25) \qquad D_i(\xi_i) = \bigcup_n D_n(\xi_n), \qquad \text{for all } n \in \{n: \xi_n \in \Gamma_i(\delta\xi_i[B_1...B_k])\}.$$

where ξ_n is a claim of type α_i, β_i, or γ_i . Let then

$$(8.26) \qquad A_i(\xi_i) = K_i(\xi_i) \bigcup D_i(\xi_i)$$

be the union without repetition of the universes of all the *justification structures* defined above at levels-k, k = 0...i.

Now consider the level-i *structure*:

$$(8.27) \qquad U_i(\delta(\xi_i)) = \; < A_i(\xi_i); \; \{R_i\}_{i \in I(\xi_i)} \; ; \; \{c_j\}_{j \in J(\xi_i)} >$$

with universe A_i defined by (8.26) and where the relation index set $I(\xi_i)$ and the constant index set $J(\xi_i)$ are the unions without repetition of the corresponding index sets of the *justification structures* $U_k(\xi_k)$ and $U_i(\xi_n)$ whose universes are defined by (8.24) and (8.25):

$$(8.28) \qquad I(\xi_i) = \bigcup_k I_k(\xi_k) \bigcup_n I_n(\xi_n),$$

$$J(\xi_i) = \bigcup_k J_k(\xi_k) \bigcup_n J_n(\xi_n),$$

$$\text{for } k \in \{0...(i-1)\}, \text{ for all } \xi_k \in \{\xi_k: \Gamma(\xi_k) \subset \delta_{k,i}(\xi_i[B_1...B_k])\}$$
$$\text{and for all } n \in \{n: \xi_n \in \Gamma_i(\delta\xi_i[B_1...B_k])\}.$$

Theorem 8.8 *(expansion structure)* The *structure* $U_i(\delta(\xi_i))$ defined by (8.27) and (8.28), is the smallest *expansion structure at level*-i, i = 1...4, for the valuation of the expansion of delegation claims $\delta_{k,i}(\xi_i[B_1...B_k])$ of the levels k = 0...(i-1) with predicate ξ_i.

Proof. The *structure* $U_i(\delta(\xi_i))$ is by construction the smallest *extension* (Definition 8.7) which satisfies the properties that are required from an *expansion structure* by Corollary 8.4 of Theorem 8.7. ∎

We will see in the next chapters that these *expansion structures* play an instrumental role in practice at the environment-system interface, architecture, design and operations control levels.

8.9.3 Internal Compound Structure and Interfaces of a Level-i, i = 1...4

In some ways, Theorem 8.8 is counter-intuitive, and was indeed somewhat unexpected. We started this investigation on the articulations between model theory and the justification of system dependability with the intuition that a single *structure* could be defined for each of the four levels of the system justification framework, with a *substructure/extension* relation "joining" these *structures* across the different levels.

Theorem 8.8 shows that matters are more complicated. Each level-i is modelled by *a set of expansion structures* $U_i(\delta(\xi_i))$, one for each delegation claim $\delta(\xi_i)$ predicated on that level by the upper levels. These expansion structures constitute the upwards interface of the level with the levels above. The complete set of *expansion structures* defined by Theorem 8.8, which valuates all the *delegation* claims addressing a given level, comprehends all the universes, relations and constants of this level.

Levels are related to each other by their *justification structures*: *justification structures* at level-i are *elementary substructures* of their *expansion structures* and also *substructures* of *expansion structures* at lower levels. These *elementary substructures* constitute the downwards interface of the level-i to the levels below.

Figure 8.3 schematically displays the various inter-dependencies between levels.

8.10 Interdependencies Between Languages of Different Levels

Because of the relations linking the underlying *structures* of different levels, the languages used at these levels to express the claim well-formed formulae are not independent. By language, we mean the set of symbols used to represent the operators, functions, predicates, variables and constants of these well-formed formulae at each level.

As we did in Section 8.8.1, let **L'** be a language which is an *expansion* of **L**. That is, every **L**-symbol is also a **L'**-symbol; and, in addition to the predicate symbols and constant symbols of **L**, **L'** contains a set $\{R_i: i \in I'\}$ of predicate symbols and a set $\{c_j: j \in J'\}$ of constant symbols.

Then, for each **L'**-*valuation* σ' (Definition 8.3), there is a unique **L**-*valuation* σ which agrees with σ' on all **L** symbols. σ is said to be the **L**-*reduction* of σ', and σ' the **L'**- *expansion* of σ. In general an **L**-*valuation* has more than one **L'**-*expansion*.

Now, let L_i be the language for expressing at level-i the predicate of a delegation claim $\delta_{ki}(\xi_i[B_1...B_k])$ and the well-formed formulae of its expansion $\Gamma_i(\delta\xi_i[B_1...B_k])$ interpreted by the L_i-*interpretation* $U_i(\delta(\xi_i))$.

$\mathbf{L_i}$ is an *expansion* of languages $\mathbf{L_k}$ used at levels k, k $\leq$ i. Indeed, a direct consequence of the *substructure* relation and of Theorem 8.8 – and in particular of the index sets defined by (8.28) – is that the sets $\{c_j : j \in J\}, \{\mathbf{R_i} : i \in I\}$ of constant and predicate symbols of $\mathbf{L_i}$ must include constant symbols and the predicate symbols of the relations restricted to the *universes* of the *justification* structures of the sets:

$$\{U_k(\xi_k): k = 0\ldots(i-1), \xi_k \in \{\xi_k : \Gamma(\xi_k) \subset \delta_{k,i}(\xi_i[B_1\ldots B_k])\}\},$$
$$\{U_i(\xi_n): \xi_n \in \Gamma_i(\delta\xi_i[B_1\ldots B_k])\}.$$

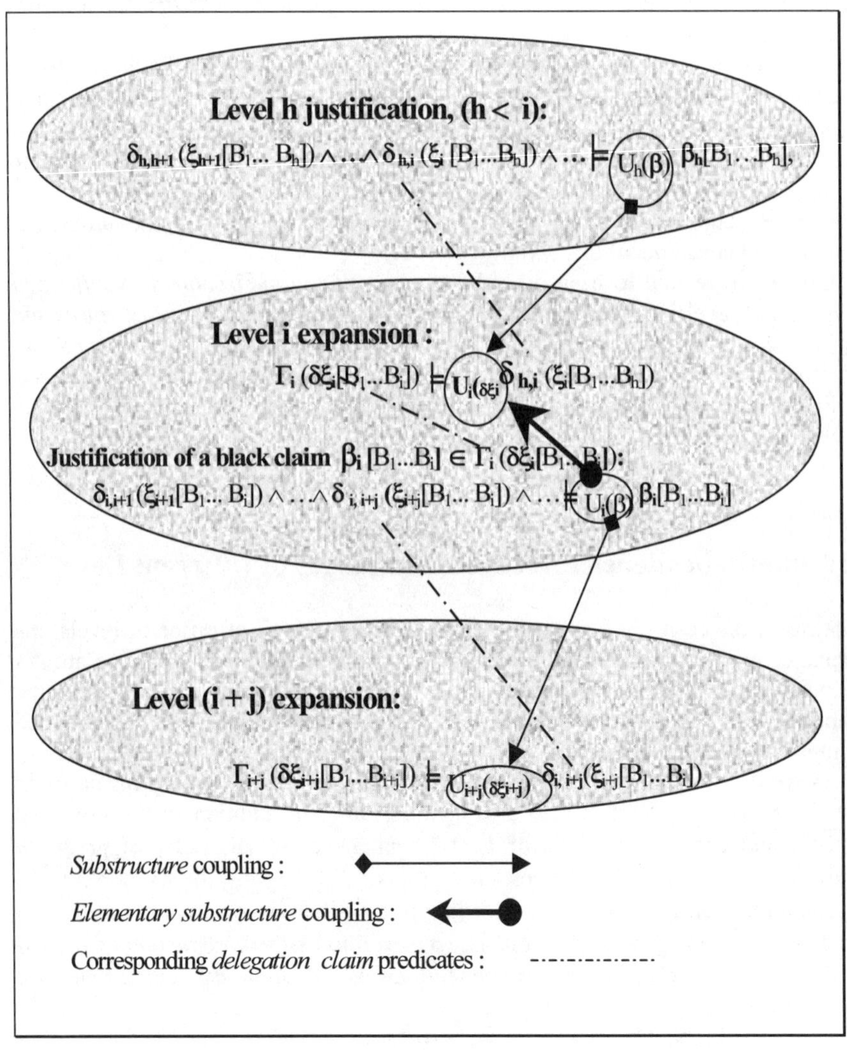

Figure 8.3. Couplings between *structures* of different levels

Besides, because of the specification role they play between *structures* of different levels, *delegation* claims have a special status in the languages used at these levels. Remember that it was stated in Section 6.6.1 that a delegation claim is a syntactic entity that, as such, can be substituted with its expansion in a argument. A distinction made by Bell and Machover [6] between two distinct expanded forms a formula can take in a language *expansion* allows us to give a more rigorous basis to this substitution:

Definition 8.14   Let φ be a formula expressed in a language **L**. A *satisfiability form* for φ in some *expansion* **L'** of **L** is a formula φ' such that any **L**-*valuation* σ satisfies φ iff σ has *a* **L'**-*expansion* σ' satisfying φ'.

Definition 8.15   A *validity form* for φ is a formula φ' in some *expansion* **L'** of **L**, such that any *L-valuation* σ satisfies φ iff *every* **L'**- *expansion* σ' of σ satisfies φ'.

Thus, φ' is a *satisfiability form* for φ iff $\neg\,\varphi'$ is *validity form* for $\neg\,\varphi$.

How do these notions apply to *delegation* claims predicates and expanded formulae? By Theorem 8.5, from every level-i, $i = 1\ldots3$, a delegation subclaim $\delta_{i,i+j}(\xi_{i+j}[B_1\ldots B_k,\ B_i])$, $j = 1\ldots(4\text{-}i)$ specifies to a lower level-$(i + j)$ a predicate ξ_{i+j} with values to free variables assigned by L_k-*valuations* at levels k, $k = 1\ldots i$. Thus, the language at level-i, $i = 1\ldots3$, must clearly include symbols for the predicate ξ_{i+j}, $j = 1\ldots(4\text{-}i)$. The delegation claim predicate $\delta_{i,i+j}(\xi_{i+j}[B_1\ldots B_i])$ is *expanded* at a lower level-$(i+j)$ where it must be the *logical consequence* of *wf's* belonging to Γ_{i+j}, the set of subclaims *wf's* at level-$(i+j)$. Any *valuation* that satisfies $\xi_{i+j}[B_1\ldots B_i]$ at level-i must also satisfy the *logical consequence* (see (8.16)):

$$\Gamma_{i+j}(\xi_{i+j}[B_1\ldots B_i]) \vdash_{U_{i+j}} \delta_{i,i+j}(\xi_{i+j}[B_1\ldots B_i])$$

at level-$(i+j)$. The *expansion* Γ_{i+j} ($\xi_{i+j}[B_1\ldots B_i]$) in a language L_{i+j} must therefore be a *satisfiability form* for the delegation claim predicate in L_i.

Moreover, remember from Section 6.5.3 that the "argument unicity" semantic property requires that a delegation claim predicate have one *expansion* only and at a single lower level. Thus, a *satisfiability form* for a delegation claim predicate is necessarily also a *validity form*.

To summarize, we therefore obtain the following:

Theorem 8.9   The language L_i for expressing at level-i the predicate of a *delegation claim* $\delta_{ki}(\xi_i[B_1\ldots B_k])$ and the well-formed formulae of its *expansion* $\Gamma_i(\delta\xi_i[B_1\ldots B_k])$ interpreted by the $L_i(\delta\xi_i)$-*interpretation* $U_i(\delta\xi_i)$ is a *language expansion* of languages L_k used at levels k, $k < i$. The sets $\{c_j : j \in J\}, \{R_i : i \in I\}$ of constant and predicate symbols of $L_i(\delta\xi_i)$ include symbols for the constant and the predicates of the *justification structures*:

$$\{U_k(\xi_k): k = 0\ldots(i-1),\ \xi_k \in \{\xi_k: \Gamma(\xi_k) \subset \delta_{k,i}(\xi_i[B_1\ldots B_k])\}\},$$
$$\{U_i(\xi_n): \xi_n \in \Gamma_i(\delta\xi_i[B_1\ldots B_k])\},$$

as well as symbols for the predicates $\delta_{i,i+j}(\xi_{i+j}[B_1\ldots B_i])$ of delegation claims to higher levels-(i+j). The unique *expansion* $\Gamma_i(\delta\xi_i[B_1\ldots B_i])$ at level-i of a delegation claim $\delta_{ki}(\xi_i[B_1\ldots,B_k])$ is a *validity form* at level-i for the predicate $\delta_{ki}(\xi_i[B_1\ldots B_k])$.

The *validity form* of the expansion of a delegation claim reflects at the language level the *logical consequence* this expansion must be at the semantic level. With regard to Section 6.6.1, the *validity form* gives a formal basis for allowing the predicate of a *delegation claim* to be substituted with its *expansion* in an argument.

A comment similar to the one made in Section (8.10) applies to Theorem 8.9. Contrary to what might *a priori* be expected, the use of a single language is not required to express all formulae of a level-i, i = 1…4; there may be as many distinct languages as there are *expansion structures* $U_i(\delta(\xi_i))$.

8.11 The Tree of Sub-structures Dependencies

It is essential and useful to pay attention to the graph which spans all the *substructures* created at the four levels of the expansion of the claims **CLM0**. Each node of this graph corresponds to a *substructure*, with an oriented edge from node *i* to node *j* if *j* is a *substructure* of *i*. As in Figure 8.3, an edge can be of two kinds, depending on whether it stands for a *substructure* or for an *elementary substructure* coupling. Figure 9.9, page 242 gives an example of such a oriented graph.

The following theorem proves that the graph is connected and has no cycle and is a tree.

Theorem 8.10 The oriented graph of *substructure* dependencies is connected and acyclic.

Proof. The graph is connected since, except for $U_1(\mathbf{CLM0})$, at every level, every *expansion* or *justification structure* is a *substructure* of another *structure* (Theorem 8.7). Because of clause (iii) of Theorem 8.7, every edge that crosses levels corresponds to a *substructure* edge from a lower level *expansion structure* to an upper level *justification structure*; hence, these crossing edges are all oriented from a lower to an upper level and no cycle can exist *between* levels. Moreover, all paths start at an *expansion structure* node and end at a *justification structure* node. Moreover, there is no cycle connecting *justification structure* nodes within a level since the *substructure/extension* relation is not commutative.∎

The tree is an instructive model of the global system design and justification; its properties provide guidance on the way and the order to follow in the elaboration of the different submodels and in the construction of the arguments of the justification. All properties holding true in a *substructure* also hold true in the parent *structures* having this *substructure* as an *elementary substructure.* Useful observations can be derived from this property.

The *elementary substructure* relation conveys a rationale to decide which claims and *substructure (interpretation)* to elaborate first. All predicates (relations) of a *substructure* must hold true as predicates (restricted relations) in the parent *extensions* (Definition 8.7); moreover a **L**-well-formed formula is not satisfied in an *interpretation structure* unless it is satisfied in the **L**-*elementary substructure(s)* (Definition 8.8) of this *interpretation*. Hence, it does not make sense to validate a *justification structure* if its *elementary substructure(s)* are not validated first. This principle figures as a construction rule (11.1.6) in Chapter 11.

Consequently, it is sensible to start by expanding and justifying the claims that must be *interpreted* (valuated) in the "most elementary" *substructures, i.e.* in those *substructures* to which point a greatest number of arrows in an oriented graph as shown in Figure 9.9; rather naturally, these *substructures* should be the first to be elaborated.

Thus, to follow this rule, the justification should start by formulating, expanding and justifying claims at level-1 and successively deal with claims of the next lower level. Within each level, the "most elementary" *substructures* should be considered first, *i.e.* those that correspond to end nodes that support the largest oriented spanning subtree of the same level. These nodes always correspond to *justification structures.*

8.12 A General Multi-level Justification Framework

Because we are primarily interested in computer-based systems, we restricted our attention to four specific levels of evidence; the pragmatic justification of this restriction was explained in the first part of this book. A generalisation of the analysis and results of this Chapter 8 to an arbitrary number of levels, substructures and associated languages is, however, instructive and rather straightforward.

Such a generalized analysis would yield the conditions under which a system property ξ_0 can be formally established as the *logical consequence* of properties satisfied at different levels of a hierarchy of nested, increasingly detailed *substructures* and language *interpretations.*

The possibility of an arbitrary number of levels immediately raises two related questions: (i) how to identify the properties that have to be predicated and justified at each level, and (ii) how many levels and how refined ?

In relation to the first question, it might be relevant to remember that many natural and artificial systems are almost *de facto* hierarchically organized [94], and that engineered systems and in particular computer systems [23] are by necessity often specified and designed with stepwise and refinement methods. The design then produces hierarchical system structures that are highly modular and exhibit dynamic time behaviours that often enjoy properties of *decomposability* [12]. These structural and behavioural properties suggest that these designs would naturally comply with the multi-level structures and interpretations studied in this chapter. The modelling and demonstration approach elaborated here should therefore be applicable to a rather wide class of systems and open potentially useful directions for further research.

Concerning the "how many levels, and how refined?" question, the theorems of Section 8.9 throw some light on this issue. More levels may offer more simplicity at each level since less distinct properties have to be justified at each level, which means a small number of *expansion structures and justification substructures* per level. But this simplicity has a price. Each property must be expanded and valuated in a *structure* which must comprehend as *substructures* all the upper level *justification substructures* in which the property is predicated; and each *justification substructure* must be *substructure* of the *expansion structures* they predicate. The more levels there are, the fewer properties per level, and the more *substructures* of distinct upper and lower levels may need to be involved and nested in the *structures* of each level. Hence, the trade-off uncovered by the theorems of Section 8.9 lies between the simplicity of the level internal *structure* and the complexity of the *substructure* connections between levels; "horizontal" simplicity at each level is obtained at the cost of a "vertical" complexity of couplings between levels.

However, whatever number of refined levels are considered, the four levels empirically introduced in Section 5.7 remain the prime constituents of the dependability justification of any computer-based system; they are necessary and sufficient for all practical purposes, simply because they correspond to the backbone of the design structure of these systems.

8.13 Design Abstractions

E.W. Dijkstra was the first, in 1968, to introduce the concept of "level of abstraction" in software design. Similar ideas had been independently developed in civil architecture [1]. Dijkstra showed how to use this concept to give a hierarchical structure to the specifications, the software and the documentation of the THE computer operating system, a system designed as a family of concurrent processes sharing common resources [23]. He considers an ordered sequence of machines M_0, M_1,..., M_l, ..., where level M_0 is the given hardware and where the software of level l, $l = 0,1,...$, defined and executed by machine M_l, transforms machine M_l into machine M_{l+1} [24]. "Abstraction is one thing that represents several real things equally well" is a definition attributed to Dijkstra in [81]. In the following years, the concept proved useful in the design of complex software for various applications but never received a true formal definition.

The bundle of *expansion* and *justification structures* introduced by Theorem 8.8 as the model of every level of the justification, and discussed in Section 8.9.3, may be regarded as a contribution of model theory to the definition. The predicate of a *delegation claim* asserts at a certain level a property which is detailed, implemented and possibly further expanded at lower levels; such a predicate can be viewed as being analogous to a procedure call at a level of a hierarchical design which calls upon and uses a service or a resource provided by a lower level while ignoring its detailed design and implementation.

If we pursue the analogy further, the results of this chapter suggest that the concept of level of abstraction has the following properties.

Contrary to what one might expect and has often been suggested, the design at a given level of abstraction cannot be described by a single model structure. A level of abstraction is defined by the set of properties predicated by its upper levels; the level must have one *expansion structure* (model) for every property and these *structures* (models) can differ from each other. Each *expansion structure* of the set embodies – as a set of *elementary substructures* – the description of the implementation (the analogue of the *justification*) of the predicated property. Every *expansion structure* of the set must also embody as *substructures* (submodels) the descriptions of the different upper level implementations (*justifications*) which predicate the property.

Hence, a level of abstraction appears to be defined as a compound set of properties with an expansion model for each property; the level relates to its upper levels by having every expansion model incorporating every upper level implementation submodel which predicates at least one of its property; the level relates to the lower levels by its implementation models that are submodels of the expansion models of the properties they predicate.

Thus, models of different levels are nested in each other; rather than being independent they are intimately coupled, contrary to what the mere viewpoint of a hierarchy of abstract machines M_0, M_1… might lead one to think. This machine hierarchy is a bottom-up view only, each machine being defined by its role to "redesign" the resources offered by the lower levels into an abstract and more user-friendly machine (see *e.g.* [12] for details). What is missing in this perspective is the top down definition of the functions expected from every level by the levels above. The approach based on model theory includes this definition.

Once more, these correspondences between the concepts of design abstraction and *expansion/justification structure* show how tenuous and easy it is to cross the border between good design and justifiability; as it should indeed be.

8.14 Recommendations for Design and Validation Models

From a pragmatic and engineering standpoint, what lessons on descriptions and models should we keep in mind from the theoretical interlude of this chapter?

Essentially that the dependability justification of a computer based system is a logical analysis based on different levels – four levels in the case of computer based systems – of nested *expansion* and *justification structures*, on their *substructure* inter-relations and their *interpretations*. *Substructures* are inter-related because a *subclaim* formula predicated and interpreted in some justification structure at some level is expanded, justified and valuated in an *expansion* and a *justification structure* at a lower level.

Each level i is described by the domains, relations and constants of a set of *expansion structures*, one for each *delegation* claim predicated on that level-i. *Justification structures* at level-i are *elementary substructures* of these *expansion structures*. Upper level *justification structures* that predicate delegation claims on level-i are *substructures* of these *expansion structures*. Lower level *expansion structures* of *delegation* claims are *extensions* of the *justification structures* that predicate these delegation claims. Thus, the description of a level and the couplings between levels are more complex and compound than the model of a single abstract or virtual machine per level may suggest.

More precisely, the justification obligations of an initial *dependability requirement* imply that these nested *substructures* preserve inter-dependencies between themselves and have the following properties:

- Every claim must hold true in a unique *justification structure* which is an *elementary substructure* (Definition 8.8) of the *expansion structure(s)* at the same level at which the claim is predicated;
- Evidence for a *white* claim is provided by the valuation of its *justification structure*;
- A *grey* claim is the *logical consequence* of *evidence* and of *delegation* subclaims to lower levels; the *evidence* is provided by the valuation of the claim *justification structure* in which the *logical consequence* must also hold true;
- A *black* claim is a *logical consequence* of *delegation* subclaims to lower levels; this logical consequence must hold true in the claim *justification structure;*
- A *delegation* subclaim predicated by - and interpreted in - a *grey* or *black* claim *justification structure,* holds true in an *expansion structure* at some lower level; the *justification structure* must be *substructure* (Definition 8.7) of the *expansion structure*;
- Each level is entirely described by *a set of expansion structures* (Theorem 8.8), one for each *delegation claim* predicated on that level by upper levels;
- The complete set of *expansion structures* which valuates all the *delegation* claims predicated on a level comprehends all the universes, relations and constants of this level;
- The domain and the constants of a *structure* include the domains and the constants of its *substructures*;
- The relations of a *substructure* are relations of its *extension* (Definition 8.7) which are restricted to its domain;
- The free variables of a delegation claim are assigned values from the union of the *domains* of the lower level grey or black *justification structures* that are *substructures* of its *expansion structure.*

The properties listed here above – despite their theoretical flavour – provide guidance for the design of a dependable system. They indicate how to specify and structure the design of a system so as to facilitate the justification of properties of its behaviour, in particular (claims on) its dependability.

For certification, licensing and other forms of *system validation*, however, the conclusion is so far less promising. *Validation* is the demonstration that system specifications are *satisfied, i.e.* that (*i*) they hold true in a model *interpretation* of the real world and (*ii*) that this *interpretation* is *valid.* Using functional relation predicates over finite or even infinite sets, one can in principle *specify any possible* acceptable behaviour of a system. However, correctness and completeness of these *specifications* can only be claimed as properties of the *interpretation* and not of the *real* system and its environment. Thus, this *interpretation* should also be proven valid, *i.e.* having a domain which is complete and relations that are true to the real world; but this is a different matter, a difficulty of an other order, in practice rarely feasible.

First, to be shown *valid*, the *interpretation* would have to include the relations and the domains of *all* relevant *substructures*, *i.e.* to be a description of the entire system behaviour (including, for instance, the detailed system-environment interface as well as the hardware behaviour). Any model which dispenses with some of the descriptions and remains based on less refined lower level *substructures* is necessarily truncated, only partially defined and thus necessarily involves some degree of indeterminacy and unverifiability.

Secondly, the *substructures* themselves must be somehow justified as acceptable representations of the perception one has of the real world; and not only of the rational and self-evident knowledge of its causal relations, but also of the right intuition and insight of its relevant underlying meaning [3].

In this sense, the validation of a system appears to be a much harder process than its *specification* or the *verification* of its properties. This difficulty should not be a real surprise to practitioners. It is the reason for the importance of the concepts of *plausibility* and *consensus* discussed in Chapter 7. But the analysis of this chapter at least provides some clarification. It shows how the justification of a system requirement is relative to a set of precisely defined *interpretations,* and how to organize these *interpretations* regardless of how disparate the domains and the sources of evidence might be. Besides, as we will see in the next chapter (Section 9.4.8), validation may be eased in certain cases when a single *interpretation* is used for the two sides of a real interface, for instance between the system and its environment.

This next chapter analyses in more detail a concrete implementation of *interpretations* for a large class of computer embedded systems.

Embedded Computer System Structures

The purpose of this chapter is to use the model structures analysed in Chapter 8 to develop concrete models that are specific to embedded computer systems. Our approach is based on the functional relations developed by D.L. Parnas[10] in [78, 79, 103]) for requirement specification. His functional relational models describe control systems, but have also been used for information systems [40]. We show in this chapter how this approach is also particularly well suited to the justification of dependability and fits well in our framework. We develop four sets of models which correspond to the levels of decomposition of the initial *dependability requirements*. These models include descriptions of the assumptions and constraints imposed on the embedded computer system at each level, *i.e.* those imposed:

- by the plant environment,
- by the computer and other existing equipment architectures,
- by the hardware and software design and technology (in particular any pre-existing hardware and software components),
- and by the operational controls and procedures in place (e.g. operator controls, maintenance, periodic tests).

These models are not specific to a particular computer system. Intended to be generic, they are not elaborated here to the degree of detail that would be required

---

[10] At the end of the eighties, D.L. Parnas published a set of seminal papers which demonstrated the great power of functional relations to specify the requirements of computer systems and their programs, and to describe their behaviour. David Parnas said to have been influenced by the work of H.D. Mills (see *e.g.* [69]) and by fruitful discussions he had at the time with the mathematician N.G. de Bruijn [18] from Eindhoven University. The previous chapter and this one throw light on the reasons why functional relations – that David Parnas wisely selected to use 20 years ago – work well for modelling computer requirements and behaviour.

in an industrial *dependability case*. However, they comprehend the essential aspects of the dependability justification and precisely define – by means of about 40 distinct embedded model structures – what has to be demonstrated at each level, what evidence is required, and how this evidence should be organized and documented.

9.1 States, Events and Other Basic Notions

The different model structures we shall deal with are the (not necessarily finite) state machines already introduced in Section 8.2.1, page 76. The universe of these machines consists of *domains* and *ranges*, (not necessarily finite) of state variable values, and of one *domain* (not necessarily finite) of time values. The values of state variables are functions of time. The behaviour of the machine is an ordinal sequence or a time sequence of states. *Relations* over the domains and ranges of values of state variables describe the possible behaviours of the machine.

Different types of state attributes may need to be considered: safe, fail-safe, fail-stop, silent-fail, and/or their opposite unsafe attributes.

It is also useful to identify *modes*. *Modes* are *equivalence classes of state values*. They are intended to make the required description of the environment and the system behaviour simpler. Examples of operational *modes* of a nuclear reactor (which are also the modes of the operation of the computer based control and protection system) are: cold shutdown, hot shutdown, hot standby, start up, power operation *modes*. A *mode* can be subdivided into *submodes*. When in a given mode, examples of *submodes* in which a system might be are: the initialisation, operating, failed, testing, or maintenance *submodes*. *Modes* and *submodes* must be defined so that the system is in one *mode* and one *submode* only at any given time.

We shall also be dealing with *events*. Events are instants in time at which state transitions occur. A "before state value, after state value" pair characterizes an event. If one knows the state transition function, an event can also be characterized by a state value and a time instant. A *class* of events is characterized by a *set* of "before state value, after state value" pairs. One is usually interested in *event classes* for which the state transition function changes value in a specific way.

Not all events are treated in the same way in a dependability justification. It is important to precisely characterize their different types, roles and significance. Standards and guidance on dependability often fail to clearly make these necessary distinctions.

Events can be external or internal with respect to the embedded computer system.

Definition 9.1 *External events* are those which are generated by the environment *and* may affect the system behaviour or need to be controlled by the system.

The detection, monitoring, control or mitigation of *all external* events and their effects is a primary – sometimes the only – function of a protection or a safety system which is specified by the level-0 *dependability requirements*. To be valid,

Table 9.1. Event Classes

	Level-1 External Events	Levels-2, 3 and 4 Internal Events
Operational Modes	PIE's	Regular Events
Failure Modes	PIE's	UE's

the *dependability requirements* must, by definition, anticipate *all* these *external events*, including failures and accidents that may occur in the environment, with their effects, desirable or not. Classical terminology refers to these *external* events as *design based events* or *postulated initiating events (PIE's)*. We shall here adopt the latter:

Definition 9.2 The *postulated initiating events (PIE's)* are the *external* events the system is expected to detect and to control.

Events that are not external are *internal events*:

Definition 9.3 All *events* generated by the system at the *architecture, design* or *operation control* levels are *internal events*.

The *operation control* level is included in Definition 9.3 since the control channel is considered as an integral part of the system. Typical *internal events* are computer, I/O devices, clock or watchdog interrupts, alarms, operator manual actions, *etc.*

Not only the normal behaviour, but also hardware, software and operation control *errors*[11] need to be anticipated by the design of a safety critical system. Software, hardware (processor, memory, input/output devices) and operation control *failures*[11] are also *internal events* that need to be considered at the four levels.

Internal events corresponding to *failures* will be coined *undesired events* because they interfere with the system functions required by the level-0 *dependability requirements*. We prefer this terminology already introduced in [40] to the term "hazards" which may take on different meanings and in particular lead to some confusion between an event and its effects. *Undesired events* may therefore occur at any of the three lower levels which were introduced in Section 5.7. At the control level they affect the state of the control channel; human failures or deficiencies in control and maintenance procedures may result in the sending of illegal or wrong control commands, in failures to issue commands in time, *etc.*

---

[11] We use the classical terminology; an *error* is the manifestation of a *defect* (also called *fault*). A *failure* is the consequence of an *error* and a deviation of the service delivered by the system to the user from the intended one [48, 57, 88].

Thus:

Definition 9.4 *Undesired events* – abbreviated *UE's* – are *internal events* occuring at the architecture, design and operation control levels, the effects of which interfere with the system functions required by the initial *dependability requirements*.

These definitions are summarized in Table 9.1 where *internal events* not *undesired* are called *regular*.

9.2 Notation

For *claims* and *delegation claims*, we shall go back to the notation of Section 6 (see Figure 6.4) rather than keep that of Chapter 7, because the semantics of the claims will matter and not only – as in Chapter 7 – their level and their white, black or grey type.

Besides, it is also convenient to adopt a notation which allows us to shorten the description of *expansions* when they are, in part or in total, common to several *delegation claims*. In such cases, we abbreviate the description, by using a standard Backus–Naur Form (BNF) like notation "|". Suppose $\xi 1$, $\xi 2$... ξk, denote the indexed predicates of *delegation claims* made at level-i, and ε, κ, ϕ, π...the formulae of the claims in which they are expanded at level-j, $(i < j)$. Then, using the notation introduced in Section 6.5.1, we shall freely use the following condensed notation for their Γ expansion at level-j:

$$(9.1) \qquad \xi 1.\text{clm}[i,j] \mid \xi 2.\text{clm}[i,j] \mid \dots \xi n.\text{clm}[i,j]$$

$$U(\xi) = \left| \begin{array}{l} \varepsilon 1.\text{clm}[j] \mid \varepsilon 2.\text{clm}[j] \mid \dots \varepsilon n.\text{clm}[j] \\ \wedge\, (\kappa.\text{clm}[j] \wedge \eta.\text{clm}[j] \wedge \dots) \\ \wedge\, (\text{nil} \mid \phi.\text{clm}[j] \mid \pi.\text{clm}[j] \mid \dots). \end{array} \right.$$

It is must be understood in (9.1) that each *delegation claim* $\delta\xi k.\text{clm}[i,j]$, $(k = 1\dots n)$, is expanded at level-j into:

$$(9.2) \qquad \xi k.\text{clm}[i,j]$$

$$U(\xi k) = \left| \begin{array}{l} \varepsilon k.\text{clm}[j] \\ \wedge\, (\kappa.\text{clm}[j] \wedge \eta.\text{clm}[j] \wedge \dots) \\ \wedge\, (\text{nil} \mid \phi.\text{clm}[j] \mid \pi.\text{clm}[j] \mid \dots), \end{array} \right.$$

where the last term of the conjunction is either empty (nil) or a claim with one of the formulae ϕ, π, ... Note that this notation does not contradict the conjunctive property introduced in 6.4, nor the argument unicity property (Section 6.5.3): (9.1) and (9.2) are generic notations and one option only in (9.1) and (9.2) is assumed to apply to every *delegation claim* of a specific system.

Finally, a *claim* or a *delegation claim* to which a set $\mathcal{A}$ of constants is associated is denoted: $\mathcal{A}\_\,\boldsymbol{\phi}.\mathbf{clm[i]}$ or $\mathcal{A}\_\,\boldsymbol{\xi}.\mathbf{clm[i,j]}$.

9.3 Level-0. Environment Requirements, Events and Constraints

As stated in Chapter 5, a dependability justification is grounded on the initial specification of two basic constituents:

- *what* has to be achieved in the environment,
- *what* the system is expected to do.

The first specification is independent of whether a system exists or not and belongs to level-0. The latter is a preliminary specification of the system-to-be; although it belongs to level-0, its pivotal and interlinking roles with the other levels make it more convenient to postpone its discussion to Section 9.4.6.

9.3.1 CLM0 Dependability Requirements

The specification of "what to achieve" is the role of application, plant and/or safety experts who must establish and validate what we collectively called in Chapter 5 the *initial dependability requirements* (**CLM0**). In the theory developed in Chapter 8, the *dependability requirements* are formalised by the well-formed formulae ξ_0. These application-specific requirements express the properties and the relations that must be maintained in the environment (*environmental relations*), independently of whether a system exists or not. *Dependability requirements* are essentially of two types (Section 5.2):

Functional dependability requirements specify – among other functions – the detection and control of *PIE's* (Definition 9.2), the protection and prevention functions, safety and security actions, inventories, alarm generation, *etc.*

Non-functional dependability requirements specify the properties that the implementation of these functions must satisfy; accuracy, stability, testability, maintainability are typical attributes that may be required. Targets on properties of the system behaviour within its environment and during operation such as timeliness, reliability, and availability are also possible requirements.

Dependability requirements may therefore be expressed as deterministic or probabilistic assertions. They may at first be expressed as goals and objectives in a natural, informal or application-oriented language. Nevertheless, they constitute, as we shall see in Section 9.4, the reference against which system specifications are to be validated at level-1. To anticipate and facilitate this validation, it is advisable to formalize the *dependability requirements* – as early and to the extent possible – in a language that can be *interpreted* in terms of constraints and relations between state variables and properties of the domains and ranges of these variables. This is an important point to which we shall come back later (in Section 9.4.7, in particular).

9.3.2 CLM0 Postulated Initiating Events and Safe Failure Modes

All *postulated initiating events* (Definition 9.2) that can be generated by the environment are part of the level-0 universe; they must be described with their consequences and effects. By definition, these events and only those are the events that have to be dealt with by the system and are assumed to affect its behaviour. The monitoring, control and mitigation of these events are primary – and often the only – functions of a protection or safety system.

Hazards, damages and failures that may occur in the environment and affect the system are included in these *postulated initiating events*; they may be generated by the failure of a motor, by a pipe break, the malfunctioning of a valve, a loss of power, or an electromagnetic or seismic interference. They may also result from conditions in the environment that are uncertain, unsafe, hard to predict or detect.

Associated with these failure PIE's, of utmost importance is the specification of *safe or preferred states* in the environment. The system should indeed be able to deal with the PIE'S as well as with the *internal undesired events* that may occur in the system. The **CLM0** requirement specifications should therefore include the precise description of the environment safe states.

9.3.3 CLM0 Environment Constraints

The system environment – and especially previously installed systems – impose constraints, restrictions and invariant relations on the values of environmental variables and on their inter-relations. Examples are:

- ranges of physical parameter values;
- relations imposed by mass, energy or flow conservation laws;
- constraints imposed by the behaviour of pre-existing hardware or software components.

For example, a typical constraint of nuclear material handling systems (Appendix B) is the need to respect conditions on mass storage and locations determined by considerations on mass nuclear criticality or by limitations on radiation release.

9.3.4 Level-0 Dependability Case Documentation

At this stage we can sum up what the initial level of the documentation of a *dependability case* should address. It should contain all the material necessary to describe the validated initial *dependability requirements* discussed above, as well as to allow the specification of a system able to deal efficiently with them.

More precisely, the level-0 documentation should at least include:

- The definition of process parameters, pre-existing equipment states, sensing and actuating devices to be monitored and controlled, with a definition of their accessible inputs and outputs;

- Parameter ranges, physical and pre-existing equipment environmental constraints;
- The external Postulated Initiating Events (PIE's), their effects and the safe environment states;
- The validated dependability *requirements* specified in terms of the above. To the extent possible, they should be amenable to an interpretation in terms of relations to be maintained between state variables, and in terms of properties of the domains and ranges of these variables.

This documentation should be available to:

- Plant/application safety experts to verify that the validated *dependability requirements* are correctly specified;
- Instrumentation engineers and computer system designers for checking intelligibility and feasibility;
- Licensors and regulators for compliance with safety requirements and regulations.

9.4 Level - 1. System-Environment Interface

To adopt a provocative tone, one might say that nothing is more central when evaluating the dependability of a system than its boundary. An adequate model must not only describe the system; it must also precisely define its interface with its environment and capture what is relevant on both sides.

We know from Chapter 7 that every initial *dependability requirement* **CLM0** must be expanded into *primary claims,* and be a logical consequence of these *primary claims.* The dependability requirement and the corresponding primary claims must be interpreted and valuated in an *expansion structure* in which the logical consequence must hold true. The *primary claims* must hold true in *justification structures* which are elementary substructures of these expansion structures. In order to define the universes and relations of these *structures*, we first examine, in this and the following sections, the different elements which constitute the system-environment interface.

9.4.1 Selection of the Entities and Relations of the Interface

As far as the environment side of this interface is concerned, the *universes* of these *structures* must include the complete set of *postulated initiating events* that can be generated by the environment, not only in normal conditions, but also in hazard, failure and accident conditions. In the example of the material-handling machine described in Appendix B, the interface includes a set of *postulated initiating events*, a set of machine components, data links, and power supplies. The relations between these elements are the mechanical relations and constraints, the real-time constraints, and the effects the events may have on components.

As a matter of fact, the functions expected from the system – including the safety functions – determine the environment characteristics that need to be taken into account and be part of the interface. For instance, a given nuclear safety system function may rely on the functioning of some power supply, and not on the coolant supply; a monitoring function may depend on the ambient radiation level, and not on the ambient temperature, *etc*. Such system considerations define the entities and the aspects of the environment that must be selected for inclusion in the *universes* of the *expansion and justification structures* describing the system-environment interface. The choice may at first rest in part on the competence, experience and creative art of modelling of experts, but must eventually be validated against the **CLM0** requirements. One of the purposes of the following sections is to show how this choice can be validated and what this validation requires.

9.4.2 Environment States

The first and essential model which is needed is a description of the time behaviour of the system environment interface. This behaviour can described by the sequence of values taken by state variables in function of time. The state variables must be a true model of the interface; their values must not only describe what happens in the environment but also how the system acts upon the environment. The mission of the system is indeed to maintain in the environment the specific relations defined by the *dependability requirements*.

Thus, following the model introduced in [79], it is useful to characterize each environmental state variable as being either *monitored* or *controlled; monitored* variables are those that need to be measured by the system; *controlled* variables are those that the system is intended to control. These variables have values that can be a function of time. We denote m(t) and c(t) the vectors of the values at time *t* of the monitored and the controlled variables respectively (see Figure 9.1). A same equipment or process parameter of the environment may in certain cases be both monitored and controlled by the system (closed loop). A pair of variables $m_i(t)$, $c_i(t)$ is in this case associated with the sensing and the actuating device of this equipment or process parameter.

Subsets of controlled values that correspond to fail-safe states of the environment with specific dependability characteristics (fail-safeness, silent-fail, or other specific safety property) are of particular interest. We shall generically denote by FS(t) the subset of such states:

Definition 9.5 FS(t) is the subset of c(t) values which correspond to safe states of the environment.

For the renovation or the replacement of a system, many monitored and controlled variables will be usually pre-determined, many of them being the same as those of the system to upgrade or replace.

9.4.3 Constraints (NAT)

The environment imposes constraints, restrictions and invariants on the values of environmental variables and on their relations. These constraints discussed in Section 9.3.3 are modelled as a binary relation between the vectors m(t) and c(t), which is called NAT in [78].

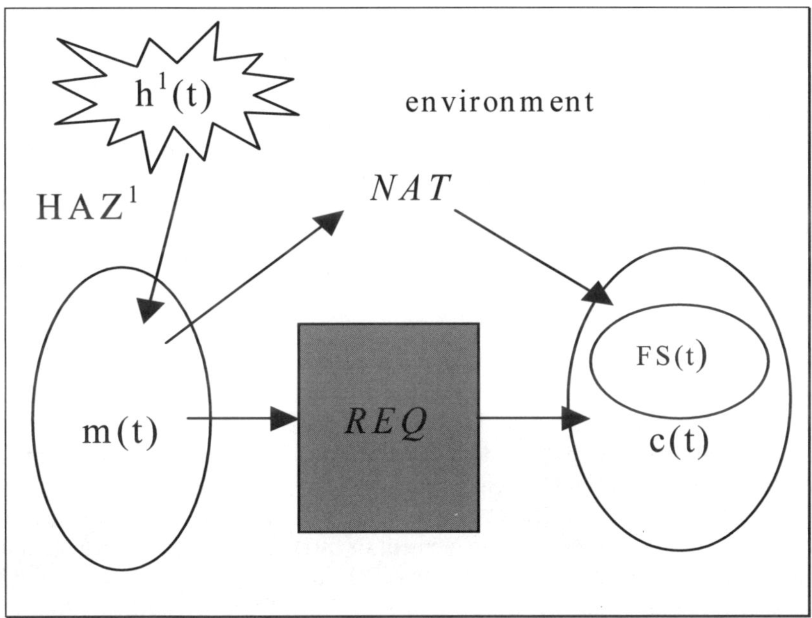

Figure 9.1. Substructure of the environment-system interface

Definition 9.6 The relation NAT is defined as follows (*cf.* Section 8.2.1):

- *Domain* (NAT) is a set of vector-valued time-functions containing all possible vector values of m(t);
- *Range* (NAT) is the set of vector valued time functions containing possible vector values of c(t);
- The *ordered pair* {m(t), c(t)} is in NAT if and only if the environment constraints allow the controlled variables to take the values c(t) when the values of the monitored variables are described by m(t).

The set of values of the *universe* of NAT and the NAT relation entirely define the environment as perceived by the system. If there is no constraint on m(t) and c(t), all *ordered pairs* of values from the *range* and the *domain* of NAT are included in NAT.

9.4.4 A Note on Environment Assumptions

The NAT *universe* must be complete. If any values of m(t) or c(t) are not included in the *universe* of NAT, it is *assumed* that these values can never occur naturally or physically in the environment. This sort of implicit, *ex absentia,* specification deserves some attention. Previous chapters stressed the need for explicitly stating assumptions. An implicit specification may be misleading in dependability assessment, especially if the environmental states left implicit are unfamiliar or are specific to another system. Examples are controlled values that are thought physically possible but cannot actually be reached because of the behaviour of a pre-existing system; or monitored values thought abnormal or impossible but actually produced by other systems. The values of such states should not be "forgotten" and absent from the *universe* of NAT so that their reachability can be "explicitly" either excluded or included by the NAT relation.

The state space of the NAT *universe* may be very large. If necessary, its set of values can be structured, for instance to reflect the fact that the environment may be in one of several *modes*, the current mode being specified by the value of one of the m(t) variables.

9.4.5 Postulated Initiating Events (Relation HAZ^1)

The **CLM0** *postulated initiating events* (*PIE's*) generated by the environment have been defined (Definition 9.2) and described in Section 9.3.2. They can be modelled as part of the system-environment universes as follows.

Let $h^1(t)$ be the vector describing the states of entities in the environment that are potential sources of *PIE's*. An element of $h^1(t)$ corresponds to such an entity; when not zero, its value is a constant that indicates (i) that a *PIE* occurred at the corresponding entity and (ii) the *type* of this *PIE*. Thus, we have the following definition:

Definition 9.7 The occurrence of a *PIE* in the time semi-open interval](t - dt), t] is defined by "states before" and "states after" the event; *i.e.* by the pair $]h^1(t - dt), h^1(t)]$. In most cases we are interested in the state resulting from the *PIE* rather than in the *PIE* transition. A *PIE* that occurred at time $t = \tau$ is then described by the pair of values $(h^1(\tau),\tau)$ where $h^1(\tau)$, the value of $h^1(t)$, is the *type* of the event occurring at time instant τ. By extension, we shall denote $(x(\tau),\tau)$ the *PIE* which corresponds to a variable x(t) taking a value $x(\tau)$ at time instant τ.

Note that simultaneous *PIE's* may occur as the result of more than one entity changing state at the same time instant.

Postulated initiating events occurring in the environment must be observable by the system. Thus, there must be a relation between these events and the variables monitored by the system.

Definition 9.8 We call HAZ^1 this binary relation:

- *Domain* (HAZ^1) is the set of vector-valued time-functions containing all possible values of the vector $h^1(t)$;

- *Range* (HAZ^1) is the set of vector valued time functions containing possible values of m(t);
- $\{h^1(t), m(t)\}$ is in HAZ^1 if and only if (m(t), t) is an event occurring whenever $(h^1(t),t)$ occurs.

As indicated in Section 9.1 and Table 9.1, there is a class of *postulated initiating events* that corresponds to the occurrences of failures or accidents in the environment. Such events need special treatment; in particular, their effects may have to be mitigated and they should lead the environment to safe states of the set FS(t). It will be convenient to characterize this class of PIE's by the following subset of $h^1(t)$ values:

Definition 9.9 F^1 denotes the subset of $h^1(t)$ values which correspond to the class of *postulated initiating events* that must result in the environment being in a safe sate of the set FS(t).

9.4.6 System Input-Output Preliminary Specifications

Together with the **CLM0** *dependability requirements*, the other foundation of the justification (*cf.* Section 5.3) is the specification of "what the system is expected to do"; that is, the preliminary specification of the relations that the system – viewed as a black box – is expected to maintain in the environment in order to satisfy these requirements, to deal with the effects (HAZ^1) of the postulated initiating events and comply with the environment constraints NAT. These *system input-output specifications* were qualified "preliminary" in Chapter 5 because at this stage of the justification they are not yet validated; their validation against the *dependability requirements* will be the first step of the justification.

The *dependability requirements* (**CLM0**) have to be refined (reified), and mapped onto functions and properties at the different levels of the architecture, design and operation control. At this early stage, these refinements – for example those implementing properties such as reliability, availability, and fail-safeness – cannot yet be specified, and need not be (*cf.* Chapter 5). The role of the computer system that must be assumed and described at this first level is that of creating and maintaining new relations between the environmental variables, in addition to – and compatible with – those of NAT and HAZ^1. We will see in the next section how these *system input-output specifications* can be expressed in terms of:

- A relation between the monitored and controlled variables m(t) and c(t),
- Properties on the domain and range of this relation,

and how this relation, domain, and range can be validated, *i.e.* shown to *satisfy* the *dependability requirements* (**CLM0**).

9.4.7 The System and the Functional Relation (REQ)

Taking the same approach as in [78], we are now in a position to give a precise definition of what we mean by *system*. The boundaries and the behaviour of the *system* are specified by a binary relation between the monitored and the controlled variables, called REQ:

Definition 9.10 REQ is defined as follows:

- *Domain* (REQ) is a set of vector-valued time-functions containing all possible values of m(t);
- *Range* (REQ) is a set of vector-valued time-functions containing the permissible values of c(t);
- {m(t), c(t)} is in REQ if and only if the computer system permits the controlled variables to take the values described by c(t) when the values of the monitored variables are described by m(t).

The functional relation REQ specifies the preliminary *input-output system specifications* that must be validated against the functional *dependability requirements*. REQ is intended to describe the relations between monitored and controlled variables that correspond to the *perfect* behaviour of the system: the values of the monitored and controlled variables are assumed to be perfect values, and REQ describes the correct – and not the possible erroneous – system behaviour that must occur at the architecture, design and operation control levels.

Remark 9.1 REQ defines not only the functions of the system; the REQ domain and range define also the *boundaries* of the system. Any variable outside this universe is not accessible to the *system*, nor to the operation control at level-4. This clear definition of the system boundaries is essential when dependability is to be justified, especially when dependability arguments involve the existence of back-up or diverse systems. Back-ups or functional diversity that may be needed to implement the REQ relations and justify the *system*, must exist within these boundaries, and not outside.

9.4.8 The Need for a Unique *Interpretation* for Requirements and System Specifications

In relation to what was said in Section 9.3.1, one can now realise that REQ is also an appropriate model of the environmental relations imposed by the functional *dependability requirements* **CLM0**. REQ can indeed be used as one single description of the *system* relations imposed by these functional *dependability requirements* and those defined by the *system specifications*. This is somehow a natural consequence of the way these requirements and specifications are obtained. Although issued from two different viewpoints and by different experts, dependability requirements and system specifications are consubstantial. The former provide the viewpoint of the application/environment and the latter of the system to be designed; both, however, have the same objective: to define relations that must exist at the environment-system interface. The two viewpoints are not superfluous. They are necessary: a correct elicitation of REQ requires a complex, intricate and fructuous alchemy of expert competences from both sides.

There is therefore great practical interest to have the same relation REQ used as the language interpretation (L_1-*interpretation*) for both the *dependability requirements* and the *system input-output preliminary specifications*. The languages used to formulate these requirements and specifications need not be strictly the same. In particular, one may be an *expansion* of the other (Section 8.10). But they should admit the same *structure* for *interpretation*. The validation of the *system input-output specifications* against the *dependability requirements* then amounts to an acceptance of this only one REQ relation and *structure* by all protagonists involved. The acceptance becomes more akin to a process of verification, alleviating somewhat the difficulties of validation discussed at the end of the previous chapter (Section 8.14). This acceptance process is the first step of the justification and is formalised by the elicitation of the *primary claims* that will be detailed in Section 9.4.10. A single *interpretation*, general and simple enough to be understandable by all parties involved will appear to be a great asset and a source of solid evidence to support these claims.

The {m(t), c(t), REQ} functional model, at the same time simple and powerful, was an inspired and enlightening stroke of David Parnas.

9.4.9 Expansion of the CLM0 Requirements into Primary Claims

Having set the scene, we now consider the justification proper.

The level-0 *dependability requirements* **CLM0** must be *expanded* into level-1 *primary claims*, which claim *properties* of the system-environment interface (cf. Sections 5.7 and 8.8.3). The specific dependability properties that must be claimed obviously depend on the application. So, the expansion at level-1 of a *dependability requirement* **CLM0** into *primary claims* will differ from one dependability case to another. However, as explained in Section 6.6.1, expansions always lead to three principal kinds of claims: *validity, dependability, and implementation*.

In a first step and in accordance with Lemma 8.8, we start by showing:

- How these different types of *primary claims* can be expressed as formulae that can be interpreted and valuated by *justification substructures*;
- How the valuations of the *justification substructures* provide the evidence that validates the *primary white* claims; and how the *black primary* claims are logical consequences of *delegation* subclaims that must hold true in these *justification substructures*.

These *justification structures* must be *elementary substructures* of the *expansion structure* in which the initial *dependability requirements* are *logical consequences* of the *primary* claims (Lemma 8.8). The level-1 expansion into *primary claims* is an initial and critical step of a justification. The comment which follows the elicitation Theorem 8.2 page 95 explains that the level-1 *expansion structure* and the *justification structures* are inter-dependent and somehow have to be determined hand in hand. For the clarity of the exposition, however, we found it logical to postpone the definition of the *expansion structure* to a later stage (Section 9.4.13), after the *justification elementary substructures* have been identified. We shall further discuss in Section 9.4.13 this important and delicate aspect of the design of justification. At this stage, we merely postulate the existence of the *expansion structure* of **CLM0**, to which we refer as the L_1-*expansion structure* U_1(**CLM0**), $\mathbf{L_1}$ being the language in which the dependability requirements are expressed.

The reader may refer to Figure 9.2, page 147 for a preview of the expansion of **CLM0** into primary claims. Evidence supporting the primary claims will be discussed in Section 9.4.14.

9.4.10 Primary Validity Claim-Justification Substructures

9.4.10.1 Environment Primary Claims
First of all, one must ensure that the relation NAT (Definition 9.6) adequately captures – *i.e.* is a *model* (Definition 8.4) of – the **CLM0** environmental constraints. That is, the relation NAT must be shown to be *a valid interpretation* of the environment constraints which are documented in the **CLM0** *dependability requirements* (Sections 9.3.3, 9.3.4). We must also ensure that this interpretation provides valid evidence.

Following the definitions and lemmas introduced in Section 8.7 and 8.8.3, let us then consider the following *structure*

$$(9.3) \qquad U_1(\mathrm{NAT}) = < \{m(t)\}\{c(t)\} \; ; \; \{\mathrm{NAT}\} \; ; \; \{k_1\}>,$$

the universe of which includes the sets $\{m(t)\}\{c(t)\}$ and the set $\{k_1\}$ which designates the constants of U_1 relevant to the environment. Valuations of U_1(NAT) provides an *interpretation* of the environmental constraints specified by the *dependability requirements* **CLM0**. This *interpretation* is obtained by the assignment of an element of the universe $\{m(t)\}\{c(t)\}$, a relation of the set NAT and a constant of $\{k_1\}$ to each variable symbol, predicate symbol, function symbol, operation symbol and parameter respectively, of the language $\mathbf{L_1}$ in which the **CLM0** environment constraints are documented.

Given this assignment, the valuations (Definition 8.2) of U_1(NAT) must *satisfy* (Definition 8.3) the **CLM0** environment constraints. If all the **CLM0** environment constraints valuate to true in U_1(NAT), U_1(NAT) *is a valid interpretation (model) for the environmental constraints specified by* **CLM0.**

Moreover, following Lemma 8.7, the evidence provided by the valuations of the *interpretation* U_1(NAT) is valid for the justification of **CLM0** if and only if U_1(NAT) is an *elementary substructure* of the L_1-*expansion structure* U_1(**CLM0**).

We therefore need the *primary claim*:

(9.4)   NAT_**constraints_validity.clm1:**

U_1(NAT) $\vdash$ **CLM0 documented environment constraints** (Section 9.3.3)

with U_1(NAT) $\prec U_1$(**CLM0**)
where the *justification substructure* U_1(NAT) is defined by (9.3).

One must also ensure that the set of environment *postulated initiating events,* defined in (9.4.5) is complete. The potential source causes of these events are described by the vector $h^1(t)$ and by the values the elements of this vector can take. This set of values is the *domain* of the HAZ^1 relations (Definition 8.1). Let us consider the *structure*:

$$(9.5) \qquad U_1(HAZ) = <\{h^1(t)\}\{m(t)\}\{c(t)\} \; ;$$
$$\{NAT\}\{HAZ^1\} \; ;$$
$$\{k_1\}>,$$

where $\{k_1\}$ includes the set of constants relevant to the relations NAT and HAZ^1. The domain $\{h^1(t)\}$ of HAZ^1 must be *complete and consistent, i.e.* must include all events anticipated by **CLM0,** compatible with the NAT constraints, and only those. In addition, the effects that these events have on the monitored variables $\{m(t)\}$ must faithfully describe the effects anticipated by the dependability requirements **CLM0.** That is $U_1(HAZ)$, as defined by (9.5) must be a *model* (Definition (8.4) of the anticipated and documented events and their effects, and an *elementary substructure* of the $\mathbf{L_1}$-*expansion structure* $U_1(\mathbf{CLM0})$. These obligations lead to the validity claim:

$(9.6) \qquad$ $h^1(t)\_\mathbf{and\_effects\_validity.clm1}$:

$\qquad\qquad U_1(HAZ) \; \vdash$ **CLM0 documented PIE'S and effects** (Section 9.3.2)

$\qquad\qquad$ with $U_1(HAZ) \prec U_1(\mathbf{CLM0})$
$\qquad\qquad$ and *justification substructure* $U_1(HAZ)$ defined by (9.5).

The set FS(t) of fail-safe state values is a subset of $\{c(t)\}$ (Definition 9.5, Section 9.4.2). FS(t) must be validated against the fail-safe modes specified by the dependability requirements **CLM0** and the NAT constraints. The system should indeed be able to deal with the *external* PIE'S and, as we shall see later, with the *internal undesired events* that may occur at the architecture, design and operation control levels. For those reasons, the **CLM0** requirements specifications have to include the definition and documentation of safe states (Sections 9.3.2, 9.3.4). Since FS(t) is within the universe of $U_1(HAZ)$ as defined by (9.5), we must have the primary validity claim:

$(9.7) \qquad$ $FS(t)\_\mathbf{validity.clm1}$:

$\qquad\qquad U_1(HAZ) \; \vdash$ **CLM0 documented safe failure modes** (Section 9.3.2)

$\qquad\qquad$ with $U_1(HAZ) \prec U_1(\mathbf{CLM0})$
$\qquad\qquad$ and *justification substructure* $U_1(HAZ)$ defined by (9.5).

The environment *primary claims* (9.4) to (9.7) assert the validity of the *justification substructures* $U_1(NAT)$ and $U_1(HAZ)$ as *models* of the environment constraints, postulated initiating events and safe states implied and documented by the *functional dependability requirements* **CLM0**. The universes and relations of these *structures* are such that the following proposition holds for the two structures being **L$_1$**-*interpretations* of the language **L$_1$** used to describe the **CLM0 documented environment constraints** (Section 9.3.3):

$$(9.8) \qquad U_1(NAT) \subseteq U_1(HAZ)$$

Since they must be *elementary substructures* of $U_1(\textbf{CLM0})$ and because of Lemma 8.5, the following proposition must also hold for the well-formed formulae satisfied by these *justification structures*:

$$(9.9) \qquad U_1(NAT) \prec U_1(HAZ).$$

Propositions (9.8) and (9.9) induce a logical sequencing amongst the corresponding justifications: because a **L**-well-formed formula with all free variables being assigned values from the universe of an *elementary substructure* is not satisfied in an **L**-*elementary extension* unless it is satisfied in the **L**-*elementary substructure* (Definition 8.8, page 89), there is no sense in validating a *justification structure* if all its *elementary substructures* cannot be validated. This ordering of justifications is recommended as the construction rule 11.1.6 in Chapter 11.

9.4.10.2 Completeness, Feasibility and Correctness Primary Claims
Once the validity of these *justification structures* is established, the validity of the *input-output system specifications* need to be explicitly claimed and valuated at level-1. The properties that must be claimed are completeness, feasibility, and correctness; they are properties of level-1 only and not of the lower level *substructures*.

The *input-output system specifications* must be *complete*; they must specify the behaviour of the system *for all cases* likely to arise in the environment. In other terms, the *input-output system specifications* must valuate to true the property:

$$(9.10) \qquad \textit{domain } (REQ) \supseteq \textit{domain } (NAT)$$

of the universe of the *structure* :

$$(9.11) \qquad U_1(REQ) = \,<\{m(t)\}\{c(t)\};\ \{NAT\}\{REQ\}\ ;\ \{k_1\}>$$

where $\{k_1\}$ includes all constants relevant to NAT and REQ.

The *input-output system specifications* must also be *feasible* with respect to the NAT constraints, *i.e.* the system behaviour modelled by REQ must be allowed by NAT. Thus, we must also have:

$$(9.12) \qquad \textit{domain } (REQ) \subseteq \textit{domain } (NAT).$$

The two sentences (9.10), (9.12) imply:

$$domain\ (REQ) \equiv domain\ (NAT).$$

Hence, the following primary claim must be *satisfied* (valuated to true) :

(9.13) REQ_completeness_and_feasibility.clm1:

$$U_1(REQ) \vdash (domain\ (REQ) \equiv domain\ (NAT))$$

with $U_1(REQ) \prec U_1(\mathbf{CLM0})$
and *justification substructure* $U_1(REQ)$ defined by (9.11).

Note that feasibility is here defined with regard to the constraints imposed by the environment. At this level, feasibility does not mean that the *input-output system specifications* involve computable functions or that an implementation is realizable. Such claims belong to and are addressed by the three lower levels.

Let us note that, since REQ is also a model of **CLM0** (Section 9.4.8), this claim (9.13) is in some way a useful "*a posteriori* verification" – involving the *input-output system specifications* – of two properties that should initially be satisfied by the *dependability requirements* **CLM0.** These requirements should indeed be complete and feasible.

Next, the *input-output system preliminary specifications* (Section 9.4.6) must be shown to comply with the perfect *system* behaviour expected by the **CLM0** *functional dependability requirements* introduced in Section 9.3.1. Specifically, the relation REQ must be shown to be a valid *interpretation* of the system input-output specifications as they are required by the documented **CLM0 functional dependability requirements**. These functional requirements must hold true in the L_1-*interpretation* $U_1(REQ)$. Again, this *interpretation* is obtained by an assignment of the elements of the *structure* $U_1(REQ)$ to the constants and variable, predicate and function symbols of L_1. Besides, following Lemma 8.7, the evidence provided by the valuations of $U_1(REQ)$ is valid for the justification of **CLM0** if and only if $U_1(REQ)$ is an *elementary substructure* of the L_1-*expansion structure* $U_1(\mathbf{CLM0})$.

Hence, the primary *claim*:

(9.14) REQ_correctness.clm1:

$$U_1(REQ) \vdash \textbf{CLM0 documented functional dependability requirements}$$
$$(\text{Section 9.3.1})$$

with $U_1(REQ) \prec U_1(\mathbf{CLM0})$
and *justification substructure* $U_1(REQ)$ defined by (9.11).

If these **CLM0** valuations hold in $U_1(REQ)$, *the relations REQ are a valid interpretation for the system input-output specifications*. If they don't, it means that the documented **CLM0** requirements cannot be satisfied by a computer system, the boundaries and the functions of which are *modelled* (Definition (8.4) by U_1(REQ). The system specifications would then have to be reconsidered, and possibly also the documented **CLM0** requirements. These initial requirements are validated against environment needs, but may not necessarily take properly into account that they must be implemented by a computer.

For reasons similar to those which lead to (9.9), one has the following relation between $\mathbf{L_1}$-*structures,* $\mathbf{L_1}$ being the language used to describe the **CLM0** documented environment constraints (Section 9.3.3) and functional dependability requirements (Section 9.3.1):

$$(9.15) \qquad\qquad U_1(NAT) \prec U_1(REQ).$$

9.4.10.3 Robustness and Functional Diversity Primary Claims
Next, safeness properties regarding the treatment of PIE's must be established at level-1.

The *postulated initiating events* described by the values of $h^1(t)$ must be unambiguously *detected, controlled* and lead to a *safe state*. The detectability, controllability and mitigation of these postulated events are essential functions of the system for which evidence must be claimed.

These claims assume that the validity claims (9.6), (9.7) are *satisfied, i.e.* that $h^1(t)$ and FS(t) are valid sets.

Detection is guaranteed if the following primary claim is *satisfied* (valuates to true) in the *justification substructure* U_1(HAZ):

(9.16) $\qquad$ $h^1(t)\_$**detectability.clm1**:

$\qquad\qquad U_1(HAZ) \vdash ((HAZ^1)^{-1}$ is a surjective function onto $h^1(t))$,

$\qquad\qquad$ with $U_1(HAZ) \prec U_1(\mathbf{CLM0})$
$\qquad\qquad$ and *justification substructure* U_1(HAZ) defined by (9.5).

where $(HAZ^1)^{-1}$ denotes the *converse* of HAZ^1 (cf. Section 8.2.1). This claim guarantees that every m(t) value within the *range* of HAZ^1 is univocally related to a unique value of $h^1(t)$, *i.e.* that $(HAZ^1)^{-1}$ is a function; and that all values of the domain of HAZ^1 are related to m(t) values, *i.e.* $(HAZ^1)^{-1}$ is a surjection.

Controllability is guaranteed if the following primary claim is *satisfied* (valuates to true) in U_1(HAZ):

(9.17) $\qquad$ $h^1(t)\_$**controllability.clm1**:

$\qquad\qquad U_1(HAZ) \vdash (range\ (HAZ^1) \subseteq domain\ (NAT))$

$\qquad\qquad$ with $U_1(HAZ) \prec U_1(\mathbf{CLM0})$
$\qquad\qquad$ and *justification substructure* U_1(HAZ) defined by (9.5).

This sub-claim guarantees that feasible values of m(t) and, by virtue of (9.4), also permissible c(t) values can be associated with every value of $h^1(t)$. Besides, because of (9.13), this claim also guarantees that the range of HAZ^1 is included in the domain of REQ.

The class $\mathcal{F}^1$ of the *postulated initiating events* introduced by Definition 9.9, page 127, corresponds to the occurrences of failures in the environment. These events need special treatment; in particular, their effects must be mitigated and they must lead to safe states of the environment. Let us consider the structure:

$$(9.18) \qquad U_1(FS) = <\{h^1(t)\,\}\{m(t)\}\{c(t)\}\{FS(t)\}\ ;$$
$$\{HAZ^1\}\{REQ\};$$
$$\{k_1\}>$$

where $\{k_1\}$ includes the constants related to HAZ^1 and REQ. *Mitigation* is guaranteed if the following primary claim is *satisfied* (valuates to true):

$$(9.19) \qquad h^1(t)\_\textbf{mitigation.clm1}:$$
$$U_1(FS) \vDash (\forall h^1(t) \mid h^1(t) \in \mathcal{F}^1: range\ h^1(t)\ (HAZ^1 \circ REQ) \subset FS(t))$$

with *justification substructure* $U_1(FS)$ defined by (9.18),
and $U_1(FS) \prec U_1(\textbf{CLM0})$,

where the symbol $\circ$ denotes the composition of relations (*cf.* Section 8.2.1, page 77), $\mathcal{F}^1 \in \{k_1\}$, and FS(t) is the sub-set of c(t) values that correspond to fail-safe states of the environment.

This claim guarantees the ability of the system to cope with postulated failures in the environment; more precisely, the set FS(t) being validated by claim (9.7), the claim guarantees that input-output specifications lead the system to a valid safe sate on occurrence of such events.

For reasons similar to those which lead to (9.9) and (9.15), the following relations hold between the **L**-*interpretations* of the **CLM0** environment constraints, functional dependability requirements, undesired events and failure modes:

$$(9.20) \qquad U_1(NAT) \prec U_1(HAZ) \prec U_1(FS)$$
$$U_1(NAT) \prec U_1(REQ) \prec U_1(FS).$$

As explained earlier – *cf.* (9.9) – these relations induce a logical ordering amongst the validations of these **CLM0** specifications. The natural constraints (NAT) should be validated first; next the detectability and controllability (HAZ^1) of the PIE's should be validated, and finally the system input-output specifications (REQ), including the proper mitigation (FS) of the PIE's by these specifications should be validated.

Finally, the **CLM0** dependability requirements may require that *postulated initiating events* be detected by more than one distinct and independent monitored variable.

There is a *functional diversity of order* d, $d \geq 2$, to monitor and to control a PIE of type h', ($h^1(t) = h'$), if and only if the following proposition is satisfied in the *structure* $U_1(HAZ)$:

$$U_1(HAZ) \vDash (\forall t : (\exists_d x : \langle h', m_x(t)\rangle \in HAZ^1))$$

where the quantifier ($\exists_d x$) stands for *"there are at least* d *distinct values of* x*"*. That is, there must exist at least d distinct monitored variables to observe the PIE h'. Hence, we associate with a PIE of type h' a *diversity set* $\mathcal{D}^1(h')$ defined as:

Definition 9.11
$$\mathcal{D}^1(h') = \{x \mid HAZ^1(h', m_x(t))\},$$
$$\|\mathcal{D}^1(h')\| = d, d \geq 1$$

The *diversity set* $\mathcal{D}^1(h')$ – sometimes abbreviated $\mathcal{D}^1$ if no confusion is possible – identifies the d distinct monitored variables that are intended to independently detect the PIE of type h'.

For any PIE h' that requires detection by d, $d \geq 1$ monitored variable(s), we therefore have the following claim:

(9.21) **(h', d)_functional_diversity.clm1:**

$$U_1(HAZ) \vDash (\|\mathcal{D}^1(h')\| = d)$$

with $U_1(HAZ) \prec U_1(\textbf{CLM0})$
and *justification substructure* $U_1(HAZ)$ defined by (9.5),
and $\mathcal{D}^1(h') \in \{k_1\}$.

In order to keep the benefits of functional diversity, these variables have to be processed independently of one another at the lower levels (*cf.* Section 9.5.12). This is why the set $\mathcal{D}^1(h')$ must be a subset of the set $\{k_1\}$ of $U_1(HAZ)$ designated constants; these values are needed at the lower levels to claim appropriate properties of independency and diversity from the architecture and design levels. $\mathcal{D}^1(h')$ must therefore be associated with delegation claims onto the lower levels in order to be assigned to free variables at those lower levels in agreement with Lemma 6.2, and Theorem 8.5. This is the role of the implementation *black claim* (9.25) introduced in the next section.

Remark 9.2 Functional diversity should not be confused with redundancy. While redundancy is an architecture and design property, functional diversity is dictated by dependability requirements which may demand that certain environment conditions be monitored by diverse physical parameters or by other diverse means. This is why all events and conditions subjected to functional diversity requirements must be described within a level-1 *justification structure* such as $U_1(HAZ)$ and subject to a *primary claim* (9.21).

9.4.11 Implementation Primary Claims – Justification Structures

The *dependability requirements* **CLM0** imply that the *system input-output specifications* are not only *valid* (*e.g.* complete, correct); they also imply that these specifications are comprehensively, correctly and consistently *implemented* by the lower levels of the architecture, design and operation control.

These properties must be the object of *primary* claims that address the implementation of the REQ relations. These primary claims are all *black* and not *grey*; they require evidence from the lower levels only. They must be expanded into delegation claims onto level 2, and from level 2 onto level 3. Some claims may even need to delegate claims on level 4 when, for instance, dependability requirements and some level 1 *postulated initiating events* cannot be dealt with by the system implementation alone and may need operator intervention.

Besides, as already noted in Remark 8.2, the set of lower levels 2, 3 and 4 is viewed from level-1 as one single aggregate, *i.e.* as a black box, so that, in practice, evidence from these levels cannot be precisely delegated to any of these levels. Only adjacent delegation subclaims are therefore formulated, *i.e.* from level 1 to level 2 (*cf.* also Section 6.6.5).

Although they may appear somewhat obvious and their supporting evidence is entirely delegated to the lower levels, these claims must be explicitly made at level 1, not only because they are part of the **CLM0** requirements but also because they are the roots of expansions at lower levels.

These implementation *primary black* claims are of three types, requiring an acceptable, safe and diverse implementation.

9.4.11.1 Acceptable Implementation
The functional relations REQ defined in 9.4.7 describe the "perfect" behaviour of the system and assume "perfect" values for the monitored and controlled variables. Therefore application-specific **CLM0** requirements usually include specifications on accuracy, stability, sensitivity and other characteristics that are indispensable for ensuring the dependability of the application; these requirements specify an acceptable implementation in terms of *measured* and *computed* values for monitoring and controlling the environment parameters.

The evidence to support such claims on the implementation comes from lower levels only and cannot be *a priori* and precisely defined at level-1. The corresponding delegation claims must therefore remain generic. The notion of *acceptable* implementation (already mentioned in Section 6.6.1) is helpful to formulate such generic claims:

Definition 9.12 *Acceptability* is defined as a generic property of an *implementation* which implies *correctness, completeness* and *accuracy*.

There may be an infinite number of ways for implementing a claim or a requirement specification, and many different properties may be required from a particular implementation. *Acceptability* as it is defined above is intended to cover the three essential implementation properties that, under various forms and to various degrees, dependability always implies; three properties that any meaningful and useful implementation cannot dispense with in practice.

With regard to *accuracy*, the **CLM0** dependability requirements may include application specific requirements on the sensitivity, precision and stability needed for monitoring and controlling the application parameters.

Thus, with *acceptability* being the generic property defined above, we have the *black primary* claim:

$$(9.22) \qquad \mathcal{A}^1\_ \text{REQ \_acceptable\_implementation.clm1}$$

where $\mathcal{A}^1$ is a set of parameter values specifying the *accuracy* required for the implementation of the *system input-output specifications*. (9.22) is a *primary black* claim which specifies evidence required exclusively from the lower levels. As said earlier, all black subclaims of level 1 require evidence of the immediate next lower level (remember comments in Section 6.6.1 and Remark 8.2). This is due to the interface between the level-1 model and the lower level models. The level-1 model is an input/output model which considers the whole system, including its operation control, as a black box the internal layers of which are yet to be defined.

Thus, following Proposition (8.13), the *justification* of the *black claim* (9.22) is the *logical consequence*:

$$(9.23) \qquad \mathcal{A}^1\_ \text{REQ \_acceptable\_implementation.clm[1,2]}$$
$$\vDash_{U_1(REQ)} \mathcal{A}^1\_ \text{REQ \_acceptable\_implementation.clm1}$$

with *justification substructure* $U_1(REQ)$ defined by (9.11),
and $U_1(REQ) \prec U_1(CLM0)$

and where $\mathcal{A}^1 \in \{k_1\}$ belongs to the set of constants of the justification structure $U_1(REQ)$.

9.4.11.2 Safe Implementation
This implementation claim is treated in a similar way to the previous one.

The mitigation claim (9.19) – validated by level-1 evidence – addresses the validity of the REQ relations to ensure the dependability required by **CLM0** and to deal with the level-1 *postulated initiating events*. This claim would remain ineffective if the implementation of the *system input-output specifications* is not itself failsafe. Hence, the *primary* claim and its *justification structure* $U_1(FS)$:

$$(9.24) \quad \text{REQ \_fail\_safe\_implementation.clm[1,2]}$$
$$\vDash_{U_1(FS)} \text{REQ \_fail\_safe\_implementation.clm1}$$

with $U_1(FS) \prec U_1(\textbf{CLM0})$,
and *justification substructure* $U_1(FS)$ defined by (9.18).

9.4.11.3 Diverse Implementation
The primary claim on functional diversity (9.21) – validated by level-1 evidence – would also remain a useless claim if it is not supported by properties of separation and independence at architectural and design levels. Thus, if the claim (9.21) is

made for a PIE h', a claim for an architecture and design that preserve the benefits of functional diversity is needed at level-1:

(9.25) $\mathcal{D}^1$(h')_**functional_diversity_implementation.clm[1,2]**

$\vDash_{U_1(HAZ)}$ (h',d)_**functional_diversity_implementation.clm1**

with $U_1(HAZ) \prec U_1(\mathbf{CLM0})$,
justification substructure $U_1(HAZ)$ being defined by (9.5),

and with $\mathcal{D}^1$(h'), Definition 9.11, being a subset of $U_1(HAZ)$ designated constants.

Of course, in addition to these three generic types of implementation claims, depending on the application, more specific properties – like real-time performances – may be required and may have to be claimed explicitly.

9.4.12 Primary Dependability Claims – Justification Substructures

9.4.12.1 The Relation $R\widetilde{E}Q$

The primary claims we have identified so far address deterministic properties. When the dependability case imposes reliability objectives, primary claims have also to specify probabilistic targets on stochastic properties of the actual behaviour of the system implementation, *e.g.* the reliability, safety and availability of this implementation. These probabilistic targets and claims are based on the two assumptions:

-   The input-output specifications of the system – as defined by REQ – are assumed to be *valid* (thanks to the validity claim (9.14)),
-   The *implementation* of these specifications may *not be perfect, i.e.* may not satisfy the specifications, as a result of random or deterministic unknown defects which have not been anticipated by the design.

These assumptions, often left vague or implicit in probabilistic evaluations of the dependable behaviour of digital systems, require some comments. The potential presence of defects in the *specifications* is sometimes considered as a possible cause of random computer system failures. Our approach takes a different view: the validity of the system requirements and of their specifications must be established by deterministic arguments at level-1, and potential random effects may occur and affect the architecture, design and control levels only. This distinction, as we shall see, is essential to allow – within a same interpretation framework – a non-ambiguous, consistent and harmonious coexistence of stochastic and deterministic claims.

The REQ relation (Definition 9.10) specifies a deterministic perfect behaviour of the system. And the proper multi-level justification of the **CLM0** dependability requirements demands (Theorem 8.7) that the relation REQ remains satisfied as a *restricted relation* in the lower level *extension structures* (Definition 8.7), and as a deterministic model of a perfect implementation. We therefore need to distinguish between the perfect REQ relation with a deterministic lower level interpretation on the one hand, and the random behaviour of a possibly imperfect implementation on the other hand.

A different relation is needed to refer to this potentially imperfect behaviour. Using a notation similar to that used for a related – albeit different – purpose in [39], we define the following:

Definition 9.13 Let $\widetilde{REQ}$ be a relation that describes the behaviour of a possibly imperfect implementation of the functional relation REQ. $\widetilde{REQ}$ has the same *domain* as REQ on the monitored variables and a *range* on controlled variables that includes *range* (REQ).

The relation $\widetilde{REQ}$ is not known with certainty: it is partially undefined. Its domain is precisely known but its range is unknown as it must include values of the controlled variables c(t) that are not permissible. If $\widetilde{REQ}$ domain and range were both exactly known, all implementation defects could be corrected and probabilistic claims would be irrelevant. Not being fully defined, $\widetilde{REQ}$ is not part of the structure U_1(REQ).

But the $\widetilde{REQ}$ relation can be valuated by probability distributions and statistical test data. Probabilistic claims can be formulated in terms of REQ and other proper relations in order to specify properties that random variables must satisfy. Hence, these claims specify the statistical evidence – *e.g.* statistical test results – required from the lower levels implementation. Later in Section 9.5.13.3 we will see how the expansions and delegations across lower levels of such probabilistic claims involve conditional probability distributions.

More precisely, let us define

$$(9.26) \qquad F_1(x,t) = \text{Prob } \{m(t) = x\}, \ x \subseteq domain \text{ (REQ)},$$

as the probability density function which describes the input profile of the *system*. Given a possibly imperfect implementation $\widetilde{REQ}$, the probability that this implementation does not provide at some time *t* the service required by the input-output specifications REQ, can then be defined as:

$$(9.27) \qquad f(t) = \text{Prob } \{\tilde{c} \ (t) \neq \text{REQ } (m(t)) \mid \widetilde{REQ} \},$$

where $\tilde{c}$ (t) denotes the actual value obtained by the implementation $\widetilde{REQ}$. The probability (9.27) is taken over the set of possible values that the monitored variables m(t) can take; m(t) is here a *random variable* with values in *domain* (REQ) and with probability defined by (9.26).

Hence, two potential sources of randomness determine the value of the probability *f(t)*:

- The probability density function $F_1(x,t)$ of the system input profile,
- And the possibly imperfect implementation $\widetilde{REQ}$ which is statistically defined and evaluable in terms of statistical data only.

If the implementation were to be "perfect", $\widetilde{REQ}$ could be replaced by the relation REQ in the conditional probability (9.27) and $f(t)$ would be equal to 0, by design one may say.

If, for reasons of simplicity, we approximate REQ as a function, then a measure of the system failure rate over a time interval $[t_2 - t_1)$ is given by:

$$(9.28) \qquad \overline{F}(t_1, t_2) = (t_2 - t_1)^{-1} \int_{t_1}^{t_2} f(t)dt \, ,$$

where the probability $f(t)$ is given by (9.27).

Now, let $h^1(t)$ be a *random variable* with values in *domain* (HAZ^1) and let us define the probability density function:

$$(9.29) \qquad H_1(x,t) = Prob\{\, h^1(t) = x \}, \ \ x \subseteq domain \ (HAZ^1).$$

Given the possibility of an imperfect implementation $\widetilde{REQ}$, the probability of this implementation being unsafe upon occurrence of some *postulated initiating event* $(h^1(t),t)$, $h^1(t) \in \mathcal{F}^1$, is then defined as:

$$(9.30) \qquad u(t) = Prob\{\, \widetilde{c}\,(t) \neq (h^1(t)\,(HAZ^1 \circ REQ)) \mid \widetilde{REQ} \, \},$$

where $(h^1(t)\,(HAZ^1 \circ REQ)) \subseteq FS(t)$.

The probability $u(t)$ is taken over the distribution of the occurrences of level-1 postulated initiating events, *i.e.* over the distribution of the values $h^1(t)$ over the set $\mathcal{F}^1$ (Definition 9.9). Again two potential sources of randomness affect the evaluation of the probability $u(t)$: the probability density function $H_1(x,t)$ of the postulated level-1 events and the possibly imperfect implementation $\widetilde{REQ}$. If this implementation were to be "perfect", $\widetilde{REQ}$ could be substituted with the relation REQ in the conditional probability (9.30) and $u(t)$ would then be equal to 0, by design.

Approximating relations by functions, a measure of the *system* unsafe failure rate, *i.e.* of the expected number of unsafe failures on occurrence of a level-1 *postulated initiating event* during a time interval $[t_2 - t_1)$ is then given by:

$$(9.31) \qquad \overline{U}(t_1, t_2) = (t_2 - t_1)^{-1} \int_{t_1}^{t_2} u(t)dt \, .$$

Other appropriate dependability probabilistic properties can be expressed in a similar way. Minimum availability, for instance, can be claimed and expressed by a generalisation of (9.27) if periods of unavailability required by repair or maintenance have to be taken into consideration.

It is important to recall that these probabilities are defined on the assumption that the set of *postulated initiating events* $h^1(t)$, the safe set $FS(t)$ and the relations

REQ and HAZ^1 are valid, *i.e.* that the level-1 *primary claims* (9.6), (9.7), (9.14) receive proper level-1 evidence.

9.4.12.2 Reliability and Safety Primary Claims
Probabilistic claims at level 1 can be expressed in terms of limiting bounds on the values of the probabilities that have been defined here above. These bounds are dictated by the dependability requirements. These requirements, however, and thus the level-1 claims on reliability, availability and safety cannot be realistically made without assumptions on the independency and the redundancy available in the implementation for processing the monitored input data and for controlling the effects of *postulated initiating events*. We have already defined (Definition 9.11) in Section 9.4.10.3 a set $\mathcal{D}^1($h'$)$ of d, (d $\geq$ 2), monitored variables that need to be processed independently to ensure functional diversity at level 1 for a PIE h'. Let us now also define the set $\mathcal{R}^1$:

Definition 9.14   With $\mathcal{R}^1 = \{< x, r >\}$, a pair $< x,r >$ is in $\mathcal{R}^1$ iff $m_x(t)$ is assumed to be processed with r redundancy, $r \geq 2$.

The sets $\mathcal{D}^1($h'$)$ and $\mathcal{R}^1$ allow us to specify and formulate the independency and the redundancy respectively, required at the levels of architecture and design. Redundancy and independency are intentionally considered as distinct properties. They must be claimed separately because their valuation requires different types of evidence. Of course, whenever redundancy requires independency at the architecture and design levels, these properties will have to be explicitly and jointly claimed at these levels.

Assuming known redundancy and independency at the lower levels, an upper bound on the failure rate, *i.e.* on the expected number of failures of the system to deliver the required service during a time interval $[t_2 - t_1)$ can be realistically determined, and formulated as the *primary* claim:

$$(9.32) \qquad \mathcal{R}^1\_\textbf{reliability.clm1:}\ \overline{F}(t_1,t_2) = (t_2 - t_1)^{-1} \int_{t_1}^{t_2} f(t)dt \leq F_{max}.$$

$\mathcal{R}^1$, the time interval $[t_2 - t_1)$, the profile probability density function $F_1(x,t)$ (9.26) and F_{max} belong to the set of constants $\{k_1\}$ of the *interpretation structure* $U_1(REQ)$. But the conditional probability density function *f(t)* is given by (9.27) and conditioned on the possibly imperfect implementation $\widetilde{REQ}$ which is unknown at this level-1. Thus, the valuation of the claim in $U_1(REQ)$ requires additional data belonging to the lower levels.

Therefore, the *justification* of the *black* primary claim (9.32) is the *logical consequence*:

(9.33) $\mathcal{R}^1\_\textbf{reliability.clm[1, 2]}$

$$\vdash_{U_1(REQ)} (\overline{F}(t_1,t_2) = (t_2 - t_1)^{-1} \int_{t_1}^{t_2} f(t)dt \le F_{max}),$$

with *justification substructure* $U_1(REQ)$ defined by (9.11),

and $U_1(REQ) \prec U_1(CLM0)$,

and where $\mathcal{R}^1$, the time interval $[t_2 - t_1)$, the profile probability density function $F_1(x,t)$ (9.26) and F_{max} belong to the set of constants $\{k_1\}$ of the justification structure $U_1(REQ)$.

In practice, the relation REQ, the domain of the *justification substructure* $U_1(REQ)$ and the profile probability density function $F_1(x,t)$, constitute the specification of random system integration tests, the results of which must provide the statistical data which valuate the system failure probability *f(t)*.

Similarly, a bound on the unsafe failure rate, *i.e.* on the probability of the system being unsafe on the occurrence of *postulated initiating events* during a time interval $[t_2 - t_1)$, given that these events are detected with some level of diversity, can be formulated and claimed as:

(9.34) $\mathcal{D}^1\_\textbf{safety.clm1}$: $\overline{U}(t_1,t_2) = (t_2 - t_1)^{-1} \int_{t_1}^{t_2} u(t)dt \le U_{max}.$

$\mathcal{D}^1$ designates the set of sets $\mathcal{D}^1(h')$ (Definition 9.11) for all $h' \in h^1$. $\mathcal{D}^1$, the time interval $[t_2 - t_1)$, the profile probability density function $H_1(x,t)$ (9.29) and U_{max} belong to the set of constants $\{k_1\}$ of the *interpretation structure* $U_1(FS)$ (9.18).

But the conditional probability $u(t)$ given by (9.30) is conditioned on the possibly imperfect implementation $\widetilde{REQ}$ which is unknown at this level. The valuation of the claim in $U_1(FS)$ requires these additional data from the lower levels.

Therefore, the *justification* of the *black primary* claim delegation claim (9.34) is the logical consequence:

(9.35) $\mathcal{D}^1\_\textbf{safety.clm[1,2]}$

$$\vdash_{U_1(FS)} (\overline{U}(t_1,t_2) = (t_2 - t_1)^{-1} \int_{t_1}^{t_2} u(t)dt \le U_{max}),$$

with $U_1(FS) \prec U_1(\textbf{CLM0})$,
and $U_1(FS)$ defined by (9.18).

In practice, the relations HAZ^1 and REQ of the *justification substructure* $U_1(FS)$, together with the undesired event profile distribution $H_1(x,t)$, are the specification of random system integration tests, the results of which must provide the statistical data which valuate the probability *u(t)* of the system being unsafe.

Let us again underline that these probabilistic delegation claims address properties of the system implementation only and not the validity of the system requirements and of the interface with the environment. More precisely, they assume that the relations REQ and HAZ^1 are models validated by the primary claims of Sections 9.4.10, 9.4.10.2 and 9.4.10.3. Thus, they are probabilistic claims requiring statistical data and evidence of the dependability of lower levels (architecture, design, operation control), and not of the validity of the requirement and undesired events specifications themselves. As said earlier, studies and evaluations of computer systems dependability often fail to clearly make this distinction.

In these probabilistic claim assertions, m(t) and h^1(t) are *random free variables*. The values of their probability density functions F_1(x,t) and H_1(h^1(t),t) must correspond to the use and environment profiles specified by the *dependability requirements* **CLM0**. Like the sets $\mathcal{R}^1$ and $\mathcal{D}^1$, these probability density distributions must be documented by the sets of constants $\{k_1\}$ of the substructures U_1(REQ) and U_1(FS). These data must indeed be transferred by the delegation claims onto the lower levels in order to be assigned to the free variables at those levels, in agreement with Lemma 6.2, and Theorem 8.5, and be available in lower level *extension structure* universes for making appropriate redundancy and independency claims.

The same applies to the time interval $[t_2 - t_1)$ selected to evaluate the integrals (9.32) and (9.34). The reliability estimation and the bounds that can be claimed are function of this interval which should correspond to periods between successive time instants at which the system can be reasonably considered as being in a state that is independent of the past. These time instants may for instance correspond to periodic maintenance activities or tests after which the system can be considered as being reset in a correct operational state. This time interval must therefore be included in the constants $\{k_1\}$ of the *justification substructures* U_1(REQ) and U_1(FS) and of their lower level *extension structure* universes, in particular at the level-4 operation level for acquiring evidence of adequate periodic maintenance and checks.

To summarize, like the implementation claims of the previous section, the probabilistic primary claims are *black subclaims* as no evidence exists at level-1 to valuate them. These claims must be the logical consequence in the level-1 *justification elementary substructures* U_1(REQ) and U_1(FS) of *delegation claims* onto the lower levels; as a matter of fact, adjacent delegation subclaims from level-1 to level-2 (*cf. Remark* 8.2 and Section 6.6.5). These probabilistic delegation claims are specifications of the statistical evidence data and statistical tests required on the lower levels on the implementation behaviour $\widetilde{REQ}$.

9.4.13 Level-1 CLM0 Expansion Structure

The *justification structures* introduced in the three previous sections must be *elementary substructures* of the *expansion structure* in which the initial *dependability requirements* are *logical consequences* of the *primary* claims (Lemma 8.8). We mentioned in Section 9.4.9 that the level-1 expansion into primary claims is an initial and creative design step of the justification, delicate in practice because the level-1 *expansion structure* and the *justification substructures* are inter-dependent. For simplification, we had postponed the definition of the *expansion structure* until the *justification elementary substructures* have been identified.

But we have now reached this stage and we can define the **L₁**-*expansion structure* $U_1($**CLM0**$)$, **L₁** being the language in which the dependability requirements are expressed. If the **CLM0** requirements and primary claims ***.clm1** are formulated in some language **L₁**, $U_1($**CLM0**$)$ is a **L₁** - *interpretation* for these *primary* claim formulae. More precisely, in accordance with Lemma 8.8:

- $U_1($**CLM0**$)$ is the single **L₁**-*expansion structure* in which initial *dependability requirements* are *logical consequences* of the *wf's* of expansions into *primary white* and *black* claims.
- The free variables of the *dependability requirements* and of the *primary* claims formulae are assigned values from the universe of $U_1($**CLM0**$)$.
- The valuations of *elementary justification substructures (submodels)* of $U_1($**CLM0**$)$ provide the evidence that validates the *primary white* claims; the *black primary* claims are logical consequences of *delegation* subclaims that must hold true in *elementary justification substructures (submodels)* of $U_1($**CLM0**$)$.

$U_1($**CLM0**$)$ must be by construction (Theorem 8.8, page 106) the smallest *structure* at level-1 for the valuation of the expansion of **CLM0** into *primary* claims; it is the *extension* (Definition 8.7) of the *justification elementary substructures* of the claims of its *expansion* (Corollary 8.4). The universe and relations are the union without repetition of the universes and relations of these *elementary substructures* $U_1($NAT$)$, $U_1($REQ$)$, $U_1($HAZ$)$ and $U_1($FS$)$.

Thus, we have:

$$(9.36) \qquad U_1(\mathbf{CLM0}) = <\{m(t)\}\,\{c(t)\}\,\{h^1(t)\}\ ;\ \{NAT\}\,\{HAZ^1\}\,\{REQ\}\ ;\ \{k_1\}>$$

For $U_1($NAT$)$, $U_1($REQ$)$, $U_1($HAZ$)$ and $U_1($FS$)$ to be its *elementary substructures*, $U_1($**CLM0**$)$ must also satisfy the Tarski–Vaught condition (Theorem 8.1, page 90).

$U_1($**CLM0**$)$ is a comprehensive description of the environment-system interface; its universe is the set of values that the variables $m(t)$, $c(t)$, and $h^1(t)$ can take; $\{k_1\}$ is the set of designated constants, *i.e.* the values of the interface parameters; in particular:

$$(9.37) \qquad \{c(t)\} \subset FS(t),$$

$$\{k_1\} \subset \{\mathcal{F}^1,\,\mathcal{D}^1(h'),\,\mathcal{A}^1,\,\mathcal{R}^1,\,[t_2-t_1),\,F_1(x,t),\,F_{max},\,H_1(x,t),\,U_{max}\}.$$

The completeness of the *expansion* of *dependability requirements* into primary claims is a property which must be *interpreted* and valuated by the *structure* $U_1($**CLM0**$)$. $U_1($**CLM0**$)$ stands as an abstraction (Section 8.13) of any lower level computer system dependable implementation of the relation REQ.

At the end of this chapter, Figure 9.9, page 242, displays the entire expansion tree of the system *substructures* and in particular the connections between $U_1($CLM0$)$ and its different level-1 *elementary justification substructures*. The required level-1 evidence and delegations to lower levels will be detailed in the next Section (9.4.14).

Some of the primary claims formulated in the previous section – such as the validity claims – must be made in every dependability case; others like the dependability claims may depend on the application and on specific **CLM0** requirements. Using the notation introduced in Section 9.2 to distinguish between optional and mandatory claims, a generic expansion of **CLM0** into primary claims is given in Figure 9.2, page 147.

Now, one can better understand why this first level expansion of the dependability requirements is a critical, creative, initial process of the justification. These requirements are given and formulated in some language (L_1) but no model (structure) is initially provided for their *interpretation* and valuation. The requirements must be logical consequences of *primary claims*, and these logical consequences have to be *interpreted* and valuated in expansion *structures* (models) that are not given at the start of the justification and have yet to be defined. Theorem 8.9 states that the language L_1 must be an *expansion* (Definition 8.5) of the languages $L_1(\alpha)$ and $L_1(\beta)$ used to formulate the white and black *primary* claims. Thus, the formulation of an initial dependability requirement and its *interpretation structure* $U_1(\textbf{CLM0})$ cannot be entirely dissociated from the elicitation and *interpretations* of the *primary* claims of its *expansion*. These initial steps of the justification have somehow to be carried out hand in hand.

Some guidance for the initial identification of *primary* claims is provided by the obligation for these claims to cover three main types of properties: validity of the *expansion structure*, correctness and dependability of the implementation. Some guidance is also dictated by the constraint for the primary claims to be interpreted and valuated in *structures* (models) that are *elementary substructures* of the *expansion structure*. This guidance, of course, leaves open many options so that experience and innovative thinking are indispensable. The selection of *primary* claims and the definition of the *expansion* and *justification* models must result from a reflexive interplay between the "prestructured understanding"[12] this guidance provides, and knowledge of the different empirical aspects and elements of the computer system interface. The primary claims *interpretations* being *elementary substructures* of the **CLM0** *interpretation*, these *interpretation* models are not mutually independent and somehow have to be jointly conceived.

Nevertheless, the three previous sections clearly indicate that an *inductive strategy* should be favoured:

1. Start with the identification of the main components (universe and relations) of the computer system interface (Sections 9.3, and 9.4.1 to 9.4.7);
2. Then identify a set of *primary* claims and define the *elementary structures* for their *interpretation* (Sections 9.4.11 and 9.4.12);

---

[12] a notion borrowed from [3].

3. Finally, derive from these *elementary structures* an *expansion structure* for verifying (*i.e. interpreting and valuating*) the logical consequences of the initial dependability requirements.

This strategy is proposed in Section 11.1 as a recommended construction rule.

Expansion and elicitation of claims at the lower levels is relatively easier as delegation claims from upper levels delineate and guide their selection.

$$
\begin{aligned}
&\textbf{expansion (CLM0, level 1)}\\[4pt]
&\underline{\textbf{begin}}\\[6pt]
&\textbf{CLM0}\ _{U_{1}(\textbf{CLM0})} =|\\
&\qquad(\ \text{NAT\_constraints \_validity.clm1}\\
&\qquad \wedge h^{1}(t)\text{\_and\_effects\_validity.clm1}\\
&\qquad \wedge FS(t)\text{\_validity.clm1}\\
&\qquad \wedge REQ\text{\_completeness\_and\_feasibility.clm1}\\
&\qquad \wedge REQ\text{\_correctness.clm1}\\
&\qquad \wedge h^{1}(t)\text{\_detectability.clm1}\\
&\qquad \wedge h^{1}(t)\text{\_controllability.clm1}\\
&\qquad \wedge h^{1}(t)\text{\_mitigation.clm1}\\
&\qquad \wedge \mathcal{A}^{1}\text{\_ REQ \_acceptable\_implementation.clm1}\\
&\qquad \wedge REQ\text{\_fail\_safe\_implementation.clm1}\,)\\[4pt]
&\qquad\qquad\quad \wedge\\[4pt]
&\qquad(\ \text{nil}\ |\ \mathcal{R}^{1}\text{\_reliability.clm1}\ |\ \mathcal{D}^{1}\text{\_safety.clm1}\\
&\qquad\qquad |\,(\,(h',d)\text{\_functional\_diversity.clm1}\ \wedge\\
&\qquad\quad \mathcal{D}^{1}(h')\text{\_functional\_diversity\_implementation.clm1}\,)\,)\\[6pt]
&\underline{\textbf{end}}\ \textbf{expansion (CLM0, level 1)}
\end{aligned}
$$

Figure 9.2. Level-1 **CLM0** expansion

9.4.14 Level-1 Evidence and Delegation Claims

Let us recapitulate and discuss some aspects of the evidence needed at this first level and the evidence that must be delegated to the lower levels. Remember that at level-1 all claims are white or black (*cf.* Section 6.6.4).

9.4.14.1 Primary White Claims Evidence
The evidence for the primary *validity claims* of Section 9.4.10 must establish the *validity* of the *expansion structure* $U_1(\textbf{CLM0})$ as a *model* of the functional relations, constraints, undesired events and safe states implied by the *dependability requirements*.

This evidence is obtained by valuations mapping each predicate symbol, each function symbol, each operation symbol and each parameter of the language $\mathbf{L_1}$ in which **CLM0** statements are expressed, to a relation on the universe and to a constant of the appropriate *elementary justification substructure* of $U_1($**CLM0**$)$ respectively; all these mappings must valuate to true. $U_1($**CLM0**$)$ *elementary substructure*s are $\mathbf{L_1}$-*interpretations* for the validity claims (Definition 8.2). The evidence thus entirely belongs to level 1; *all primary validity claims are white.*

Next, a number of other *primary* claims are also *white* because they strictly specify and claim no other properties than properties of $U_1($**CLM0**$)$ *elementary substructures.* These claims and their valuations in $U_1($**CLM0**$)$ are (9.13), (9.16), (9.17), (9.19), (9.21):

- REQ_**completeness_and_feasibility.clm1**:
  $$domain\ (\mathrm{REQ}) \equiv domain\ (\mathrm{NAT}),$$
- $h^1(t)$_**detectability.clm1**:
  $$(\mathrm{HAZ}^1)^{-1}\ \text{is a surjective function onto } h^1(t),$$
- $h^1(t)$_**controllability.clm1**: $range\ (\mathrm{HAZ}^1) \subseteq domain\ (\mathrm{NAT}),$
- $h^1(t)$_**mitigation.clm1**:
  $$(\forall h^1(t)\,|\,h^1(t) \in \mathcal{F}^1 : range\ h^1(t)\ (\mathrm{HAZ}^1 \circ \mathrm{REQ}) \subset \mathrm{FS}(t)),$$
- (h',d)_**functional_diversity.clm1**:
  $$U_1(\mathrm{HAZ}) \vDash (\,\|\,\mathcal{D}^1\,(h')\,\| = d\,).$$

These claims are *white* because they are exclusively interpreted by NAT, HAZ, FS(t) and the domain and relations of REQ. They are interpreted as properties of the structure $U_1($**CLM0**$)$ which is assumed to be the valid $\mathbf{L_1}$-*interpretation* of the requirements **CLM0**, thanks to the validity claims.

Except for the **mitigation.clm1,** these claims make no reference to the range of REQ, *i.e.* to values of the controlled variables c(t). Hence, they are independent from properties of the implementation of REQ. The claim **mitigation.clm1** may be perceived as an exception but is not. This claim addresses the *existence* of REQ relations able to lead the computer system into a state of FS(t) from a state in the range of HAZ1; it does not address properties of the implementation of REQ; these properties are claimed by (9.23).

Environment-system interface and dependability experts are responsible for producing the evidence of these *white claim primary claims*, thus for making the required valuation mappings between the **CLM0** initial *dependability requirements* and the valuations of the *elementary justification substructures (submodels)* of the $\mathbf{L_1}$-*interpretation* $U_1($**CLM0**$)$.

9.4.14.2 Primary Black Claim Delegations

All primary claims that address properties interpreted by the *range* of REQ controlled variables imply properties related to the implementation of the REQ relations, *i.e.* to the lower levels of the architecture, design and operation control. These primary claims must be interpreted and satisfied by evidence at lower levels; these claims and their valuations in *elementary justification substructures* of $U_1($**CLM0**$)$ are (9.23), (9.24), (9.25), (9.33), (9.35):

- $\mathcal{R}^1$_**reliability.clm1**: $\overline{F}(t_1,t_2) = \int\limits_{t_1}^{t_2} f(t)\,dt \leq F_{max}$,

- $\mathcal{D}^1$_**safety.clm1**: $\overline{U}(t_1,t_2) = \int\limits_{t_1}^{t_2} u(t)\,dt \leq U_{max}$,

- $\mathcal{A}^1$_ REQ _**acceptable_implementation.clm1**,
- REQ _**fail-safe_implementation.clm1**,
- $\mathcal{D}^1$(h')_**functional_diversity_implementation.clm1**.

More precisely, these six claims are all *black* and not *grey*; they require evidence from the lower levels only. They do not require other evidence from level 1 than the validity of the *interpretation structure* $U_1(\mathbf{CLM0})$, which is ensured by the *primary* validity claims. This is a confirmation of the absence of grey claim at level 1, already underlined in Section 6.6.4.

These claims must be expanded into delegation claims onto level 2 and, as we shall see, from level 2 onto level 3. Some claims may ultimately also need evidence from level 4 when, for instance, dependability claims and some level-1 *postulated initiating events* cannot be dealt with by the system alone and may need operator intervention. In accordance with Lemma 8.9 and Theorem 8.5, the predicates of the black claims listed above are transferred to level-2 with proper assignments of constant values from the set $\{k_1\}$ of $U_1(\mathbf{CLM0})$.

9.4.14.3 Adjacent Delegation Subclaims
Let us underline again that all black subclaims of level 1 require evidence of the immediate next lower layer. This is due to the interface between the level-1 model and the other lower level models. The level-1 model is an input/output model which considers the whole system, including its operation control as a black box, the internal layers of which are yet to be defined. Thus all delegation subclaims from level 1 are formulated as adjacent subclaims (see comments in Section 6.6.1 and Remark 8.2).

9.4.15 Level-1 Argument

To recapitulate, using the notation introduced in Section 9.2, the level-1 main structure of *arguments* (Definition 6.3) consists of the expansions, justifications and delegations (9.38) which follow here below. *White primary claims* are identified by the references to their formulae and their *justification elementary substructures*; *black primary delegation claims* are merely identified by their predicates as they are all adjacent single claims; the relations between the different *elementary justification structures* of Proposition (9.20) induce a logical ordering between the primary white claim justifications of this argument.

(9.38) **argument (CLM0, level-1)**

<u>begin</u>

    expansion (CLM0, level 1);   /**cf.* Figure 9.2 in Section 9.4.13) */

    */ <u>**primary white claims justifications by**</u> U_1**(CLM0)** <u>**valuations:**</u> */

    U_1**(CLM0)** $\models$ **validity claims**; /* (9.4) (9.6) (9.7)*/

    U_1**(CLM0)** $\models$ **completeness and correctness claims**; /* (9.13) (9.14)*/

    U_1**(CLM0)** $\models$ **robustness claims**; /* (9.16) (9.17) (9.19)*/

    U_1**(CLM0)** $\models$ **(h',d)_functional_diversity claim**; /* (9.21)*/

    */ <u>**black primary claims delegation predicates:**</u> */

    $\mathcal{A}^1$**_REQ_acceptable_implementation.clm[1,2]**;
    REQ_fail_safe_implementation.clm[1,2];
    $\mathcal{D}^1$**(h')_functional_diversity_implementation.clm[1,2]**;
    $\mathcal{R}^1$**_reliability.clm[1,2]**;
    $\mathcal{D}^1$**_safety.clm[1,2]**;

<u>end</u> argument(CLM0, level-1)

9.4.16 Level -1 Documentation

The analysis of this Section 9.4 allows us to specify the documentation which is required to expand and justify level-1 claims. This material should necessarily be part of the level-1 documentation of a *dependability case* supporting the **CLM0** *initial dependability requirements*. Since the level-1 *primary claims* address essential system properties, this material should also be part of the documentation of the requirement specifications of a dependable system.

From the previous sections, we can conclude that the material documenting level-1 should necessarily include:

1.   The description of a **L_1**-*interpretation* and *structure* U_1**(CLM0)** (9.36) which defines:

    -    The functional relations REQ between the environmental variables $m(t)$ and $c(t)$, with the assumptions discussed in 9.4.4 and the relations NAT.

    -    The *postulated initiating events* (domain of $h^1(t)$ and the event class $\mathcal{F}^1$), the relations HAZ^1 and the set of failsafe states $FS(t)$.

    -    The set of parameter values $\{k_1\}$ which specifies the redundancy and the independency required from the implementation, the probability distributions of the monitored input data and of the *postulated*

2. *initiating events* occurrences, and the reliability assessment time interval $[t_2 - t_1)$.

2. The arguments (9.38) detailing the *expansion* into primary claims, the *justifications* and the *delegations*.

3. The description of the level-2 *justification elementary substructures* of $U_1(\mathbf{CLM0})$, on the interpretation of which the arguments are based: $U_1(NAT)$ (9.3), $U_1(REQ)$ (9.11), $U_1(HAZ)$ (9.5), and $U_1(FS)$ (9.18).

4. The documented evidence which validates – by valuation of these *elementary justification substructures* – the primary white claims against the **CLM0** dependability requirements; that is, the mappings of the **CLM0** dependability requirements onto these *substructures* which valuate to true (i) the *expansion* of the *dependability requirements* **CLM0** into primary claims and (ii) the white primary claims, (see 9.4.13). These valuations allow, in particular, the relations REQ to be used as a valid *interpretation* of the system input-output specifications.

Recalling Section 9.4.8, let us again underline that these valuations will be easier to produce and to document if the initial *dependability requirements* **CLM0** are formulated in terms of $U_1(\mathbf{CLM0})$ universe and relations.

The documentation should be made accessible to parties such as:

- Safety experts of the application domain in order to approve the **CLM0** requirements and the *interpretation structure* $U_1(\mathbf{CLM0})$, ensuring that the documented behaviour is adequate in all eventualities.
- Teams of independent reviewers of the system input-output specifications.
- Computer system designers for approval with respect to feasibility and readability.
- Licensing authorities for compliance with plant safety requirements.

9.5 Level-2. Computer System Architecture

Level-1 considers the computer system as a black box. Level-2 opens this black box and focuses attention on its internal organization and main structure. This structure is the arrangement of the physical and functional inter-relations between all hardware, software and communication units, and is what we call the computer system architecture.

A precise description of the hardware and software architecture is essential; many dependability claims address properties of the architecture. Section 5.7 discussed the necessity and the specificity of the evidence provided by the architecture. And yet, as noted in [30], there are no tools or representations today that can precisely and efficiently represent the full scope of a system architecture and of its interface with the environment. Wiring diagrams and structural charts may represent physical hardware parts but not the software, nor the interactions with the environment, nor the behaviours resulting from undesired external events. Besides, typically, these representations do not support rigorous analysis techniques. This level-2 description offers an approach to make up for this deficiency.

9.5.1 Elements of the Architecture

The model of the architecture must provide the evidence needed to expand and justify the delegation claims of the level-1 argument (9.4.15); these delegation claims define which truth the level-2 *subtructures* must valuate and justify. Redundancy, availability and diversity are characteristic properties of dependable architectures. Redundant and diverse parallel processing channels as well as their communication links and interconnection logic (*e.g.* voting and priority mechanisms)[13] are therefore the main constitutive parts of the architecture (see Figure 9.3). Physically, each channel basically consists of one or more interconnected computers, data acquisition and output registers. Channels may also need to exchange data for validating and prioritising their input signals, and for validating their outputs.

We assume at this level that the *expansion structure* $U_1(\textbf{CLM0})$ and its *elementary substructures* are validated; in particular that REQ, the model of the system input output specifications, is validated against the **CLM0** dependability requirements and is feasible with respect to NAT.

9.5.2 Input and Output Variables

For each channel, two sets of variables must be identified: a set of input variables whose digital values are read and processed by the computers of the channel, and a set of output variables whose digital values are produced by the computers of the channel. These digital values are stored in an input and an output register of the channel. Sensing and analogue-to-digital conversion devices acquire the values of the monitored variable m(t) values and load the input register. Likewise, output devices convert the contents of the output register into the values of the controlled variables c(t). The behaviour of these acquisition and conversion devices is modelled by relations detailed hereafter.

Some of the digital input and output values are accessible to the operators, to the control room and the maintenance team for monitoring, testing and interacting with the channels, but these accesses and functions are part of the level-4 model and will be discussed in Section 9.7.

---

[13] For clarity, these links are merely symbolized on Figure 9.3 by single lines between processing units.

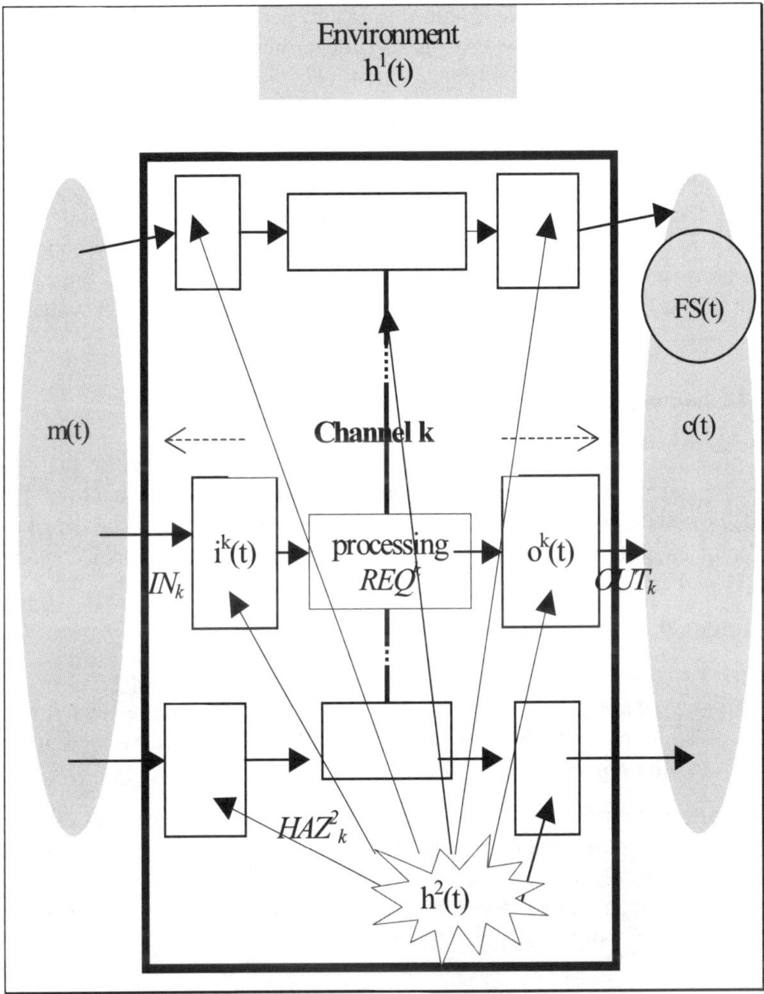

Figure 9.3. Structures at the architecture level

9.5.3 Channel k Data Acquisition (Relations IN_k)

Let $i^k(t)$ denote the vector containing one element for each of the input variables of channel k, k=1, 2, 3,…

Definition 9.15 A relation IN_k defined as follows describes the behaviour of the input devices of channel k:

- *domain* (IN_k) is a set of vector-valued time functions containing possible values of a sub-vector of m(t);
- *range* (IN_k) is a set of vector-valued time functions containing the possible values of $i^k(t)$;

- $(m(t), i^k(t))$ is in IN_k if and only if $i^k(t)$ describes values taken by the input variables of channel k when $m(t)$ describes the values taken by monitored variables.

IN_k specifies the behaviour of the channel k data acquisition and input devices (*e.g.* intelligent sensing devices, analogue to digital converters). The IN_k relations are an abstraction of these devices and of the protocol – polling, multiplexing, sharing – used to access the channel input register and load the digital values $i^k(t)$.

IN_k is a relation and not a function because of imperfections in the measurements. Different $m(t)$ values may be converted to a same $i^k(t)$ value (imprecision) or a same $m(t)$ value may be captured by distinct $i^k(t)$ values (drift).

9.5.4 Channel Assignment Relation (IN)

We observed that level-1 claims on reliability (9.32) may require that more than one channel redundantly process a same monitored variable $m_i(t)$ (*cf.* Definition 9.14). In this case the value of a monitored variable $m_i(t)$ must be distributed over different channel data acquisition input registers. The relation IN specifies this distributed assignment:

Definition 9.16 The relation IN is defined as follows:

- *domain* (IN) is the set of monitored variables $m_i(t)$;
- *range* (IN) is the set of channel data acquisition registers, $k = 1, 2,...$;
- $< m_i(t), k >$ is in IN if and only if the monitored variable $m_i(t)$ is read by register k, $k = 1, 2,...$

It is worth noting that IN is a relation between identifiers (names) of architecture elements, while the relations IN_k are relations between the values of these elements. The relation IN will be instrumental in Section 9.5.12.2 to discuss the problem of assigning monitored variables to channels in order to satisfy constraints imposed by diversity.

9.5.5 Channel k Output Device (Relations OUT_k)

Let $o^k(t)$ denote the vector containing one element for each of the output variables of channel k

Definition 9.17 A relation OUT_k, defined as follows, describes the behaviour of the output devices of channel k:

- *domain* (OUT_k) is a set of vector-valued time-functions containing all the possible values of $o^k(t)$;
- *range* (OUT_k) is a set of vector-valued time-functions containing possible values of a sub-vector of $c(t)$;
- $<o^k(t), c(t)>$ is in OUT_k if and only if $c(t)$ describes possible values of the controlled variables when $o^k(t)$ describes the values of the output variables.

The OUT_k relations specify the behaviour of the channel k data conversion and output devices (*e.g.* digital to analogue converters, actuator drivers). OUT_k is an abstraction of the protocol – polling, multiplexing, sharing – used to access the channel output register, read the digital values $o^k(t)$ and deliver c(t) control values.

OUT_k is a relation – rather than a function – into the set of possible values of c(t) because of the imperfections of the output devices (*e.g.* digital to analogue converters, actuator drivers). Different values of an element of $o^k(t)$ may correspond to a given c(t) value (imprecision), or a given value of an $o^k(t)$ element can be converted to different c(t) values (drift).

9.5.6 Voting Relations (OUT)

Level-1 claims on reliability (9.32) may require that a pair of monitored/controlled variables be processed in distinct redundant channels. In this case, the jth output variables $o^k_j(t)$, from different channels k = 1, 2,... are used to compute the value of a same controlled variable $c_i(t)$. The relation OUT is the specification of the computation of the controlled variable values in function of channel output redundant variable values:

Definition 9.18 The relation OUT is defined as follows:

- *domain* (OUT) is the set of output registers, k = 1, 2...;
- *range* (OUT) is the set of controlled variables $c_i(t)$;
- <k, $c_i(t)$> is in OUT if and only if the controlled variable $c_i(t)$ is accessed by the output register k, k = 1, 2 ...

OUT is a relation between names of architecture elements, while the relation OUT_k is a relation between the values of these elements.

When *undesired events* affect the values of some of the output variables, we shall see how the computation (voting) defined by OUT can be "safety-oriented" or "availability-oriented", depending on whether OUT privileges or not values of the controlled variables that are fail-safe, *i.e.* that belong to FS(t).

9.5.7 Channel k Requirements (REQ_k)

The functional relations between the input and the output register of channel k are specified by a relation REQ_k.

Definition 9.19 REQ_k is defined as follows:

- *domain* (REQ_k) is a set of vector-valued time functions containing the possible values of $i^k(t)$,
- *range* (REQ_k) is a set of vector-valued time functions containing values of $o^k(t)$,
- <$i^k(t)$, $o^k(t)$> is in REQ_k if and only if the channel assigns the values $o^k(t)$ to its output register when the input register takes the value $i^k(t)$.

9.5.8 Level-2 Undesired Events (HAZ^2)

Not only the normal behaviour of the computer system but also its possible failures have to be considered. Several safety guides and standards, *e.g.* [48], require from safety-critical systems the ability to detect and, to the extent possible, to correct their own failures. These *undesired events* (UE's) must be anticipated; they must be part of the model and of the conceptual design basis.

UE's that may occur and affect the architecture must therefore be postulated. Let $h^2(t)$ (see Figure 9.3) be the vector of architecture entities that are potential sources of UE's; these sources may be a failing sensor, a failing input or output register, a communication line rupture, a communication protocol, a voting device. It is important to underline that only operational UE's have to be considered; the possibility of events that would result from deficiencies of the architecture design itself (*e.g.* a missing unit or channel) has to be excluded by claims on the validity of the architecture; in particular, claims on architecture completeness.

Each element of the vector $h^2(t)$ corresponds to an entity of the architecture which is a potential source of *undesired events*. When different from zero, the current value of the element indicates that the corresponding entity has caused an *undesired event* (*e.g.* has failed), and the type of the event. The occurrence of an *undesired event* at time t is thus described by the pair $(h^2(t), t)$. An event may be caused by (the failure of) a single entity, or by (the simultaneous failure of) more than one entity.

The effects UE's may have on the contents of the channel input/output registers are described by relations between the values taken by the registers of the channels and the values of these events:

Definition 9.20 For every channel k, we call HAZ^2_k the relation defined as follows:

- *Domain* (HAZ^2_k) is the set of vector-valued time-functions containing the possible values of $h^2(t)$;
- *Range* (HAZ^2_k) is the set of vector valued time functions containing possible values of $i^k(t)$ and of $o^k(t)$;
- $<h^2(t), i^k(t)>$ is in HAZ^2_k if and only if $(i^k(t), t)$ is an event occurring whenever $(h^2(t),t)$ occurs; and $< h^2(t), o^k(t)>$ is in HAZ^2_k if and only if $(o^k(t), t)$ is an event occurring whenever $(h^2(t),t)$ occurs.

Remark 9.3 The sources of *undesired events* are described by a unique vector $h^2(t)$, and not by separate vectors associated with channels. One must indeed take into account that some UE's may simultaneously affect more than one channel, causing coincident failures in these channels. Such a common cause $(h^2(t),t)$ event value belongs to the domain of several channel HAZ^2_k relations. In contrast, a failing sensing device for instance can affect a single or a few channels only. For these $h^2(t)$ source of UE non-zero values, the *range* of HAZ^2_k includes, for the affected channel(s) k only, special $i^k(t)$ values that unambiguously reflect the nature of the failure (input value missing, out of range, corrupted, *etc.*).

expansion (CLM0, level 2)

<u>**begin**</u>

$\mathcal{A}^1$_REQ_**acceptable_implementation.clm[1,2]**

$\quad$ $U_2 (\delta \mathcal{A}^1 REQ)$ $\sqsupset$

$\qquad$ **architectural_feasibility_and_completeness.clm2**
$\qquad$ $\wedge$ ($\forall$k: **channel_k _correctness.clm2**
$\qquad\qquad$ $\wedge$ $\mathcal{A}^2$_**channel_k_accuracy.clm2**
$\qquad\qquad$ $\wedge$ $\mathcal{A}^2$_**channel_k _acceptable_implementation.clm2**)

REQ_fail_safe_implementation.clm[1,2]

$\quad$ $U_2 (\delta \, FS)$ $\sqsupset$

$\qquad$ ($\forall$k: h^2(t)_k_**effects_validity.clm2**
$\qquad\qquad$ $\wedge$ h^2(t)_k_**detectability.clm2**
$\qquad\qquad$ $\wedge$ h^2(t)_ k_**mitigation.clm2**
$\qquad\qquad$ $\wedge$ **channel_k_fail_safe_implementation.clm2**)

$\mathcal{D}^1$(h')_**functional_diversity_implementation.clm[1,2]**

$\quad$ $U_2(\delta \, \mathcal{D}^1 div)$ $\sqsupset$

$\qquad$ **h'_channel_functional_diversity.clm2**
$\qquad$ $\wedge$ ($\forall$ pair (i,j) $\in$ $\mathcal{D}^2$(h'):
$\qquad\qquad$ (i,j)_**channel_ independency.clm2**)

$\mathcal{R}^1$_**reliability.clm[1,2]**

$\quad$ $U_2 (\delta \mathcal{R}^1 rel)$ $\sqsupset$

$\qquad$ ($\forall$k:**channel_k_reliability.clm2**)
$\qquad$ $\wedge$ ($\forall$(i,r) $\in$ $\mathcal{R}^1$:
$\qquad$ m_i(t)_**r_redundancy.clm2**
$\qquad\qquad$ $\wedge$ ($\forall$ pair (j,k) $\in$ I^2(i,r): (j,k)_**channel_ independency.clm2**))

$\mathcal{D}^1$_**safety.clm[1,2]**

$\quad$ $U_2 (\delta \, \mathcal{D}^1 saf)$ $\sqsupset$

$\qquad$ ($\forall$k: **channel_k_safety.clm2**)
$\qquad$ $\wedge$ ($\forall$h': **h'_channel_functional_diversity.clm2**)

<u>**end**</u> **expansion (CLM0, level 2)**

Figure 9.4. Level-2 Expansions

9.5.9 Level-2 Expansion of Level-1 Delegation Claims

The delegation claims of the level-1 argument (Section 9.4.15) have to be expanded into level-2 claims, formulated in some architecture descriptive language L_2. As for level-1, these *expansions* and the level-2 claims *justifications* must be interpreted and valuated by L_2- *substructures*.

However, at level-2, as well as at levels 3 and 4, these *expansion substructures* must satisfy an essential property that was not explicitly required at level-1. A level-2 *expansion structure* – denoted $U_2(\delta\xi_2)$ (see Definition 8.11) – which valuates a delegation claim $\delta_{1,2}\xi_2$ must be an *extension* (Definition 8.7) of the L_1-*justification structures* of the *primary black* claims β which predicate this delegation claim (Proposition (iii) in Theorem 8.7); and the language $L_2(\delta\xi_2)$ must be an *expansion* (Definition 8.5) of the languages $L_1(\beta)$.

This obligation provides additional guidance on the elicitation and interpretation of the level-2 expansions, in contrast with level-1 where *justification structures* for **CLM0** *dependability requirements* are in general not available.

Just as for level-1, *justification structures* at level-2 must be *elementary substructures* of the *expansion structures*; as a consequence, and for the sake of clarity, it is therefore helpful to postpone the definition of the level-2 *expansion structures* to a later stage (Section 9.5.14), after the *justification elementary substructures* have been identified. The reader may find a preview of the expansions at level-2 in Figure 9.4.

Claims at level 2 result from expansions of the level-1 black claim delegations specified in the level-1 argument (Section 9.4.15, page 149). The following sections develop these expansions and elicit the level-2 claims. As explained in Section 6.6.1, these expansions produce claims of three different kinds: architecture validity, acceptable and fail-safe implementation, and architecture dependability.

9.5.10 Expansion of REQ Acceptable Implementation Delegation Claim[1,2]

The level-1 delegation claim $\mathcal{A}^1\_REQ\_\textbf{acceptable\_implementation.clm[1,2]}$ (see the argument in 9.4.15) is an important delegation claim to be expanded at level 2. This claim – see (9.23) – was predicated in the level-1 *justification structure* $U_1(REQ)$ defined by (9.11), and in that structure alone:

$$(9.39) \qquad U_1(REQ) = <\{m(t)\}\{c(t)\}; \{NAT\}\{REQ\} ; \{k_1\}>$$

where the set $\{k_1\}$ includes all level-1 constants relevant to NAT and REQ, and in particular the set $\mathcal{A}^1$ of parameter values specifying the *accuracy* required for the implementation of the *system input-output specifications*.

According to Proposition (iii) in Theorem 8.7, page 103, $U_1(REQ)$ has to be a *substructure* (Definition 8.7) of the level-2 *expansion structure* of this delegation claim $\mathcal{A}^1\_REQ\_\textbf{acceptable\_implementation.clm[1,2]}$. In agreement with the notation introduced in Section 8.8.4, we call $U_2(\delta\mathcal{A}^1REQ)$ this level-2 *expansion structure*; thus, we must have:

$$(9.40) \qquad U_1(REQ) \subseteq U_2(\delta\mathcal{A}^1REQ).$$

The components of $U_2(\delta\mathcal{A}^1REQ)$ will be identified in Section 9.5.10.5. Let us recall that *acceptability* was defined at level-1 (Definition 9.12) as a generic property requiring a comprehensive, correct and accurate implementation. This delegation claim must therefore be expanded and valuated at level-2 by claims on (i) architecture completeness, (ii) channel correctness, (iii) channel accuracy and (iv) channel acceptable implementation.

9.5.10.1 Architecture Completeness and Feasibility

The set of channel data acquisition devices must be able to transmit some value for every condition that can occur in the environment; thus one must necessarily have for the *whole set* of channels k:

$$(9.41) \qquad \{\cup_k: domain\ (IN_k)\} \supseteq domain\ (NAT),$$

where *domain* (NAT) is given by Definition 9.6.

One must also have:

$$(9.42) \qquad \{\cup_k:\ range\ (OUT_k)\} \supseteq range\ (REQ)$$

where *range* (REQ) is given by Definition 9.10, since the *set* of output devices of the channels must be able to transmit control data for all effects specified by REQ.

Clearly, the Propositions (9.41), (9.42) must valuate to true in the *justification structure*:

$$(9.43) \qquad U_2(REQ) = \ <\{m(t)\}\{c(t)\}\{\cup_k:\{i^k(t)\}\{o^k(t)\}\}\ ;$$
$$\{NAT\}\{REQ\}\{\cup_k:\{IN_k\}\{OUT_k\}\}\ ;$$
$$\{k_2\}>$$

where $\{k_2\}$ includes the constants related to all channels.

Besides, the values produced by a channel output device must be "feasible", *i.e.* compatible with the environment constraints specified by NAT; thus:

$$(9.44) \qquad \forall_k: (range\ (OUT_k) \subseteq range\ (NAT))$$

must also hold in U_2 (REQ).

Moreover, for every pair of monitored–controlled variable vector values that satisfy the *system input-output specifications* REQ, there must exist at least one complete channel with acquisition input register, processing unit and delivering output register which implements this specification. This condition implies that the following *logical consequence* valuates to true in U_2 (REQ):

(9.45) $\forall$ m(t), $\forall$ c(t):

$$U_2(\text{REQ}) \models (\text{REQ}(m(t),c(t)) \Rightarrow (\exists k: IN_k(m(t),i^k(t)) \cap OUT_k(o^k(t), c(t))\,)).$$

Last, following Lemma 8.7, page 92, the evidence provided by the valuations of the *interpretation* U_2(REQ) is valid (correct, complete and consistent) if and only if U_2(REQ) is an *elementary substructure* of the $\mathbf{L_2}$-*expansion structure* $U_2(\delta \mathcal{A}^1\text{REQ})$ of the delegation claim $\mathcal{A}^1\_\text{REQ}\_\textbf{acceptable\_implementation.clm}\,[\,1,2\,]$.
  In conclusion, we have the following claim:

(9.46) **architectural_feasibility_and_completeness.clm2:**
  Propositions (9.41),(9.42), (9.44), (9.45) are *satisfied* (valuate to true) in the *justification substructure* U_2 (REQ) defined by (9.43),
  and U_2 (REQ) $\prec U_2(\delta \mathcal{A}^1\text{REQ})$.

Thus, Propositions (9.40) and (9.46) imply the following obligations on *structures*:

$$U_1(\text{REQ}) \subseteq U_2(\delta \mathcal{A}^1\text{REQ})$$
$$U_2(\text{REQ}) \prec U_2(\delta \mathcal{A}^1\text{REQ}).$$

9.5.10.2 Channel Correctness

The functional relations between the input and output registers of any channel k must comply with the *system input-output specifications*. Let us consider, for every channel k, the *structure*:

(9.47) $\qquad U_2(\text{REQ}_k) = <\ \{m(t)\}\,\{c(t)\}\,\{i^k(t)\}\,\{o^k(t)\}\ ;$
$$\qquad\qquad\qquad \{\text{REQ}\}\,\{IN_k\}\,\{OUT_k\}\,\{\text{REQ}_k\}\ ;$$
$$\qquad\qquad\qquad \{k_2\} >$$

where $\{k_2\}$ is the set of constants of channel k. Correctness of channel k implies that the following logical consequence holds in the structure $U_2(\text{REQ}_k)$ for any value m(t) of the monitored variables such that IN_k (m(t), i^k(t)):

(9.48) **channel_k_correctness.clm2:**
  $\forall$ m(t) | IN_k (m(t), i^k(t)):
  REQ(m(t), c(t))
$$\underset{U_2(\text{REQ}_k)}{\models} (IN_k(m(t),i^k(t)) \circ \text{REQ}_k(i^k(t),o^k(t)) \circ OUT_k(o^k(t),c(t))),$$

  with *justification substructure* U_2 (REQ_k) defined by (9.47)
  and U_2 (REQ_k) $\prec U_2(\delta \mathcal{A}^1\text{REQ})$.

The *logical consequence* (9.48) constrains the channel specifications REQ_k. Every valuation of $U_2(REQ_k)$ which valuates to true the closed well-formed formula (*sentence*) on the right hand side must also valuate to true the predicate $REQ(m(t), c(t))$, (*cf.* (8.11)).

Remark 9.4 The *logical consequence* (9.48) implies that the relation $REQ(m(t),c(t))$ in $U_1(REQ)$ remains a *restricted* relation (Definition 8.7) in $U_2(REQ_k)$. If one or two of the predicates IN_k, OUT_k are false, then the behaviour of the channel is not constrained by the claim. For example, if a given value of $c(t)$ is not in the range of OUT_k, any channel behaviour would be considered acceptable.

The domains and relations of the substructures are such that:

$$U_2 (REQ) \supseteq U_2 (REQ_k)$$

and since both structures are elementary substructures of $U_2(\delta\mathcal{A}^{\,1}REQ)$, Lemma 8.5, page 90, implies also:

$$(9.49) \qquad\qquad U_2 (REQ_k) \prec U_2 (REQ).$$

which induces a logical sequence between the corresponding justifications that must be made in these two substructures.

9.5.10.3 Channel Accuracy

Accuracy (Definition 9.12), is an essential property of the behaviour of a channel. The accuracy of the A/D and D/A conversions and of the estimation of the validity of parameter value ranges should be commensurate with the functions to be served. This accuracy is specified by IN_k and OUT_k, which define the relations between the monitored and controlled variables and their value estimations and representations in the input and output registers. Precision and sensitivity can be interpreted at level-2 by upper bounds on the cardinality (denoted $\| \; \|$) of the ranges and domains of these relations.

Consider the following structure:

$$(9.50) \qquad U_2(\mathcal{A}^{\,2}I/O_k) = \;<\{m(t)\}\{c(t)\}\{i^k(t)\}\{o^k(t)\}\;;$$
$$\{IN_k\}\{OUT_k\}\;;$$
$$\{k_2\}>$$

the universe and constants of which are included in those of $U_2(REQ)$ (9.43). And also because of its relations, $U_2(\mathcal{A}^{\,2}I/O_k)$ is $\mathbf{L_2}$-*substructure* of $U_2(REQ)$ and of $U_2(REQ_k)$ for the interpretation of the language $\mathbf{L_2}$ describing the input and output devices of channel k:

$$(9.51) \qquad\qquad U_2(\mathcal{A}^{\,2}I/O_k) \subseteq U_2(REQ)$$

$$U_2(\mathcal{A}^{\,2}I/O_k) \subseteq U_2(REQ_k).$$

Within this structure $U_2(\mathcal{A}^2 \, I/O_k)$, level-2 specifications on accuracy of channel k can be formulated and *interpreted* as:

$$(9.52) \qquad (\forall m_i(t): \quad \| \, range \, (IN_k \, (m_i(t), \, i^k(t))) \, \| \leq \bar{u}_2 \, (m_i(t))$$
$$(\forall c_i(t): \quad \| \, domain \, (OUT_k \, (o^k(t), \, c_i \, (t))) \, \| \leq \bar{u}_2 \, (c_i(t))$$

$$\text{with} \qquad \bar{u}_2 \, (m_i(t)), \, \bar{u}_2 \, (c_i \, (t)) > 0$$
$$\bar{u}_2 \, (m_i(t)), \, \bar{u}_2 \, (c_i \, (t)) \in \mathcal{A}^2 \subset \{k_2\}.$$

The set $\mathcal{A}^2$ of upper bound values $\bar{u}_2$ is included in the set of constants $\{k_2\}$ of $U_2(\mathcal{A}^2 \, I/O_k)$; these upper bounds are derived from the accuracy and tolerance specifications of the set $\mathcal{A}^1$ of $U_1(REQ)$ parameters specified by the level-1 implementation black claim (9.23), a set of parameters itself derived from the **CLM0** requirements.

Accuracy is determined by the implementation of the relations IN_k, OUT_k and REQ_k. The values $m_i(t)$ and $c_i(t)$ in (9.52) designate perfect values for the corresponding monitored and controlled environment parameters, as defined by the domain and range of the relation REQ. The bounds $\bar{u}_2$ are specified by the application. These bounds have to be satisfied by the channel implementation at level-3. However, the channel design alone, even if assumed to be properly and correctly implemented at level-3, may not necessarily be able to enforce and guarantee these bounds. Re-calibration and/or periodic test procedures may be necessary to protect against undesired events, for instance, aging effects on sensing and actuating devices, electromagnetic or other external interferences and instabilities of the input-output devices. Intervention by the operation control is needed in those situations, and claims on channel accuracy require further reification and evidence from level-4.

Hence, the level-2 claim on accuracy is *grey* and its justification is the *logical consequence*:

$$(9.53) \qquad \mathcal{A}^2\_k\_hw\_sw\_accuracy.clm[2,4]$$
$$\wedge \, (\forall m_i(t) \mid IN(m_i(t),k) : \| \, range \, (IN_k \, (m_i(t), \, i^k(t))) \, \| \leq \bar{u}_2(m_i(t))$$
$$\wedge \, (\forall c_i(t) \mid OUT(k, c_i(t)) : \| \, domain \, (OUT_k \, (o^k(t), \, c_i(t))) \, \| \leq \bar{u}_2(c_i \, (t))$$

$$\vdash_{U_2(\mathcal{A}^2 \, I/Ok)} \mathcal{A}^2\_channel\_k\_accuracy.clm2$$

$$\text{with } U_2(\mathcal{A}^2 \, I/O_k) \prec U_2(\delta \mathcal{A}^1 REQ)$$
$$\text{and } U_2(\mathcal{A}^2 \, I/O_k) \text{ defined by (9.50).}$$

Relations in (9.46) (9.48) (9.51) and Lemma 8.5, page 90, imply that:

$$(9.54) \qquad U_2(\mathcal{A}^2 \, I/O_k) \prec U_2(REQ)$$
$$U_2(\mathcal{A}^2 \, I/O_k) \prec U_2(REQ_k).$$

And by (9.49):

$$(9.55) \qquad U_2(\mathcal{A}^2 I/O_k) \prec U_2(REQ_k) \prec U_2(REQ).$$

Similarly to the previous orderings (9.9) and (9.20), (9.55) suggests a logical sequence for justifying the claims that must valuate in these different structures: the channel accuracy and correctness being justified first, before the justification of the completeness of the architecture.

9.5.10.4 Channel Acceptable Implementation

The level-2 evidence provided by $U_2(REQ)$ and $U_2(\mathcal{A}^2 I/O_k)$ valuations to justify (9.46), (9.48) and (9.53) is not sufficient. The *acceptable* implementation of the *system input-ouput specifications* implies that the channel input, output and requirement specifications (IN_k, OUT_k, REQ_k) not only are validated against **CLM0** and the level-1 *primary* claims, but are also comprehensively, correctly and accurately implemented by the lower levels. Accuracy, in particular, is determined by the implementation of the relations IN_k, OUT_k and REQ_k, and especially by the software.

Thus, the *acceptability* of the channels' design requires further evidence from the lower levels; a *black* claim $\mathcal{A}^2$**_channel_k_acceptable_implementation.clm2** requesting an *acceptable* channel implementation at lower levels is necessary in the expansion of the delegation claim **REQ_acceptable_implementation. clm [1,2]**. Although it is somewhat obvious and the supporting evidence is entirely delegated to the lower levels, this claim must be explicitly made at level 2, as part of the delegation claim expansion and as a root of expansions at lower levels. Thus, for every channel k, the *justification* of this *black* claim is the logical consequence, *cf.* (8.13), page 91:

$$(9.56) \qquad \mathcal{A}^2\_\textbf{channel\_k \_acceptable\_implementation.clm[2,3]}$$
$$\vdash_{U_2(REQ_k)} \mathcal{A}^2\_\textbf{channel\_k \_acceptable\_implementation.clm2}$$

> *is satisfied in justification substructure* $U_2(REQ_k)$ *defined by* (9.47) *with* $U_2(REQ_k) \prec U_2(\delta\mathcal{A}^1 REQ)$ *and* $\mathcal{A}^2 \in \{k_2\}$,

$\mathcal{A}^2$ being the set of parameter values specifying the accuracy and tolerance bounds defined by (9.52).

9.5.10.5 The Expansion Structure $U_2(\delta\mathcal{A}^1 REQ)$

We now have all necessary information to identify the elements of the *expansion structure* of the level-1 delegation claim $\mathcal{A}^1$_REQ_acceptable_implementation.clm[1,2]. According to Corollary 8.4, page 105, an *expansion structure* is the smallest *extension* (Definition 8.7) of the upper level *justification structures* in which the delegation claim is predicated and of the *elementary substructures* of the claims of the expansion. Thus, $U_2(\delta \mathcal{A}^1 REQ)$ must satisfy the Propositions (9.40), (9.46), (9.48), (9.53), (9.56):

$$U_1(REQ) \subseteq U_2(\delta \mathcal{A}^1 REQ)$$
$$U_2(REQ) \prec U_2(\delta \mathcal{A}^1 REQ)$$
$$\forall k: U_2(REQ_k) \prec U_2(\delta \mathcal{A}^1 REQ)$$
$$\forall k: U_2(\mathcal{A}^2 I/O_k) \prec U_2(\delta \mathcal{A}^1 REQ)$$

Since the following Propositions (9.51)

$$\forall k: U_2(\mathcal{A}^2 I/O_k) \subseteq U_2(REQ)$$
$$\forall k: U_2(\mathcal{A}^2 I/O_k) \subseteq U_2(REQ_k)$$

hold between these different structures, a necessary and sufficient condition for (9.40), (9.46), (9.48), and (9.53) to hold is:

$$U_1(REQ) \subseteq U_2(\delta \mathcal{A}^1 REQ)$$
$$U_2(REQ) \prec U_2(\delta \mathcal{A}^1 REQ)$$
$$\forall k: U_2(REQ_k) \prec U_2(\delta \mathcal{A}^1 REQ).$$

Following the construction rules of Theorem 8.8, page 106, the "smallest" *expansion structure* which satisfies these propositions is:

$$(9.57) \qquad U_2(\delta \mathcal{A}^1 REQ) = <\{m(t)\}\{c(t)\}\{\cup_k: \{i^k(t)\}\{o^k(t)\}\} ;$$
$$\{NAT\}\{\cup_k:\{IN_k\}\{OUT_k\}\{REQ_k\}\} ;$$
$$\{k_1\},\{k_2\}>,$$

where $\{k_1\}$ are the constants of $U_1(REQ)$ and $\{k_2\}$ includes the constants related to all channels, in particular the set $\mathcal{A}^2$ of parameter specifying the accuracy and tolerance bounds that was defined by (9.52).

Moreover, $U_2(REQ)$ and $U_2(REQ_k)$ must be *elementary substructures* of $U_2(\delta \mathcal{A}^1 REQ)$ which, therefore must satisfy the Tarski–Vaught condition (Theorem 8.1, page 90). Note also that, although REQ is a relation of the *substructure* $U_1(REQ)$, REQ need not be an explicit relation in (9.57) since, thanks to the claims (9.48) on channel correctness and Remark 9.4, REQ is a restricted relation of $U_2(REQ)$.

9.5.11 Expansion of REQ Fail-Safe Implementation Delegation Claim[1,2]

The level-1 delegation claim $\mathcal{A}^1$_REQ_**fail_safe_implementation.clm[1,2]** (see the argument (9.4.15)) was predicated – *cf.* (9.24) – in the level-1 *justification structure* $U_1(FS)$ defined by (9.18), and only in that structure:

$$(9.58) \qquad U_1(FS) = <\{h^1(t)\}\{m(t)\}\{c(t)\}\{FS(t)\} ; \{HAZ^1\}\{REQ\} ; \{k_1\}>$$

where $\{k_1\}$ includes the constants related to HAZ^1 and REQ and, in particular, the class $\mathcal{F}^1$ of the *postulated initiating events* introduced by Definition 9.9, page 127, which must result in the system environment being in a safe state.

Hence, because of Proposition (iii) in Theorem 8.7, page 88, $U_1(FS)$ has to be a *substructure* (Definition 8.7) of the level-2 *expansion structure* of the delegation claim $\mathcal{A}^1$_**REQ_fail_safe_implementation.clm[1,2]**. In agreement with the notation introduced in Section 8.8.4, we call $U_2(\delta FS)$ this level-2 *expansion structure*; thus, we must have:

$$(9.59) \qquad\qquad U_1(FS) \subseteq U_2(\delta FS).$$

$U_2(\delta FS)$ must valuate the *expansion (logical consequence)* of the above delegation claim into claims on the level-2 postulated undesired events validity, detection, mitigation and implementation mechanisms. $U_2(\delta FS)$ constituents include those of these *elementary justification substructures* which will be identified in the following sections.

9.5.11.1 Level-2 Undesired Events Validity

As we did for the level-1 *postulated initiating events*, one must ensure at level-2 that the set of *postulated undesired events* that may affect the architecture is complete. The sources of these events are described by the vector $h^2(t)$ and by the values the elements of this vector can take. This set of values is the *domain* of the HAZ^2_k relations (Definition 9.20).

The domain $\{h^2(t)\}$ of the relations HAZ^2_k must be *complete and consistent*, *i.e.* it must include all architecture *undesired events* of the kind discussed in Section 9.5.8. In addition, HAZ^2_k range must faithfully describe the effects of these *undesired events* on the variables $i^k(t)$ and $o^k(t)$. Let us then consider the *structure*:

$$(9.60) \qquad U_2(HAZ^2_k) = <\{h^2(t)\}\{i^k(t)\}\{o^k(t)\};$$
$$\{HAZ^2_k\};$$
$$\{k_2\}>$$

where $\{k_2\}$ includes the constants relevant to $\{HAZ^2_k\}$. $U_2(HAZ^2_k)$ must be a valid $\mathbf{L_2}$-*interpretation* for potential architecture defects and their effects on channel k, *i.e.* a *model* – Definition (8.4 – of the level-2 anticipated and documented *undesired events* and of their effects on channel k.

Hence the validity claim for every channel k:

$(9.61) \qquad h^2(t)$_**k_effects_validity.clm2**:

$\qquad\qquad U_2(HAZ^2_k) \vdash \mathbf{L_2}\text{-}\textbf{postulated architecture UE's and effects,}$

$\qquad\qquad$ with *justification substructures* $U_2(HAZ^2_k)$ defined by (9.60)
$\qquad\qquad$ and $U_2(HAZ^2_k) \prec U_2(\delta FS)$.

9.5.11.2 Channel Robustness

The adequate detection and mitigation of the level-2 *undesired events* by the channels must be claimed. The architecture must be designed in such a way that *all undesired events* $(h^2(t),t)$ that may correspond to single or multiple failures are unambiguously *detected, controlled* and lead to a *safe state*.

Detection by channel k is guaranteed if the level-2 claim

(9.62) $h^2(t)\_k\_\textbf{detectability.clm2:}$

$$U_2(HAZ^2_k) \vdash (\forall h^2(t) \,|\, (HAZ^2_k (h^2(t), i^k(t)) \vee HAZ^2_k (h^2(t), o^k(t)):$$

$$(HAZ^2_k)^{-1} \text{ is a surjective function } into \, h^2(t))$$

$$\text{with } U_2(HAZ^2_k) \text{ defined by (9.60) and } U_2(HAZ^2_k) \prec U_2(\delta FS).$$

This claim guarantees that every $i^k(t)$, $o^k(t)$ value in the *range* of HAZ^2_k is related to a unique value of $h^2(t)$, *i.e.* that $(HAZ^2_k)^{-1}$ is a function, and that all values of the domain of HAZ^2_k are related to $i^k(t)$ or $o^k(t)$.

With regard to *mitigation*, let us consider the *substructure*:

(9.63) $$U_2(FS;k) = < \{FS(t)\} \,\{h^2(t)\}\,\{i^k(t)\}\,\{o^k(t)\};$$
$$\{IN_k\}\,\{OUT_k\}\,\{REQ_k\}\,\{HAZ^2_k\} \,;$$
$$\{k_2\}>$$

where $\{k_2\}$ includes the constants relevant to channel k and to $\{HAZ^2_k\}$. The channel k is fail-safe if the level-2 claim holds in this *justification substructure*:

(9.64) $h^2(t)\_\, k\_\textbf{mitigation.clm2:}$

$$U_2(FS;k) \vdash$$

$$((\forall h^2(t) \,|\, HAZ^2_k (h^2(t), i^k(t)):$$
$$range\,(HAZ^2_k \circ IN_k \circ REQ_k \circ OUT_k) \in FS(t))$$

$$\wedge (\forall h^2(t) \,|\, HAZ^2_k (h^2(t), o^k(t)):$$
$$range\,(HAZ^2_k \circ OUT_k) \in FS(t)))$$

$$\text{with } justification\ structure \ \ U_2(FS;k) \text{ being defined by (9.63)},$$
$$\text{and } U_2(FS;k) \prec U_2(\delta FS).$$

The claim (9.64) requires from the system the ability to cope with the level-2 *undesired events*. Events of the first kind in (9.64) affect the input devices and can be controlled and mitigated by the channel processor, while the second kind needs to be dealt with by the channel output devices. Given that the fail-safe set FS(t) is validated by the claim (9.7), the claim (9.64) requires that IN_k, REQ_k, and OUT_k include relations that lead the system to a valid safe state on occurrence of an *undesired event*.

Because of their universes and relations, the *structures* $U_2(HAZ^2_k)$ and $U_2(FS;k)$ are such that the following proposition holds for the *interpretation* of the language $\mathbf{L_2}$ used to describe the undesired events and their effects:

$$(9.65) \qquad U_2(HAZ^2_k) \subseteq U_2(FS;k).$$

Since these two *justification structures* must be *elementary substructures* of $U_2(\delta FS)$, and because of Lemma 8.5, page 90, the following proposition must also hold:

$$(9.66) \qquad U_2(HAZ^2_k) \prec U_2(FS;k).$$

Again – as it was already the case for level-1, *cf.* (9.9) and (9.20) – this relation induces a logical sequencing for the justifications that take place in these *substructures*. Quite naturally, the detectability and effects of the level-2 undesired effects (HAZ^2_k) should be validated first, before the validation of their mitigation by the channel specifications (IN_k, REQ_k, OUT_k), so as to lead the system to the safe state subset FS of the environment.

Remark 9.5 It is worth noting that there is a difference with level-1. In contrast with level-1 where $U_1(REQ) \prec U_1(FS)$ (9.20), here one can verify that $U_2(REQ_k)$ is not a *substructure* of $U_2(FS;k)$. Channel correctness with respect to the level-1 system input-output specification REQ (claim (9.48)), and level-2 undesired event mitigation (claim (9.64)) are at level-2 independent properties while, at level-1, the mitigation of the postulated initiating events is part of the system input-output specifications.

9.5.11.3 Channel Fail-Safe Implementation

The level-2 evidence provided by the valuations of the justification *substructures* $U_2(HAZ^2_k)$ and $U_2(FS;k)$ is not sufficient to ensure the fail-safeness of channel k. Detection and mitigation of undesired events must also be ensured at the design level-3. Moreover, assistance of the operation control may be needed also; especially for *undesired events* affecting output devices: because these devices are downstream the channel processor, they cannot be controlled by the processor. Delegation claims to level 3 are therefore necessary.

Although they are somewhat obvious and with evidence entirely delegated to the lower level, such claims must be explicitly made at level 2, as they are part of the primary claim expansions and the roots of expansions at lower levels. Thus, we have:

$$(9.67) \qquad \textbf{channel\_k\_fail\_safe\_implementation.clm2}.$$

For every channel k, the *justification* of this *black* claim is the logical consequence, *cf.* (8.13):

(9.68)　　　　**channel_k _ fail_safe _implementation.clm[2,3]**

　　　　　　$\vDash_{U_2(FS;k)}$ **channel_k _ fail_safe _implementation.clm2**

　　　　　　is satisfied in justification substructure $U_2(FS;k)$ defined by (9.63), with $U_2(FS;k) \prec U_2(\delta FS)$.

9.5.11.4 *The Expansion Structure* $U_2(\delta FS)$

All elements are now available to define the expansion structure of the delegation claim $\mathcal{A}^1$_REQ_**fail_safe_implementation.clm[1,2]**. According to Corollary 8.4, page 105, an *expansion structure* is the smallest *extension* (Definition 8.7) of the upper level *justification structures* in which the delegation claim is predicated and of the *elementary substructures* of the claims of the expansion. Thus, $U_2(\delta FS)$ must satisfy the propositions (9.59), (9.61), (9.62), (9.64) and (9.68), or:

(9.69)　　　　　　　　　　$U_1(FS) \subseteq U_2(\delta FS)$

　　　　　　　　　　　　$\forall k: U_2(HAZ^2_k) \prec U_2(\delta FS)$

　　　　　　　　　　　　$\forall k: U_2(FS;k) \prec U_2(\delta FS);$

these different structures are related by (9.66):

$$\forall k: U_2(HAZ^2_k) \prec U_2(FS;k),$$

so that the second Proposition in (9.69) is implied by the third. Let us then consider the *structure*:

(9.70)　　　　$U_2 = \ <\ \{h^1(t)\}\,\{h^2(t)\}\,\{FS(t)\}\,\{\cup_k: \{i^k(t)\}\,\{o^k(t)\}\};$

　　　　　　　　$\{HAZ^1\}\,\{\cup_k:\{IN_k\}\,\{OUT_k\}\,\{REQ_k\}\,\{HAZ^2_k\}\}\ ;$

　　　　　　　　$\{k_1\}\,\{k_2\}>.$

Every structure $U_2(FS;k)$ is a *reduction* (Definition 8.5) of (9.70); moreover, by virtue of the claim (9.48) and Remark 9.4, page 161, the relation REQ is a *restricted* relation of (9.70); hence, the structure $U_1(FS)$ (9.58) is a *substructure* of (9.70). Therefore, (9.70) is the smallest structure which satisfies

$$\forall k: U_2(FS;k) \prec U_2$$
$$U_1(FS) \subseteq U_2$$

and we have:

$$(9.71) \qquad U_2(\delta FS) = U_2 = <\{h^1(t)\}\{h^2(t)\}\{FS(t)\}\ \{\cup_k: \{i^k(t)\}\ \{o^k(t)\}\};$$
$$\{HAZ^1\}\{\cup_k:\{IN_k\}\{OUT_k\}\ \{REQ_k\}\{HAZ^2_k\}\}\ ;$$
$$\{k_1\}\{k_2\}>$$

where $\{k_1\}$ includes the constants related of $U_1(FS)$, in particular, the class $\mathcal{F}^1$ of the *postulated initiating events* introduced by Definition 9.9 in Section 9.4.5; $\{k_2\}$ includes the constants relevant to channels and relations $\{HAZ^2_k\}$. Besides, for every k, $U_2(FS;k)$ must be an *elementary substructure* of $U_2(\delta FS)$; thus, $U_2(\delta FS)$ must satisfy the Tarski–Vaught condition (Theorem 8.1, page 90).

9.5.12 Expansion of Functional Diversity Delegation Claim[1,2]

The level-1 delegation claim (9.25):

$$\mathcal{D}^1(h')\_\textbf{functional\_diversity\_implementation.clm[1,2]}$$

(see the argument (9.4.15)) was predicated in the level-1 *justification structure* $U_1(HAZ)$ defined by (9.5), and only in that structure:

$$U_1(HAZ) = <\{h^1(t)\}\{m(t)\}\{c(t)\}\ ; \{NAT\}\{HAZ^1\}\ ; \{k_1\}>$$

where $\{k_1\}$ includes the constants related to NAT and HAZ^1 and, in particular, the set of constants $\mathcal{D}^1(h')$ introduced by Definition 9.11 which identifies the distinct monitored variables that are intended to independently detect the PIE of type h'.

Because of Proposition (iii) in Theorem 8.7, page 103, $U_1(HAZ)$ has to be a *substructure* (Definition 8.7) of the level-2 *expansion structure* of the delegation claim (9.25). In compliance with the notation introduced in Section 8.8.4, we call $U_2(\delta\mathcal{D}^1 div)$ this level-2 *expansion structure*; thus, we must have:

$$(9.72) \qquad\qquad U_1(HAZ) \subseteq U_2(\delta\mathcal{D}^1 div).$$

$U_2(\delta\mathcal{D}^1 div)$ must valuate the *expansion (logical consequence)* of the delegation claim from level-1 into claims on architecture, as well as on independency, an indispensable ingredient of functional diversity. These claims are detailed in the next sections. The components of $U_2(\delta\mathcal{D}^1 div)$ will be given in Section 9.5.12.4.

9.5.12.1 Preservation of Functional Diversity at Architecture Level
The *functional diversity* being considered here must not be confused with the hardware and software *Design Diversity* of level 3, different by its nature and purpose and discussed in Section 9.6.7.2.

The different monitoring variables $m_x(t)$ of the diversity set $\mathcal{D}^1(h')$, $d \geq 2$, (Definition 9.11 in Section 9.4.10.3):

$$\mathcal{D}^1 (h') = \{x \mid HAZ^1 (h', m_x(t))\},$$
$$\|\mathcal{D}^1 (h')\| = d, d \geq 1$$

are needed at level-1 to independently detect – and react upon – a level-1 *postulated initiating event* of type h', $h' \in \{h_1\}$. Several independent variables are needed so as to tolerate one or more variables failing to detect the PIE. The system architecture must preserve this independency and, in particular keep the ability to cope with the PIE in situations where parts of the computer architecture could be affected by a failure. A necessary condition for this independency and "graceful degradation" is to have the different variables of $D^1(h')$ independently processed by distinct channels.

For a PIE $h^1(t)$ of type h', the channels that acquire the values of the relevant monitored variables $m_x(t)$ are the channels j of the set:

(9.73) $$\{j \mid IN (m_x(t), j) \; and \; x \in \mathcal{D}^1(h')\}$$

where the channel assignment relation IN is given by Definition 9.16. Let us define the set $\mathcal{D}^2 (h')$ of channels j which process the variables monitoring h':

Definition 9.21 $\qquad \mathcal{D}^2 (h') = \{j \mid IN (m_x(t), j) \; and \; (x \in \mathcal{D}^1(h'))\}.$

$\mathcal{D}^1 (h')$ must be a *one-one injection* ($\rightarrow$) into $\mathcal{D}^2 (h')$ if every monitored variable of $\mathcal{D}^1 (h')$ is allocated to a distinct channel of $\mathcal{D}^2 (h')$, and both sets must have same cardinality: $\|\mathcal{D}^2 (h')\| = \|\mathcal{D}^1 (h')\|$.

Then, let us consider the *substructure* $U_2 (\mathcal{D}^2 div)$:

(9.74) $$U_2(\mathcal{D}^2 div) = <\{h^1(t)\}\{m(t)\}\{j\};$$
$$\{HAZ^1\}\{IN\};$$
$$\{k_2\}>,$$

where $\{j\}$ is the set of input channels:

$$\{j\} = \{j \mid j \in \mathcal{D}^2(h') , h' \in h^1(t)\}.$$

The one-one injection claim can be formulated and valuated in $U_2(\mathcal{D}^2 div)$:

(9.75) **h'_channel_functional_diversity.clm2:**
$$U_2(\mathcal{D}^2\text{div}) \models (\mathcal{D}^1(\text{h'}) \rightarrow \mathcal{D}^2(\text{h'}));$$
with $U_2(\mathcal{D}^2\text{div}) \prec U_2(\delta\mathcal{D}^1\text{div})$
and *justification structure* $U_2(\mathcal{D}^2\text{div})$ defined by (9.74).

The consequence of the channel allocation claimed by (9.75) is that any pair (i,j) of channels belonging to a same set $\mathcal{D}^2$ (h') have different input domains (*domain* (IN_i)), and therefore different functional requirements REQ_i and different ranges of output variables (*range* (OUT_i)) to control actuators:

(9.76) $\forall (i,j) \mid i,j \in \mathcal{D}^2(\text{h'})$ and $(i \neq j)$:
$$((domain\ (\text{IN}_i\ ) \neq domain(\text{IN}_j\ )$$
$$and\ (\text{REQ}_i \neq \text{REQ}_j\ )$$
$$and\ (range\ (\text{OUT}_i) \neq range(\text{OUT}_j)).$$

Channels with such different structures are unlikely to be affected by the same level-2 undesired event and with the same consequences at the same time (claims on independency are also necessary and discussed in Section 9.5.12.3 later on); hence, they offer less possibility of "losing" more than one variable monitoring a same PIE being at the same time.

9.5.12.2 Channel Allocation (Relation IN)

However, the channel allocation scheme claimed by (9.75) raises a practical difficulty and, for economical reasons, is seldom implemented in real systems. Care must be taken that two monitored variables associated with the same postulated initiating event h', h'$\in$ h^1(t), are not allocated to the same channel, even though a same physical parameter, and thus a same monitored variable may have to be associated with different events h'.

Hence, the problem of allocating monitored variables to channels – *i.e.* defining the relation IN – while keeping the number of necessary channels to a minimum, is equivalent to a problem of graph colouring.

In its simplest form, this problem is the search for an optimal way of colouring the nodes of a graph so that no two adjacent nodes share the same colour. In our case, the nodes are the monitored variables; they are adjacent, *i.e.* connected by an edge, if they are associated with the same PIE h'; and their colour is the channel they are allocated to.

The problem of finding the minimum number of colours (channels) needed is known to be NP-hard. The corresponding decision problem (is there an allocation which uses at most k channels?) is NP-complete, and was one of Karp's 21 NP-complete problems [54].

Heuristic algorithms, however, exist. The Welsh–Powell algorithm for colouring a graph uses a simple heuristic improvement to a naive greedy algorithm.

It is easy to show that this heuristic uses at most H + 1 colours (channels), where H is the maximum degree of the graph, *i.e.* the maximum number of distinct PIE's monitored by a same variable.

The algorithm runs as follows:

1. Sort the variables in decreasing order of number of events monitored, and initially have every variable unallocated.
2. Allocate the variables in this decreasing order, allocating a variable channel 1 if it is unallocated and if there is not yet a variable monitoring the same event and allocated to channel 1.
3. Repeat this process with channels 2, 3, *etc.* until no variable is unallocated.

This algorithm can be used to determine the allocation scheme defined by the relation IN, but does not necessarily find a minimum number of channels.

In practice, the average number of events monitored by a same variable is such that the number of distinct independent channels required to satisfy the constraint would be too high. Compromises have to be made. For instance, a channel may be allocated to more than one set $\mathcal{D}^2$ (h'). Or, instead of allocating a full channel – with input, processor and output stage – to a set $\mathcal{D}^2$ (h'), a more economical possibility is to have one channel only with a few independent processor units, each unit processing the variables of one or a few sets $\mathcal{D}^2$ (h'). For example, the monitored variables of the SIP system described in Appendix A are distributed over four so-called "functional groups", each functional group being processed by a separate processing unit in each redundant train. The architecture of some French nuclear reactors consists of five [4] or seven [9] such functional units. These various compromises necessarily imply that some postulated initiating events are monitored by variables sharing some common equipment. They must result from a comparative evaluation of the likelihood and consequences of the simultaneous occurrence of an event h', h'$\in$ h^1(t), and of a system failure causing the loss of all variables of the set $\mathcal{D}^2$(h').

9.5.12.3 Architecture Independency

Functional diversity and, as we shall see in Section 9.5.13, redundancy between channels require independency. Mutual independency between functionally diverse or redundant channels is an essential precondition for the ability of the system architecture to gracefully sustain *undesired events* at levels 2 and 3, events that otherwise would affect more than one – or even all – diverse or redundant channels at the same time.

Channels which require mutual independency are those which process monitored variables assumed to be processed redundantly by the *primary dependability claims* of Section 9.4.12 as well as those which process the functionally diverse monitored variables that detect level-1 *postulated initiating events* specified by the *primary claim on functional diversity* (9.21).

These monitored variables were identified by the sets $\mathcal{R}^1$ (Definition 9.14) and $\mathcal{D}^1$ (Definition 9.11) *respectively*, and documented by the sets of constants $\{k_1\}$ of the substructures $U_1(REQ)$ and $U_1(FS)$.

Hence, the corresponding sets of r channels (k) that must be mutually independent are the sets defined by:

Definition 9.22 $I^2(x,r) = \{k \mid IN\ (m_x(t), k)\ and < x, r > \in \mathcal{R}^1\},$

as well as the sets $\mathcal{D}^2$ (h') (Definition 9.21).

Like diversity, independency is not a transitive property. In order to preserve the independency assumed at level 1 by the *primary claims*, independency between *all* pairs of channels (j,k) in the set I^2 and between all pairs in the sets $\mathcal{D}^2$ must be claimed.

Level-2 independency between two channels j and k requires that both channels cannot be simultaneously affected by a *undesired event* $h^2(t)$ that would cause a coincident failure mode. Let us then consider the *substructure*:

(9.77)$$U_2\ (ind(j,k)) = < \{h^2(t)\}\,\{i^j(t)\}\,\{i^k(t)\}\,\{o^j(t)\}\,\{o^k(t)\}\ ;$$
$$\{HAZ^2_j\}\,\{HAZ^2_k\}\ ;$$
$$\{k_2\}>$$

where $\{k_2\}$ includes the constants of channels j,k.

One can define two channels j and k as being mutually independent at any time *t* with respect to a postulated level-2 *undesired event* $(h^2(t) = h')$ if the following property holds in U_2 (ind(j,k)), *i.e.* if the following intersection of relations is empty:

(9.78)$$\forall\ t:\quad \{[HAZ^2_j(h'(t), i^j(t)) \vee HAZ^2_j(h'(t), o^j(t))]$$
$$\wedge\ [HAZ^2_k(h'(t), i^k(t)) \vee HAZ^2_k(h'(t), o^k(t))]\} = 0.$$

Remember from Section 9.5.8 that failures of the processing unit are not considered as failures of the architecture; these *undesired events* are taken into account at the design level (level-3). The evidence provided by the valuation of (9.78) is therefore not sufficient to guarantee independency between two channels j,k at the lower levels. Common cause failures may affect the processors and/or the software in both channels. Additional evidence will therefore be needed from, and must be delegated to, level-3.

Hence, the level-2 claim on the independency between two channels j and k is *grey* and its justification is the *logical consequence*:

$$(9.79) \qquad \forall (i,r) \in \mathcal{R}^1, \forall h' \in h^1(t): \ \forall\, (j,k) \in I^2(i,r) \cup \mathcal{D}^2(h'):$$

$$\textbf{(j, k)\_channel\_ independency.clm[2,3]}$$
$$\wedge\, (\, U_2\,(\mathrm{ind}(j,k)) \models (\, (\forall h' \in h^2(t), \forall t:$$
$$\{[HAZ^2{}_j(h'(t), i^j(t)) \vee HAZ^2{}_j(h'(t), o^j(t))]$$
$$\wedge\, [HAZ^2{}_k(h'(t), i^k(t)) \vee HAZ^2{}_k(h'(t), o^k(t))]\} = 0)\,)$$

$$U_2(\mathrm{ind}(j,k)) \models \textbf{(j, k)\_channel\_independency.clm2}.$$

with *justification structure* $U_2(\mathrm{ind}(j,k))$ defined by (9.77)
and $U_2(\mathrm{ind}(j,k)) \prec U_2(\delta \mathcal{D}^1 \mathrm{div})$.

This is the first occurrence of a *grey* claim we meet in the **CLM0** expansion; its justification implies a valuation of the *white claim* specification (9.78) in the level-2 structure $U_2(\mathrm{ind}(j,k))$ and a *delegation claim* for evidence of a proper reification of this specification at level-3.

To sum up, channels that need to be mutually independent result from the dependability requirements **CLM0** and the dependability and functional diversity *primary claims* at level-1; $I^2(i,r)$ and $\mathcal{D}^2(h')$ are the sets which specify these designated channels.

As redundant channels are not protected against a common cause failure if they are not independent according to (9.79), the expansions of the level-1 dependability delegation claims must explicitly associate claims on redundancy with claims on independency and appropriate delegation claims; this association will be explicitly made in the expansion detailed later on in Section 9.5.14, and Figure 9.4.

9.5.12.4 The Expansion Structure $U_2(\delta \mathcal{D}^1 \mathrm{div})$

As we did for the previous level-1 delegation claims in Sections 9.5.10.5 and 9.5.11.4, we can now proceed to define the elements of the expansion structure $U_2(\delta \mathcal{D}^1 \mathrm{div})$ for the level-1 delegation claim (9.25):

$$\mathcal{D}^1(h')\_\textbf{functional\_diversity\_implementation.clm[1,2]}.$$

Recall that, according to Corollary 8.4, page 105, an *expansion structure* is the smallest *extension* (Definition 8.7) made of the upper level *justification structures* in which the delegation claim is predicated and of the *elementary substructures* of the claims of the expansion.

Thus, $U_2(\delta \mathcal{D}^1 \text{div})$ must satisfy the following Propositions (9.72), (9.75), (9.79):

$$U_1(\text{HAZ}) \subseteq U_2(\delta \mathcal{D}^1 \text{div})$$
$$U_2(\mathcal{D}^2 \text{div}) \prec U_2(\delta \mathcal{D}^1 \text{div})$$
$$\forall h' \in h^1(t) \colon \ \forall \ (i,j) \in \mathcal{D}^2(h') \colon U_2(\text{ind}(i,j)) \prec U_2(\delta \mathcal{D}^1 \text{div}).$$

The smallest *expansion structure* (Theorem 8.8) which satisfies these Propositions is:

$$(9.80) \qquad U_2(\delta \mathcal{D}^1 \text{div}) = \ < m(t)\} \{c(t)\} \ \{h^1(t)\} \ \{h^2(t)\} \ \{\cup_k \colon \{i^k(t)\} \ \{o^k(t)\}\} \ ;$$
$$\{\text{NAT}\}\{\text{HAZ}^1\}\{\text{IN}\}\{\cup_k \colon \{\text{HAZ}^2_k\}\} \ ;$$
$$\{k_1\}\{k_2\}>$$

where $\forall h' \in h^1(t) \colon \mathcal{D}^2(h') \in \{k_2\}$; $\{k_1\}$ is the set of constants of $U_1(\text{HAZ})$ and, in particular, the set of constants $\mathcal{D}^1(h')$ introduced by Definition 9.11, Section 9.4.10.3, which identifies the distinct monitored variables that are intended to independently detect the PIE of type h'.

Besides, $U_2(\mathcal{D}^2 \text{div})$ and $U_2(\text{ind}(i,j))$, for $(i,j) \in \mathcal{D}^2(h')$ and $h' \in h^1(t))$, must be *elementary substructures* of $U_2(\delta \mathcal{D}^1 \text{div})$; thus $U_2(\delta \mathcal{D}^1 \text{div})$ must satisfy the Tarski–Vaught condition (Theorem 8.1, page 90).

9.5.13 Expansion of Reliability and Safety Delegation Claims[1,2]

The delegation claim $\mathcal{R}^1\_\text{reliability.clm[1,2]}$ **(9.33)** is predicated in the *justification substructure* $U_1(\text{REQ})$ defined by (9.11), and the delegation claim $\mathcal{D}^1\_\text{safety.clm[1,2]}$ (9.35) in *justification substructure* $U_1(\text{FS})$ defined by (9.18).

Reliability and safety were explicitly claimed at level-1 on the assumption that some degrees ($\mathcal{R}^1$) of redundant processing and diversity ($\mathcal{D}^1$) are available at level-2 (*cf.* Section 9.4.12.2). The delegation claims **(9.33)** and (9.35) must therefore be expanded into level-2 claims requiring not only channel reliability and safety - but also architecture independency, diversity and redundancy.

Let $U_2(\delta \mathcal{R}^1 \text{rel})$ and $U_2(\delta \mathcal{D}^1 \text{saf})$ be the *expansion structures* in which the expansions (*logical consequences*) of **(9.33)** and (9.35) must hold respectively. According to Proposition (iii) in Theorem 8.7, page103, these structures have to be *extensions* (Definition 8.7) of the level-1 *justification substructures* in which of the delegation claims are predicated:

$$(9.81) \qquad\qquad U_1(\text{REQ}) \subseteq U_2(\delta \mathcal{R}^1 \text{rel})$$
$$U_1(\text{FS}) \subseteq U_2(\delta \mathcal{D}^1 \text{saf});$$

and the two **CLM0** level-1 *interpretations* are related by the Proposition (9.20) established at level-1:

$$(9.82) \qquad\qquad U_1(REQ) \prec U_1(FS).$$

The following sections successively detail the expansions of **(9.33)** and (9.35) into claims on architecture redundancy, channel reliability and safety.

9.5.13.1 Architecture Redundancy and Independency

Which monitored variables need to be processed redundantly is determined – or more precisely "assumed" by the primary dependability claims on reliability and safety.

$\mathcal{R}^1$ (Definition 9.14) is the set of those designated monitored variables with their required redundancy r. This set belongs to the set of constants of the *substructure* $U_1(REQ)$, given by (9.11). And the relation IN (Definition 9.16) specifies how the monitored variable values are distributed over different channels.

A given environmental variable $m_i(t)$ is monitored and correctly processed (9.48) by at least r distinct identical channels at any time t if the Proposition (9.48) *holds* for each of these channels in a structure $U_2(REQ_k)$ (9.47) which valuates the channel correctness.

The following *white* claim must therefore hold:

$(9.83) \quad \forall\, (i,r) \in \mathcal{R}^1 : m_i(t)\_\boldsymbol{r}\_\textbf{redundancy.clm2}:$

$$(\exists_r k: IN_k(m_i(t), i^k(t)) \circ REQ_k(i^k(t), o^k(t)) \circ OUT_k(o^k(t), c(t))$$
$$\underset{U_2(REQ_k)}{\dashv} REQ(m_i(t), c(t)\,)$$

with *justification structure* $U_2(REQ_k)$ defined by (9.47),
and $U_2(REQ_k) \prec U_2(\delta\mathcal{R}^1 rel)$,

and where the quantifier $\exists_r k$ must be interpreted as "*there exists at least r values k such that*". If *satisfied,* this claim guarantees that the monitored variable $m_i(t)$ are processed according to REQ by at least r distinct channels.

As stated in Section 9.5.12.3 on architecture independency, these r redundant channels must also be mutually independent and satisfy claim (9.79). The set of redundant channels processing the monitored variable $m_i(t)$ is (Definition 9.22) the set

$$I^2(i,r) = \{k \mid IN\,(m_i(t), k) \text{ and} < i, r > \in \mathcal{R}^1\}.$$

Thus, we have also here the claim, similar to (9.79):

(9.84) $(\forall\ (i,r) \in \mathcal{R}^1 : \forall j,k \in I^2(i,r):$

$(j,\ k)\_\textbf{channel\_ independency.clm2}$

with *justification structure* $U_2(\text{ind}(j,k))$ defined by (9.77)

and $U_2\ (\text{ind}(j,k)) \prec U_2(\delta\mathcal{R}^1\text{rel})$,

where the sets $\mathcal{R}^1$ and $I^2(i,r)$ are included in the set $\{k_2\}$ of $U_2(\delta\mathcal{R}^{-1}\text{rel})$. Hence, structures $U_2(\text{ind}(j,k))$ are *elementary substructures* of the two *expansion structures* $U_2(\delta\mathcal{R}^1\text{rel})$ and $U_2(\delta\mathcal{D}^1\text{div})$.

9.5.13.2 The Relations $\widetilde{IN}_k$, $\widetilde{REQ}_k$ and $\widetilde{OUT}_k$

The dependability *delegation* claims $\mathcal{R}^1\_\textbf{reliability.clm}[1,2]$ and $\mathcal{D}^1\_\textbf{safety.clm}[1,2]$ are based on the assumption that there is a non-zero probability that the implementation of the REQ relation is flawed and affected by defects that have not been anticipated; these probabilistic claims assume that the $\widetilde{REQ}$ implementation may not satisfy the *primary delegation claims* of the level-1 argument (Section 9.4.15):

$$\mathcal{A}^1\_\text{REQ\_acceptable\_implementation.clm}[1,2]$$
$$\text{REQ\_fail\_safe\_implementation.clm}[1,2].$$

In other words, the dependability delegation claims on reliability and safety assume a non-zero probability that the level-2 claims on channel completeness, correctness, accuracy or robustness we have expanded in the previous section are not satisfied. The dependability *delegation claims* must be expanded into appropriate probabilistic level-2 claims which require level-2 statistical evidence on the dependable behaviour of a possibly imperfect architecture implementation $\widetilde{REQ}$.

Probabilistic claims on the dependability of the architecture can be formulated, interpreted and valuated by imposing bounds on reliability, and safety probabilistic measures similar to those of level-1. The assumptions are thus of the same nature as those made and discussed in Section 9.4.12 for the probabilistic *primary claims*.

Assuming that the sets of *undesired events* and of relations HAZ^2_k, IN_k, REQ_k, OUT_k are validated by the previous level-2 claims for channel k, and given possibly imperfect implementations $\widetilde{IN}_k$, $\widetilde{REQ}_k$ and $\widetilde{OUT}_k$ of the channels' input, output and processing devices, the probability that these implementations do not provide at some time t the service required by the input-output specifications REQ can then be defined as:

(9.85) $f_k(t) = \text{Prob}\ \{\widetilde{c}\ (t) \neq m(t)\ (IN_k \circ REQ_k \circ OUT_k)\ |\ \widetilde{IN}_k, \widetilde{REQ}_k\ \widetilde{OUT}_k\ \}$,

where $\widetilde{c}\ (t)$ is the c(t) vector value obtained by these implementations. The probability $f_k(t)$ is taken over the set of possible values of the monitored variables m(t); m(t) is here a *random variable* with values in *domain* (REQ) and with probability defined by (9.26):

$$F_1(x,t) = \text{Prob } \{m(t) = x\}, \ x \subseteq domain \text{ (REQ)}.$$

Let $h^2(t)$ be a *random variable* with values in *domain* (HAZ^2) and let us define the probability density function:

(9.86) $\qquad H_2(x,t) = \text{Prob}\{ h^2(t) = x\}, \ x \subseteq domain \text{ }(HAZ^2).$

The probability of the implementations $\widetilde{IN}_k$, $\widetilde{REQ}_k$ and $\widetilde{OUT}_k$ being unsafe upon occurrence of some *undesired event* $(h^2(t),t)$ is then defined as:

(9.87) $u_k(t) =$

$$\text{Prob}\{ (\tilde{c}(t) \neq (h^2(t)(HAZ^2_k \circ IN_k \circ REQ_k \circ OUT_k)))$$
$$\cup (h^2(t)(HAZ^2_k \circ OUT_k)) \notin FS(t)) \mid \widetilde{IN}_k ; \widetilde{OUT}_k ; HAZ^2_k \}.$$

Probabilities (9.85) and (9.87) are models of two types of randomness: the randomness captured by the probability distributions $F_1(x,t)$, $H_2(x,t)$ and the randomness of failure occurrences caused by the imperfections of the implementations described by $\widetilde{IN}_k$, $\widetilde{REQ}_k$ and $\widetilde{OUT}_k$.

These imperfect implementations are of the same nature as $\widetilde{REQ}$ (Definition 9.13); they are not only uncertain, but also to a certain degree undefined. The domain of $\widetilde{IN}_k$ is the same as the *domain* (IN_k) of monitored variables, but the range of $\widetilde{IN}_k$ and the domains and ranges of $\widetilde{REQ}_k$ and $\widetilde{OUT}_k$ are unknown as they include values of the input, output and controlled variables that are not permissible. Should these implementations be "perfect", then $\widetilde{IN}_k$, $\widetilde{REQ}_k$ $\widetilde{OUT}_k$ could be replaced by their corresponding relations in the conditional probabilities (9.85), (9.87), which would then be necessarily equal to 0.

Since they are undefined, $\widetilde{IN}_k$, $\widetilde{REQ}_k$ and $\widetilde{OUT}_k$ do not belong to level-2 deterministic *interpretations*. They have to be valuated by statistical numerical test data. For the implementations $\widetilde{IN}_k$ and $\widetilde{OUT}_k$, the statistical data are part of the evidence to be provided at level 2. For $\widetilde{REQ}_k$, the data have to be provided by level-3, since the dependability of the implementation of the REQ_k relations is a level-3 property of the hardware and software design of the channel k processing unit. In particular, this explains why the expression of $u_k(t)$ takes into account *undesired events* affecting the channel input and output devices only, *undesired events* affecting the processing unit being modelled at the design level.

The vectors m(t) and $h^2(t)$ in the Expressions (9.85) and (9.87) are vectors of *random variables* and the probabilities are taken over their possible values, *i.e.* the *domains* of REQ and HAZ^2_k. The probability distribution $F_1(x,t)$ of the free variable m(t) corresponds to the use and environment profiles specified by the *dependability requirements* **CLM0**; $F_1(x,t)$ data are included in the set of level-1 constants $\{k_1\}$ of the structure $U_1(REQ)$, *cf.* claim **(9.33)**. The $h^2(t)$ probability distribution $H_2(x,t)$ is part of the evidence to be provided at level 2 and must be included in the set of constants $\{k_2\}$ of an appropriate level-2 *structure.*

9.5.13.3 Channel Reliability and Safety

The probability of failure:

$$f(t) = \text{Prob} \{ \widetilde{c}\,(t) \neq REQ\,(m(t)) \mid \widetilde{REQ} \}$$

the valuation of which is claimed at level-1 by the delegation claim $\mathcal{R}^1\_\textbf{reliability.clm}[1,2]$ **(9.33)**, and the probability:

$$u(t) = \text{Prob}\{ \widetilde{c}\,(t) \neq (h^1(t)\,(HAZ^1 \circ REQ)) \mid \widetilde{REQ} \}$$

the valuation of which is claimed at level-1 by the delegation claim $\mathcal{D}^1\_\textbf{safety.clm}[1,2]$ (9.35), are computed from the sets of channel probabilities $f_k(t)$ (9.85) and $u_k(t)$ (9.87), respectively; the computation must take into account the level-2 architecture redundancy implemented between the channels and the voting strategies (relation OUT) implemented to process the channel outputs.

The Delegation Claims **(9.33)** and (9.35) have to be expanded into level-2 claims specifying bounds for $f_k(t)$ and $u_k(t)$. These claims have to take into account the bounds specified by the probabilistic claims of level-1 and the redundancy and diversity implied and specified by the parameter sets $\mathcal{D}^1(h',d)$ and $\mathcal{R}^1$ (*cf.* Section 9.4.12.2) of the structures $U_1(FS)$ and $U_1(REQ)$.

If we approximate relations by functions, a probabilistic claim on an upper bound for the channel failure rate, *i.e.* for the expected number of failures of channel k to deliver the required service during an interval $[t_2 - t_1)$ can be *formulated* as :

$$(9.88) \quad \textbf{channel\_k\_reliability.clm2}: \quad \overline{F}_k(t_1,t_2) = (t_2 - t_1)^{-1} \int_{t_1}^{t_2} f_k(t)\,dt \leq F_{kmax},$$

and valuated in the interpretation structure $U_2(REQ_k)$ defined by (9.47).

F_{kmax} belongs to the set of constants $\{k_2\}$ of $U_2(REQ_k)$. The interval $[t_2 - t_1)$ is specified at level-1 as the interval over which the system is expected to deliver a correct service (Claim (9.32) page 142);this interval and the profile probability density $F_1(x,t)$, (9.26) belong to the set $\{k_1\}$ of the *structure* $U_1(REQ)$ (*cf.* claim **(9.33)** in Section 9.4.12.2), which is a *substructure* of $U_2(\delta\mathcal{R}^1 rel)$; they are also included in the constants of $U_2(\delta\mathcal{R}^1 rel)$ and of its elementary substructure $U_2(REQ_k)$.

Likewise, a claim on an upper bound for the expected number of a channel k unsafe failures on occurrence of a level-2 *undesired event* during a time interval $[t_2 - t_1)$ can be *formulated* as:

(9.89) **channel_k_safety.clm2:** $\qquad \overline{U}_k(t_1,t_2) = (t_2 - t_1)^{-1} \displaystyle\int_{t_1}^{t_2} u_k(t)\,dt \ \leq U_{kmax}\,,$

and valuated in the interpretation structure $U_2(FS;k)$ defined by (9.63). The bound U_{kmax}, the constants relevant to $\{HAZ^2_k\}$, the interval $[t_2 - t_1)$ and the density function $H_2(x,t)$ (9.86) of the probability of a level-2 undesired event belong to the set of constants $\{k_2\}$ of $U_2(FS;k)$. Note that Proposition (9.65) holds:

$$U_2(HAZ^2_k) \prec U_2(FS;k).$$

Like level-1 probabilistic claims, the claims (9.88), (9.89) address the channel implementations and not the validity of the channel requirements and *undesired events* specifications; they assume that the validity claims of Sections 9.5.10.1 and 9.5.10.2 are satisfied.   But, contrary to level-1, these level-2 claims are probabilistic *grey claims*; they require level-2 statistical evidence on the dependable behaviour of the $\widetilde{IN}_k$ and $\widetilde{OUT}_k$ input-output device implementations as well as lower level (design) evidence on the dependability of the $\widetilde{REQ}_k$ implementation.

Moreover, the time intervals for the integrals (9.88), (9.89) have to be selected as being the time instants at which the behaviour of the system can be considered as being independent of the past. The times at which periodic maintenance, re-calibrations and periodic functional tests of the channels are scheduled are examples of such regeneration times. Evidence of such regeneration points has to be provided by the operation control level.

The claims (9.88), (9.89) need therefore to be expanded into – and inferred from – sub-claims at the *design* and the *operation* levels; they are *grey* claims requiring evidence from levels-2, 3 and 4. For every channel k, their justifications are the following two logical consequences:

(9.90) $\qquad$ **channel_k_reliability [2,3]** $\wedge$ **channel_k_reliability [2,4]**

$$\wedge\ (\overline{F}_k(t_1,t_2) = (t_2 - t_1)^{-1} \int_{t_1}^{t_2} f_k(t)\,dt \ \leq F_{kmax})$$

$$\vDash_{U_2(REQ_k)}\ \textbf{channel\_k\_reliability.clm2},$$

with *justification structure* $U_2(REQ_k)$ defined by (9.47) and $U_2(REQ_k) \prec U_2(\delta\mathcal{R}^1\text{rel})$.

(9.91)   **channel_k_safety [2,3]** $\wedge$ **channel_k_safety [2,4]**

$$(\overline{U}_k(t_1,t_2) = (t_2 - t_1)^{-1} \int_{t_1}^{t_2} u_k(t)\,dt \ \leq U_{k\max}$$

$$\models_{U_2(FS;k)} \ \textbf{channel\_k\_reliability.clm2}$$

with *justification structure* $U_2(FS;k)$ defined by (9.63) and $U_2(FS;k) \prec U_2(\delta\mathcal{D}^1\text{saf})$.

In practice, the *justification substructures* $U_2(REQ_k)$, $U_2(FS;k)$, together with the profile distributions $F_1(x,t)$, $H_2(x,t)$, are the specification of the random channel tests, the results of which provide the statistical data which valuate the probabilities $f_k(t)$ and $u_k(t)$ of the channel failing or being unsafe.

Remark 9.6 Under dependability requirements more stringent than the single failure criterion, the architecture may be required to sustain the occurrence of concomitant *undesired events* both at the environment and architecture levels; then, joint distributions of the random variables $h^1(t)$ and $h^2(t)$ would have to be considered instead of the distribution $H_2(x,t)$, (9.86).

9.5.13.4 *The Expansion Structures* $U_2(\delta\mathcal{R}^1\text{rel})$ *and* $U_2(\delta\mathcal{D}^1\text{saf})$
We can now proceed to define the components of the expansion structures introduced in Section 9.5.13 for the level-1 delegation claims **(9.33)**, (9.35):

$$\mathcal{R}^1\_\textbf{reliability.clm[1,2]}$$
$$\mathcal{D}^1\_\textbf{safety.clm[1,2]}.$$

According to Corollary 8.4 the *expansion structures* are the smallest *extensions* (Definition 8.7) of the upper level *justification structures* in which the delegation claim is predicated and of the *elementary substructures* of the claims of the expansion. Thus, $U_2(\delta\mathcal{R}^1\text{rel})$ and $U_2(\delta\mathcal{D}^1\text{saf})$ must satisfy the following Propositions (9.81), (9.83), (9.84), (9.90), (9.91):

$$U_1(REQ) \subseteq U_2(\delta\mathcal{R}^1\text{rel})$$
$$U_1(FS) \subseteq U_2(\delta\mathcal{D}^1\text{saf})$$
$$\forall k: U_2\,(REQ_k) \prec U_2(\delta\mathcal{R}^1\text{rel})$$
$$\forall (i,r) \in \mathcal{R}^1 : \forall\, (j,k) \in I^2(i,r): U_2\,(\text{ind}(j,k)) \prec U_2(\delta\mathcal{R}^1\text{rel})$$
$$\forall k: U_2(FS;k) \prec U_2(\delta\mathcal{D}^1\text{saf}).$$

The smallest *expansion structures* (Theorem 8.8) which satisfy these Propositions are:

$$(9.92) \qquad U_2(\delta \mathcal{R}^1 \text{rel}) = <\{m(t)\}\{c(t)\}\{h^2(t)\}\{\cup_k: \{i^k(t)\}\ \{o^k(t)\}\} ;$$
$$\{NAT\}\{\cup_k:\{IN_k\}\{OUT_k\}\{REQ_k\}\{HAZ^2_k\}\} ;$$
$$\{k_1\}\{k_2\}>$$

where $\{k_1\}$ is the set of constants of $U_1(REQ)$ and $\{k_2\}$ includes the sets $I^2(i,r)$, $(i,r) \in \mathcal{R}^1$, and the set of constants of the *elementary substructures* U_2 (REQ_k), $\forall k$; and the structure:

$$(9.93) \qquad U_2(\delta \mathcal{D}^1 \text{saf}) = <\{m(t)\}\{c(t)\}\ \{h^1(t)\}\{h^2(t)\}\ \{FS(t)\}\{\cup_k: \{i^k(t)\}\ \{o^k(t)\}\} ;$$
$$\{REQ\}\ \{HAZ^1\};\{\cup_k:\{IN_k\}\{OUT_k\}\{REQ_k\}\{HAZ^2_k\}\}$$
$$\{k_1\}\{k_2\}>$$

where $\{k_1\}$ is the set of constants of $U_1(FS)$ and $\{k_2\}$ includes the sets of constants of the *elementary substructures* $\forall k$: $U_2(FS;k)$.

Besides, for all k, U_2 (REQ_k), $U_2(FS;k)$ and for $(i,r) \in \mathcal{R}^1$, and $(j,k)\in I^2(i,r)$, $U_2(\text{ind}(j,k))$ must be *elementary substructures* of $U_2(\delta \mathcal{R}^1 \text{rel})$ and $U_2(\delta \mathcal{D}^1 \text{saf})$; these *expansion structures* have therefore to satisfy the Tarski–Vaught condition (Theorem 8.1, page 90).

9.5.14 Level-2 Expansions

Contrary to level-1, level-2 (as well as lower levels) has more than one *expansion structure*; the level-2 *expansion structures* that have been constructed are: $U_2(\delta \mathcal{A}^1 REQ)$ (9.57), $U_2(\delta FS)$ (9.71), $U_2(\delta \mathcal{D}^1 \text{div})$ (9.80), $U_2(\delta \mathcal{R}^1 \text{rel})$ (9.92) and $U_2(\delta \mathcal{D}^1 \text{saf})$ (9.93).

As explained in Section 8.9.3, these *expansion structures* constitute the upwards interface of level-2 to level-1; each *expansion structure* is the smallest *extension* (Definition 8.7) of upper level *justification structures* in which the delegation claim being extended is predicated, and of the *elementary substructures* of the claims of the expansion (Corollary 8.4, page105).

Figure 9.9, page 242, gives a comprehensive view of these *expansion* structures and of their interdependencies with their *elementary substructures* and their *substructures* of the level-1.

Claims at level-2 are generated by the expansions of the *black delegation claims* of level-1 specified in the level-1 argument (Section 9.4.15, page 149). These expansions are the first step of the argument at level-2; they are given in Figure 9.4, page 157, with the notation introduced in Section 9.2. As said in Section 6.6.1, the claims ***.clm2** in these expansions are basically claims on validity, dependability and implementation. They are the most typical of the architecture level; obviously, they do not exclude other possible claims. Depending on the application, other claims may need to address properties required from multiplexing, voting and timer mechanisms (*e.g.* external watchdogs); typical claims of this kind are claims on the absence of spurious trips.

Another observation is that, as opposed to the level-1 expansions, the level-2 expansions – as well as the lower level expansions – are less specific to the application domain. Since they address the computer system architecture and design, these expansions do not vary as much from one safety case to the other; hence they are representative of a relatively wide family of systems.

9.5.15 Level-2 Evidence and Delegation Claims

Let us now summarise what evidence is needed at this second level and what evidence has to be delegated to the lower levels. Level-2 claims are white, grey or black.

9.5.15.1 Level-2 White Claims Evidence

There is a number of level-2 claims that are exclusively *interpreted* and *valuated* at level-2. They *hold* true in level-2 *elementary substructures*; their evidence entirely belongs to level-2; all these claims are *white*.

These *white* claims are the *validity* claims on the architecture completeness (9.46), and channel correctness (9.48); the claims requiring architecture redundancy (9.83) and diversity (9.75); and the claims on channel robustness (9.61), (9.62), (9.64).

The *justification elementary substructures* $U_2(REQ)$ (9.43), $U_2(REQ_k)$ (9.47), $U_2(\mathcal{D}^2 div)$ (9.74), $U_2(HAZ^2_k)$ (9.60) and $U_2(FS;k)$ (9.63) are the *L_2-interpretation* for valuating these claims (Definition 8.2, page 85). Evidence is obtained by valuations mapping each predicate symbol, each function symbol, each operation symbol and each parameter of the language $\mathbf{L_2}$ in which the architecture is described, to a relation and a constant of these justification structures, and by showing that all these mappings valuate to true.

9.5.15.2 L-2 Black Claims

The level-2 is an abstraction of the lower levels of design and operation and control. Claims on the channels' implementation *acceptability* (9.56) and *safeness* (9.67) must hold true in lower level *justification structures*. Their evidence is exclusively provided by the valuation of these structures. In particular, detection and mitigation of level-2 channel *undesired events* (9.64) may require assistance by the design and operation control. This level-2 and 3 evidence is at level-2 required by the black claim on *safeness* (9.67).

9.5.15.3 L-2 Grey Claims

We observed that claims on accuracy (9.53), independency (9.79) and on probabilistic dependability properties (9.88), (9.89), must not only *hold* at level-2 but also at lower levels. These claims must be inferred not only from level-2 evidence but also from delegation subclaims onto the lower levels.

Evidence on channel accuracy is obtained by validation and testing of the ADC and DAC converters. Testing these converters is discussed in the next Section

9.5.16. As explained in section 9.5.10.3, a complement of evidence is required by a *delegation claim* on level-4 operation control.

Additional evidence on independency is required by a *delegation claim* on design independency at level-3.

The completeness, correctness and accuracy of the level-3 channel design is implied by the *black* claim on acceptable implementation (9.56).

The statistical evidence needed to valuate the dependability claims is discussed in Section 9.5.13. Evidence for the claims (9.90), (9.91) on channel reliability and safety is provided at level-2 by the valuations of the structures $U_2(REQ_k)$ (9.47) and $U_2(FS;k)$ (9.63); evidence on the behaviour of the implementation $\widetilde{REQ}_k$ is delegated to level 3. Evidence on the existence of regeneration points for the estimation of reliability and safety is delegated to level 4.

9.5.16 A Digression on Testing

On the beginning, the introduction called attention on the difficulties of testing digital systems, as compared to analogous systems. Testing an ADC module for obtaining evidence of its accuracy is a good example of the problem.

Consider an ADC module with an input analogue signal varying between 0 and 100% and a 8 bits parallel output. Considering the module as a black box and ignoring its digital implementation, a classical "hardware" verification would involve testing 5 points, at 0%, 25%, 50%, 75%, and 100% of the input range; interpolation and continuity arguments would justify that if the module is correct at these points, it is also correct in between. The input value corresponding to the output bit of weight x, $0 \leq x \leq 7$, is equal to $(100 \times 2^x/256)\%$. Hence, the three hardware verifications 25%, 50%, and 75% exercise the two bits $x = 6$ and 7 only, and it is the 100% test alone which exercise the other six bits, and all at once. A minimum of 8 distinct tests is necessary to flip every output bit individually, but many more test cases are clearly required to check all relevant bit combinations, depending on the specific internal design of the module.

9.5.17 Level-2 Argument

A generic structure of level-2 *arguments* (Definition 6.3, page 45) is given hereafter; references in parentheses point to the text occurrences of *white* claim *justifications, grey* claim *expansions* and *black* claim *assertions*:

(9.94) **argument (CLM0, level-2)**

   begin

   expansion (CLM0, level-2); /*see Figure 9.4*/

   /* **white claim justifications:** */

   $U_2(REQ) \models$ **architectural_feasibility_and_completeness.clm2;** /*(9.46)*/

$\forall k: U_2(REQ_k) \models$ **channel_k_correctness.clm2**; /*(9.48)*/

$\forall k: U_2(HAZ^2_k) \models h^2(t)\_k\_$**effects_validity.clm2**; /*(9.61)*/
$\forall k: U_2(HAZ^2_k) \models h^2(t)\_k\_$**detectability.clm2**; /*(9.62)*/
$\forall k: U_2(HAZ^2_k) \models h^2(t)\_k\_$**mitigation.clm2**; /*(9.64)*/

$\forall h' \in h^1(t): U_2(\mathcal{D}^2 div) \models h'\_$**channel_functional_diversity.clm2**; /* (9.75)*/
$\forall (i,r) \in \mathcal{R}^1 : (\exists_r k: U_2(REQ_k) \models (m_i(t)\_r\_$**redundancy.clm2**);/* (9.83)*/

*/ <u>**grey claims evidence and delegations**</u>: */

$\forall k: U_2(\mathcal{A}^2 I/O_k) \wedge \mathcal{A}^2\_$**channel_k_accuracy.clm[2,4]**

$\qquad \models_{U_2(\mathcal{A}^2 I/Ok)} \mathcal{A}^2\_$**channel_k_accuracy.clm2**; /*(9.53)*/

$\forall (i,r) \in \mathcal{R}^1, \forall h' \in h^1(t): \forall (j,k) \in I^2(i,r) \cup \mathcal{D}^2(h'):$
$\quad U_2(ind(j,k)) \wedge (j,k)\_$**channel_independency[2,3]**
$\qquad \models_{U_2(ind(j,k))} (j,k)\_$**channel_independency.clm2**; /*(9.79)*/

$\forall k: U_2(REQ_k) \wedge$ **channel_k_reliability [2,3]** $\wedge$ **channel_k_reliability [2,4]**
$\qquad \models_{U_2(REQ_k)}$ **channel_k_reliability.clm2**; /*(9.90)*/
$\forall k: U_2(FS;k) \wedge$ **channel_k_safety [2,3]** $\wedge$ **channel_k_safety [2,4]**
$\qquad \models_{U_2(FSk)}$ **channel_k_safety clm2**; /*(9.91)*/

*/ <u>**black claims delegation predicates**</u>: */

$(\forall k: \mathcal{A}^2\_$**channel_k_acceptable_implementation.clm[2,3]**
$\qquad \models_{U_2(REQ_k)} \mathcal{A}^2\_$**channel_k_acceptable_implementation.clm2**; /*(9.56)*/

$(\forall k:$ **channel_k_fail_safe_implementation.clm[2,3]**
$\qquad \models_{U_2(FS;k)}$ **channel_k_fail_safe_implementation.clm2**); /*(9.68)*/

<u>**end**</u> **argument (CLM0, level-2)**

9.5.18 Level -2 Documentation

An outcome of practical interest of this Section 9.5 analysis is to precisely and comprehensively delineate the information which is required to expand and justify the claims made at level-2. This information should necessarily be part of the level-2 documentation of a safety case supporting the **CLM0** claims. To the extent that these claims address essential properties of the architecture, this material should also be part of the documentation of any dependable system architecture.

The level-2 documentation should include:

1. For each *primary black* claim from level-1, the description of the universe and relations of the corresponding expansion model (**L$_2$**-*interpretation* and *expansion structure)*; these *expansion structures* $U_2(\delta\mathcal{A}^1 REQ)$ (9.57), $U_2(\delta FS)$ (9.71), $U_2(\delta\mathcal{D}^1 div)$ (9.80), $U_2(\delta\mathcal{R}^1 rel)$ (9.92) and $U_2(\delta\mathcal{D}^1 saf)$ (9.93) include:

   - for every channel k, the sets of input variables $i^k(t)$ and output variables $o^k(t)$, the relations IN_k, OUT_k, REQ_k, the channel assignment and voting relation IN and OUT;
   - the set of *undesired events* $h^2(t)$ and the relations HAZ^2_k;
   - the parameters $\{k_2\}$ which include:
     - the set $\mathcal{A}^2$ of channel accuracy bounds,
     - the sets $\mathcal{D}^2(h')$ and $I^2(i,r)$ of the channels claimed diverse, redundant and mutually independent,
     - the $i^k(t)$, $o^k(t)$ and $h^2(t)$ probability distributions and time intervals that valuate reliability and safety claims at level 2.

2. The arguments (9.94) detailing the *expansion* of every black primary claim into level-2 claims, the level-2 *justifications* and the *delegations* to lower levels required for these claims.
3. The description of the level-2 *justification elementary substuctures* on the interpretation of which the argument is based: $U_2(REQ)$ (9.43), $U_2(REQ_k)$ (9.47), $U_2(\mathcal{A}^2 I/O_k)$ (9.50), $U_2(ind(i,j))$ (9.77), $U_2(\mathcal{D}^2 div)$ (9.74), $U_2(HAZ^2_k)$ (9.60) and $U_2(FS;k)$ (9.63).
4. The documented evidence which validates – by valuation of these *justification structures* – the level-2 *white, grey, delegation and black* claims; this evidence is the set of mappings onto these *justification elementary structures* which valuate these claims to true.

This documentation, essential to the validation of the architecture of a dependable system, must be accessible to:

- System architects and designers, software and hardware engineers; before final release, they should approve the documentation with respect to feasibility and readability.
- Safety experts and licensing authorities, for compliance with the system dependability requirements and their implications on architecture diversity, redundancy, independency, reliability, and safety.

9.6 Level-3. Design

9.6.1 Complexity and Elements of the Design

The level-3 addresses the innermost part of the system: the internal structure of the channel(s). The level-3 *substructures* must provide the representations of the channels' hardware and software behaviour that are needed to valuate the evidence required by the upper level claims. The *domain* of these representations may have to cover for each channel as much as the processing and the diagnostic hardware, the hardware clocks and interrupt mechanisms, the operating system, the application software development platform, the software libraries, and the application specific software. Hardware and software *undesired events* must also be described, with the mechanisms provided to control them: hardware and software auto-tests, watchdogs, timers, *etc*.

These different hardware and software elements can make the *substructures* at this level quite complex, probably more than at any of the three other levels. In practice, and for a real system, it might therefore be useful to distinguish in level-3 several inner levels of abstraction and to organize the corresponding *substructures* (*interpretations*) as a hierarchy of abstractions, along principles similar to those suggested in [12, 23, 74, 89]. In any case, the degree of detail and the complexity of the representations should remain limited to what is necessary to formulate and expand the claims that are delegated onto this level and to valuate the evidence needed to support them. The ability of circumscribing the scope of the representations to what is necessary and objectively dictated by specified valuations is a powerful weapon against complexity which a claim oriented scheme only can offer.

This ability is exploited here. What is given below is what appears to be a minimal set of *substructures* to valuate the evidence required by the delegation claims introduced in the two previous sections. The *domain* includes the states of the I/O registers, the processor registers, the states of clocks, watchdogs and timers as well as the *undesired events* that can affect these states.

The *relations* of the level-3 *substructures* correspond to the input data conversion and validation checks, the software functional relations, the software library functions and the impact of *undesired events*.

9.6.2 The Design Relations (SOF$_k$)

At level-3, the level-1 and level-2 *substructures* can be assumed to be valid representations of the environment system interface and the architecture. In particular, the REQ system input-output relations are assumed to be feasible with respect to the NAT constraints, the variables $m(t)$, $c(t)$, $i^k(t)$ and $o^k(t)$ being defined as in the previous sections.

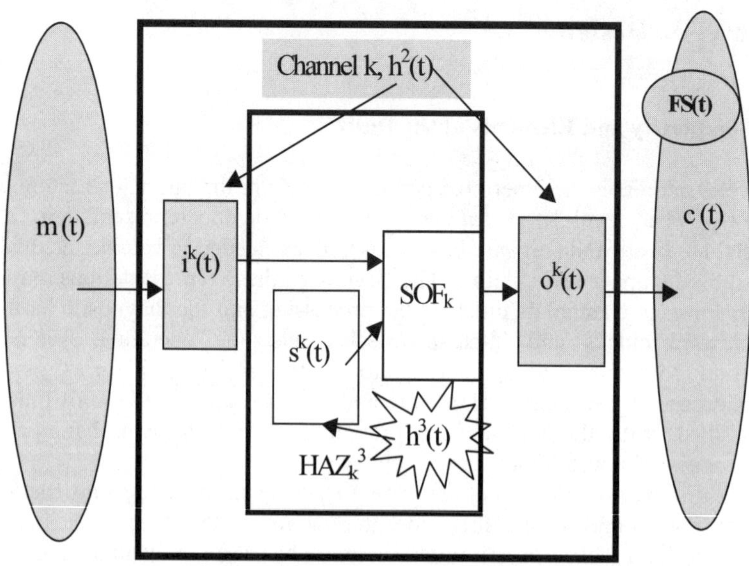

Figure 9.5. Structures at design level

As already mentioned in Sections 8.2.1 and 9.1, *substructures* at the design level are (not necessarily finite) *state machines*. The entire system implementation rests on these level-3 *substructures*, at the heart of which are the states of the channel processing units. In essence, the behaviour of a hardware and software processing unit is an ordinal sequence or a time sequence of states. The universes of the *substructures* include sets (not necessarily finite) of states and time values; the possible behaviours of the processing units are described by *relations* over these state and time values.

Definition 9.23 Let the vector $s^k(t)$ define the *state* at time t of the channel k processing unit; the elements of the vector describe at time t the contents of the processor registers, the value of the clock, the state of the watchdog, and the contents of the memory cells in such a way that a time sequence $s^k(0)$, $s^k(1)$,... $s^k(t)$, $s^k(t+1)$... uniquely and completely defines the behaviour of the processing unit.

Following an approach similar to [79], the input-output behaviour of the software and of the underlying processing hardware of a channel k is described by a relation, called SOF$_k$. SOF$_k$ describes the permissible new values at time (t+1) of the *state* $s^k(t+1)$ and the permissible values $o^k(t)$ of the output register if, at time t, the *state* is $s^k(t)$ and $i^k(t)$ is the value of the input data acquisition register.

Definition 9.24 The relation SOF$_k$ is defined by:

- *Domain* (SOF$_k$) is a set of vector-valued time functions containing the possible values of the vectors $s^k(t)$ and $i^k(t)$, t = 0,1,...;

- *Range* (SOF_k) is a set of vector-valued time functions containing values of $s^k(t)$ and $o^k(t)$, $t = 0,1,\ldots$;
- $(i^k(t)\ s^k(t), o^k(t)\ s^k(t+1))$ is in SOF_k if and only if the pair $(i^k(t)\ s^k(t))$ describes the values of the state and the input variables at time t, and the pair $(o^k(t)\ s^k(t+1))$ describes the permissible values of the output variables of channel k at time t and the new values of the state at time $(t+1)$.

In this definition, the grain of the time t is determined by the successive time instants at which the value of $s^k(t)$ is modified and depends on the level of detail at which the hardware and the software need to be described. For a detailed description of the software behaviour, the successive time instants of interest correspond to the incrementations of the processor instruction counter. For some properties of the channel behaviour, it may be sufficient to consider the channel internal state at successive end of cycles when output values are updated. For even more macroscopic properties, it may be sufficient to consider the values $i^k(t)$ and $o^k(t)$, $t = 0,1,\ldots$ only, ignoring the state $s^k(t)$, as done in [79], for instance.

The hardware and software behaviour defined by SOF_k is deterministic if the relation SOF_k is a function.

9.6.3 Design Level Undesired Events

As for the architecture level-2, it is essential that the design of a dependable system anticipates the *undesired events* (UE's) that may occur at the level-3 of the hardware/software implementation. These events must be postulated in a dependability case, and their consequences must be evaluated; hence, they are an essential element of the level-3 *substructures*.

But contrary to level-2, two kinds of *undesired events* need to be considered at level-3. Some events directly affect the hardware, these being caused by physical effects like aging, maintenance, electromagnetic or seismic interferences, variations of environmental conditions (temperature, pressure...). These events are to a great extent unpredictable and must therefore be considered as being random events. Other events are the consequence of defects remaining in the hardware or the software design. When triggered by an appropriate input profile, these defects systematically – and reproducibly if the system is deterministic – affect the system behaviour.

These two types of *undesired events* are thus of a quite different nature and their effects on the system behaviour can be quite different. They require different treatment and need to be modelled differently. They are analysed in the two following sections.

9.6.3.1 Random Undesired Events (Relation HAZ^3*)*
Potential causes of random *hardware undesired events* are a processor arithmetic or logic unit failure, a failing register, a failing memory location, a watchdog run-out time, *etc.* Let $h^3(t)$ be the vector of these potential sources. When not zero, the value of an element of $h^3(t)$ indicates that the corresponding source of a hardware *undesired event* has been triggered and the value indicates the type of

event. The occurrence of a hardware *undesired event* at time t is described by the pair $(h^3(t),t)$. Such an event may be caused by a single or more than one source (multiple failure).

Random hardware *undesired events* affect the *state* of the channel processing unit. The channel hardware/software implementation of a dependable system should be designed so as to detect and deal with these events by means of self-diagnostics, correcting or fault-tolerance mechanisms, in an attempt either to continue processing or to assign the control variables with safe state values. SOF_k relations model these mechanisms.

The impact of random *undesired events* is modelled by a relation between these events and the *states* of the channel processing unit, regarded as a state machine.

Definition 9.25: For channel k, we call this relation HAZ^3_k:

- *Domain*(HAZ^3_k) is the set of vector-valued time-functions containing the possible values of $h^3(t)$;
- *Range*(HAZ^3_k) is the set of vector valued time functions containing values of $s^k(t)$;
- $<h^3(t), s^k(t)>$ is in HAZ^3_k if and only if $(s^k(t), t)$ is an event occurring whenever $(h^3(t),t)$ occurs.

9.6.3.2 Design Defects and Exceptions

The processing unit and its software may exhibit *undesired* run-time behaviours which do not comply with the input-output specifications REQ and/or the channel specifications REQ_k. These undesired behaviours are the consequences of errors caused by design flaws in the hardware or by programming and coding anomalies of the software, which escaped reviews, verifications and tests.

These hardware design and software defects are of two kinds: those that can be detected and mitigated by the system and those that cannot.

Design and software defects of the first kind cause what are usually called *exceptions. Exceptions* are detectable by the software; for instance, a memory depletion or an abnormal result of calculation – *e.g.* a zero divisor – may be observed and detected by a module or a subroutine expecting and not receiving an input with given properties. Other examples are a numerical or memory under/overflow, a buffer or stack over/underflow, an interrupt caused by a software check or an invariant violation. Such *exceptions* affect the channel processing unit state $s^k(t)$. When detected, they can be handled by exception handling and recovery software, which attempts to continue processing or to assign the control variables with safe state values. SOF_k relations also model this recovery software. However, contrary to the *random undesired events* discussed earlier, exceptions of this kind are likely to be caused by *systematic* errors or *systematic* misuses of a software module. They are likely to be recurrent, eventually forcing the channel to a stop, hopefully but not certainly in a safe state; a situation which, in any case, can be tolerated by the system for one or a few redundant channels only, but not for all of them.

Hardware design and software defects of the second kind are those that are not anticipated and thus not observable. They result in channel behaviour SOF_k that is not *acceptable*, *i.e.* that does not comply with the relations REQ_k and cause the system to have behaviour different from the behaviour specified by REQ. We will see later how evidence on the acceptability of the relations SOF_k is claimed and defined (acceptability claim (9.100)). And we shall discuss in Section 9.6.7.2 how design diversity can mitigate the consequences of design defects.

9.6.4 Level-3 Expansion of Level-2 Delegation Claims

Claims at level 3 result from the expansions of the *grey* and *black delegation* claims of level 2 specified in the level-2 argument (9.94). As for level-2, these claims essentially address validity, dependability and requirements on the lower level, in this case the control of operation.

Level-3 expansions are interpreted and valuated by L_3-*expansion structures*. A level-3 *expansion structure* $U_3(\delta\xi_3)$ (see *e.g.* (8.16)) – which valuates the logical sequence of the expansion of a delegation claim $\delta_{23}(\xi_3)$ – must be an *extension* (Definition 8.7, page 88) of the L_2-*justification structure* in which this delegation claim is predicated (Proposition (iii) in Theorem 8.7, page 103).

As explained in Section 6.6.1, Level-3 expansions are performed with regard to criteria of design validity, implementation acceptability and robustness, and dependability. The following sections develop the expansions of the upper level delegation claims and elicit the level-3 claims. An overview of these expansions is given on Figure 9.6.

9.6.5 Expansion of Channel Acceptable Implementation Delegation Claim[2,3]

The delegation claim from level 2:
$\mathcal{A}^2$_**channel_k_acceptable_implementation.clm[2,3]**, in argument (9.5.17), is a central claim on level-3, predicated at level-2 in the *justification structure* $U_2(REQ_k)$, (9.56). This claim is expanded at level-3 into claims on *acceptability* (Definition 9.12). The expansion is a logical consequence valuated in an *expansion structure* which, to be consistent with the notation of Section 8.8.4 (page 96), we will call $U_3(\delta\mathcal{A}^2REQ_k)$. According to Proposition (iii) in Theorem 8.7, $U_2(REQ_k)$ must be a *substructure* (Definition 8.7) of this level-3 *expansion structure*:

$$(9.95) \qquad \forall k: U_2(REQ_k) \subseteq U_3(\delta\mathcal{A}^2REQ_k)$$

The constituents of $U_3(\delta\mathcal{A}^2REQ_k)$ must include those of its *elementary justification substructures* which are identified in the following sections.

Acceptability was defined at level-1 as a generic property requiring (Definition 9.12) a comprehensive, correct and accurate implementation. The delegation claim above must therefore be expanded and valuated at level-3 by claims on a (i) complete, (ii) correct and (iii) accurate hardware and software implementation of the specifications valuated at level-2 by the IN_k, REQ_k, and OUT_k relations.

---

expansion (CLM0, level 3)

<u>begin</u>

$\forall$k: $\mathcal{A}^2$_**channel_k_acceptable_implementation.clm[2,3]**
    $U_3(\delta\mathcal{A}^2REQ_k) \sqcap$
        $\forall$k: k_hw_sw_**completeness.clm3**
           $\wedge$ (k_hw_sw_**correctness.clm3**
           $\wedge$ $\mathcal{A}^3$_k_hw_sw_**accuracy.clm3**

$\forall$k: **channel_k_fail_safe_implementation.clm[2,3]**
    $U_3(\delta FS_k) \sqcap$
        ($\forall$k: $h^3(t)$_k_**effects_ validity.clm3**
          $\wedge$ $h^3(t)$_k_**detectability.clm3**
          $\wedge$ $h^3(t)$_k_**mitigation.clm3**
          $\wedge$ **channel_k_fail_safe_control.clm3**

($\forall$ pair (j,k) $\in$ (I^2 $\cup$ ($\forall$h' $\in$ $h^1(t)$: $\mathcal{D}^2$(h')))):
    (j,k)_**channel_independency[2,3]**
    $U_3(\delta ind(i,j)) \sqcap$
        ($\forall$h' $\in$ $h^3(t)$: (j,k)_**hardware_h'_independency.clm3**
            $\wedge$ (nil | (j,k)_**design_diversity.clm3**)

channel_k_reliability [2,3]
    $U_3(\delta rel_k) \sqcap$
      software_k_reliability.clm3

channel_k_safety [2,3]
    $U_3(\delta saf_k) \sqcap$
      software_k_safety.clm3

<u>end</u> expansion (CLM0, level 3)

Figure 9.6. Level-3 expansions

9.6.5.1 Hardware and Software Completeness

Completeness of the hardware-software implementation implies that the universe of the relation SOF_k includes the universe of REQ_k, for every channel k:

(9.96) $\qquad$ $(\forall k: \textit{domain}\ (SOF_k) \supseteq \textit{domain}\ (REQ_k)).$

$(\forall k: \textit{range}\ (SOF_k) \supseteq \textit{range}\ (REQ_k))$

Moreover, every channel specification $REQ_k(i^k(t), o^k(t))$ constrained by the level-2 claim (9.45) on architectural completeness, and satisfying the channel correctness claim (9.48), must entail the existence of a hardware-software implementation, that is a SOF_k relation and a state such that:

(9.97) $\qquad$ $\forall\ i^k(t),\ o^k(t) \mid REQ_k(i^k(t),\ o^k(t)) :$

$$(\exists s^k(t): (SOF_k(i^k(t)\ s^k(t), o^k(t)\ s^k(t+1))$$
$$\underset{U_3(SOF_k)}{\models} REQ_k\ (i^k(t),\ o^k(t));$$

hence, the following claim :

(9.98) $\qquad$ **k_hw_sw_completeness.clm3:**
(9.96) and (9.97) are *satisfied* in the *structure* $U_3(SOF_k)$
with $U_3(SOF_k) \prec U_3(\delta \mathcal{A}^2 REQ_k)$,

and where the *elementary justification substructure* is:

(9.99) $\qquad$ $U_3(SOF_k) = <\{i^k(t)\}\ \{s^k(t)\}\ \{o^k(t)\};$
$\{SOF_k\}\ \{REQ_k\}\ ;$
$\{k_3\}>;$

$\{k_3\}$ is the set of constants related to the implementation described by SOF_k.

9.6.5.2 Correctness
The software of channel k and its processing hardware must comply with the functional relations between the input and output registers of that channel. More precisely, the following claim must be *satisfied* (valuate to true) in the structure $U_3(SOF_k)$ for any monitored variable $m(t)$ such that $IN_k\ (m(t),\ i^k(t))$:

(9.100) $\qquad$ **k_hw_sw_correctness.clm3:**

$$(\forall s^k(t) : REQ_k\ (i^k(t),\ o^k(t))$$
$$\underset{U_3(SOF_k)}{\models} SOF_k(i^k(t)\,s^k(t),\ o^k(t)\,s^k(t+1)),$$

with $U_3(SOF_k)$ defined by (9.99)
and $U_3(SOF_k) \prec U_3(\delta \mathcal{A}^2 REQ_k)$.

In other words, every valuation of the structure $U_3(SOF_k)$ *satisfying, i.e.* valuating to true, the right hand side closed well formed formula of (9.100) must also *satisfy* the left hand side predicate: the channel input-output specification predicate must be the *logical consequence* of the channel input-output and software relations.

Remark 9.7 The channel specifications REQ_k ($i^k(t)$, $o^k(t)$) are validated against the system input-output specifications $REQ(m(t),c(t))$ by the level-2 claim on channel correctness (9.48). This claim and the *logical consequence* (9.100) constrain the acceptability of the software and its supporting hardware, and bear a special meaning. If one or two of the predicates IN_k, OUT_k are false, then any software behaviour is considered *acceptable*. For example if a given value $i^k(t)$ is not in the range of IN_k, the behaviour of acceptable software is not constrained by (9.100). This is nothing else but a formal instantiation of the familiar statement: "If a program has not been specified, it cannot be incorrect". These faulty situations for which the software remains here unspecified are dealt with at level 4 by claim (9.157) and Remark 9.12, page 228.

Remark 9.8 (9.97) and (9.100) imply that for all pairs $<(i^k(t), o^k(t))>$, the relation $REQ_k(i^k(t),o^k(t))$ in $U_2(REQ_k)$ is a *restricted* relation of the relation $SOF_k(i^k(t), o^k(t))$ in $U_3(SOF_k)$ (Definition 8.7). The relations SOF_k and REQ_k are indeed not identical. In particular, SOF_k additionally includes the pairs $<(i^k(t), o^k(t))>$ resulting from the detection and mitigation of the level-3 *undesired events* (see below). However, because of its wider set of relations, $U_2(REQ_k)$ is not a *substructure* of $U_3(SOF_k)$.

9.6.5.3 Accuracy
Accuracy is the third attribute of *acceptability* required from level 2 onto level 3 by the delegation claim $\mathcal{A}^2$_**channel_k_acceptable_implementation.clm[2,3]** in the level-2 argument (9.5.17).   Errors resulting from chopping number machine representations, rounding errors, error propagation and algorithm instabilities must be kept within limits. Let us consider the following substructure:

$$(9.101) \qquad U_3(\mathcal{A}^3 I/SOF_k) = \; <\{m(t)\}\{c(t)\}\{i^k(t)\}\{s^k(t)\}\{o^k(t)\} \; ;$$
$$\{IN_k\}\{SOF_k\};$$
$$\{k_3\}>.$$

Within this structure, software accuracy of channel k is specified for each monitored variable $m_i(t)$ by:

$$(9.102) \qquad \| \, range \, (IN_k(m_i(t), i^k(t)) \circ SOF_k(i^k(t) \; s^k(t), o^k(t) \, s^k(t+1))) \, \| \leq \varepsilon_3 \, (m_i(t)),$$
$$\text{with } \varepsilon_3 \, (m_i(t)) \in \mathcal{A}^3 \subset \{k_3\},$$

where $\varepsilon_3(m_i(t))$ is the vector of bounds on absolute errors on the computed values of $o^k(t)$ when processing the monitored variable $m_i(t)$. $\varepsilon_3(m_i(t))$ is derived from the accuracy bounds $\bar{u}_2(m_i(t))$ of the set $\mathcal{A}^2$, themselves derived from the bounds of the set $\mathcal{A}^1$ of $U_1(REQ)$ parameters specified by the level-1 implementation black claim (9.23), a set of parameters itself derived from the **CLM0** requirements.

Thus, the level-3 claim on accuracy is a *white* claim:

$$(9.103) \quad (\forall m_i(t) \mid IN(m_i(t), k) :$$
$$\| \mathit{range} \; (IN_k(m_i(t), i^k(t)) \circ SOF_k(i^k(t) \; s^k(t), o^k(t) \; s^k(t+1))) \| \leq \varepsilon_3 (m_i(t)))$$
$$\vDash_{U_3(\mathcal{A}^3 I/SOF_k)} \; \mathcal{A}^3\_k\_\mathbf{hw\_sw\_accuracy.clm3}$$

with $U_3(\mathcal{A}^3 I/SOF_k)$ defined by (9.101)
and $U_3(\mathcal{A}^3 I/SOF_k) \prec U_3(\delta \mathcal{A}^2 REQ_k)$.

9.6.5.4 The Expansion Structure $U_3(\delta \mathcal{A}^2 REQ_k)$

The elements of the *expansion structure* of the delegation claim $\mathcal{A}^2\_\mathbf{channel\_k\_acceptable\_implementation.clm[2,3]}$ are now identified. By Corollary 8.4, this *expansion structure* is the smallest *extension* (Definition 8.7) of (i) the upper level *justification structures* in which the delegation claim is predicated and (ii) of the *justification elementary substructures* of the claims of this *delegation* claim expansion. $U_3(\delta \mathcal{A}^2 REQ_k)$ must satisfy the Propositions (9.95), (9.98), (9.100), (9.103), or equivalently:

$$\forall k: \quad U_2(REQ_k) \subseteq U_3(\delta \mathcal{A}^2 REQ_k)$$
$$U_3(SOF_k) \prec U_3(\delta \mathcal{A}^2 REQ_k)$$
$$U_3(\mathcal{A}^3 I/SOF_k) \prec U_3(\delta \mathcal{A}^2 REQ_k).$$

Thus, following Theorem 8.8 in Section 8.9.2, the "smallest" *expansion structures* which satisfy these Propositions are, for every channel k:

$$(9.104) \quad U_3(\delta \mathcal{A}^2 REQ_k) = \; <\{m(t)\} \, \{c(t)\} \, \{i^k(t)\} \, \{s^k(t)\} \, \{o^k(t)\} \; ;$$
$$\{IN_k\} \, \{REQ_k\} \, \{SOF_k\};$$
$$\{k_2\} \, \{k_3\}>$$

where $\{k_2\}$ are the constants of $U_2(REQ_k)$, (9.47); $\{k_3\}$ includes the constants related to the implementation SOF_k and the set $\mathcal{A}^3$ specifying the accuracy and tolerance bounds introduced in (9.103). Moreover, $U_3(SOF_k)$ and $U_3(\mathcal{A}^3 I/SOF_k)$ must be *elementary substructures* of $U_3(\delta \mathcal{A}^2 REQ_k)$; $U_3(\delta \mathcal{A}^2 REQ_k)$ should therefore satisfy the Tarski–Vaught condition (Theorem 8.1, page 90).

9.6.6 Expansion of Channel Fail-Safe Implementation Claim[2,3]

For every channel k, the level-2 delegation claim **channel_k_fail_safe_implementation.clm[2,3]** (in argument (9.94)) is predicated in the level-2 *justification structure* $U_2(FS;k)$ defined by (9.63), and only in this *structure*:

$$U_2(FS;k) = < \{FS(t)\} \{h^2(t)\} \{i^k(t)\} \{o^k(t)\};$$
$$\{IN_k\} \{OUT_k\} \{REQ_k\} \{HAZ^2_k\} ;$$
$$\{k_2\}>$$

where $\{k_2\}$ includes the constants relevant to channel k and $\{HAZ^2_k\}$.

Thus, because of Proposition (iii) in Theorem 8.7, $U_2(FS;k)$ has to be a *substructure* (Definition 8.7) of the level-3 *expansion structure* of this delegation claim. In agreement with the notation introduced in Section 8.8.4, we call $U_3(\delta FS;k)$ these level-3 *expansion structures*; and we must have:

$$(9.105) \qquad \forall k: U_2(FS;k) \subseteq U_3(\delta FS;k).$$

$U_3(\delta FS;k)$ valuates the *expansion (logical consequence)* of the delegation claim into claims on level-3 postulated undesired events validity, detection, and mitigation mechanisms. $U_3(\delta FS;k)$ constituents include those of its *elementary justification substructures* which are identified in the following three sections.

9.6.6.1 Level-3 Undesired Random Events Validity
As for the previous levels, one must ensure the validity of the sets of undesired events that are postulated. However, as explained in Section 9.6.3, because of their very nature, undesired events caused by design defects cannot be anticipated. *Random undesired events* of the kind defined in Section 9.6.3.1 only can be postulated. Their domain $\{h^3(t)\}$ and the sets of relations HAZ^3_k must be *complete and consistent*. The HAZ^3_k range must also be a valid description of the effects of these *undesired events* on the variables $s^k(t)$. That is, the level-3 *structure reduction* (Definition 8.5):

$$(9.106) \qquad U_3(HAZ_k^3) = <\{h^3(t)\} \{s^k(t)\};$$
$$\{HAZ^3_k\};$$
$$\{k_3\}>$$

must be a *model*, in the sense of (8.4), of level-3 anticipated and documented *random undesired events* and of their effects on the processing hardware and software of channel k; $\{k_3\}$ is the set of designated constants relevant to HAZ^3_k.

These obligations lead for a channel k to the validity claim:

(9.107) $\qquad$ $h^3(t)\_k\_$**effects_validity.clm3**:

$$U_3(HAZ_k^3) \vDash \textbf{L}_3\textbf{-postulated random UE's and effects,}$$

with *justification substructure* $U_3(HAZ_k^3)$ defined by (9.106)
and $U_3(HAZ_k^3) \prec U_3(\delta FS;k)$.

9.6.6.2 Robustness

All random *undesired events* $(h^3(t),t)$ discussed in Section 9.6.3.1 must be unambiguously *detected* and *controlled* by the processing hardware and software design, and must lead to a *safe state*.

Detection of *undesired events* by the processing hardware and software of channel k is required by the level-3 claim

(9.108) $\qquad$ $h^3(t)\_k\_$**detectability.clm3**:

$$U_3(HAZ_k^3) \vDash (\forall\ h^3(t) \mid (HAZ^3_k\ (h^3(t),\ s^k(t)):$$
$$(HAZ^3_k)^{-1} \text{ is a surjective function } \textit{into } h^3(t\ )),$$

with *justification substructure* $U_3(HAZ_k^3)$ defined by (9.106)
and $U_3(HAZ_k^3) \prec U_3(\delta FS;k)$.

This claim requires that every $s^k(t)$ value in the *range* of $HAZ^3_k(h^3(t))$ be univocally related to a unique value of $h^3(t)$, *i.e.* that $(HAZ^3_k(h^3(t),s^k(t)))^{-1}$ be a function and all values of the domain of HAZ^3_k be related to $s^k(t)$.

As for the *mitigation* of level-3 random undesired events, let us consider the *substructure*:

(9.109) $\qquad$ $U_3(FS;k) = <\ \{FS(t)\}\ \{h^3(t)\}\ \{i^k(t)\}\ \{s^k(t)\}\ \{o^k(t)\}\ ;$
$$\{IN_k\}\ \{OUT_k\}\ \{SOF_k\}\ \{HAZ^3_k\}\ ;$$
$$\{k_3\}>$$

where $\{k_3\}$ includes the constants of $U_3(HAZ^3_k)$. Channel k is fail-safe under the condition that the following condition holds true in this *substructure*:

(9.110) $\quad$ $h^3(t)\_k\_$**mitigation.clm3**:

$$U_3(FS;k) \vDash (\forall h^3(t) \mid HAZ^3_k(h^3(t),s^k(t)): \textit{range}(HAZ^3_k \circ SOF_k \circ OUT_k) \in FS(t))$$

with $U_3(FS;k)$ defined by (9.109)
and $U_3(FS;k) \prec U_3(\delta FS;k)$.

Because of their universes and relations, the *structures* $U_3(HAZ^3_k)$ and $U_3(FS;k)$ satisfy the following Proposition:

$$(9.111) \qquad U_3(HAZ^3_k) \subseteq U_3(FS;k).$$

Since these two *justification structures* are *elementary substructures* of $U_3(\delta FS;k)$, and because of Lemma 8.5, page 90, the following Proposition also holds:

$$(9.112) \qquad U_3(HAZ^3_k) \prec U_3(FS;k).$$

Again – as for the upper levels, *cf.* (9.9), (9.20) and (9.66) – this relation induces a logical sequencing for the justifications that take place in these *substructures*. Quite naturally, the detectability and effects of the undesired effects (HAZ^3_k) should be validated first, before their mitigation by the channel software specifications (SOF_k).

Remark 9.9 There is the same difference with level-1 as that noted in Remark 9.5. By opposition to level-1 where $U_1(REQ) \prec U_1(FS)$ (9.20), one can here verify that $U_3(SOF_k)$ is not a *substructure* of $U_3(FS;k)$. Channel software correctness with respect to the level-2 channel input-output specification REQ_k (claim (9.100)) and the level-3 undesired event mitigation (claim (9.110)) are two properties that must be independently valuated while, at level-1, the mitigation of the postulated initiating events is part of the system input-output specifications.

9.6.6.3 Channel Fail-Safe Control

The level-3 evidence provided by the valuations of the *justification substructures* $U_3(HAZ^3_k)$ and $U_3(FS;k)$ is not sufficient to ensure the fail-safeness of channel k design. The consequences of some failures cannot be entirely mitigated by the channel processing hardware and software; in certain conditions, assistance of the operation control may be needed; especially for *undesired events* affecting output devices which, being downstream, cannot be controlled by the channel processor. Delegation claims to level-4 are therefore necessary.

Although the required evidence is entirely delegated to the lower level, a *black* claim *justified* by a *delegation* claim to level-4 must be explicitly made at level 3, so as to be the root of expansions at level-4:

$$(9.113) \qquad \textbf{channel\_k\_fail\_safe\_control.clm[3,4]}$$

$$\vdash_{U_3(FS;k)} \textbf{channel\_k\_fail\_safe\_control.clm3}$$

with *justification substructure* $U_3(FS;k)$ defined by (9.109) , with $U_3(FS;k) \prec U_3(\delta FS;k)$.

9.6.6.4 *The Expansion Structures* $U_3(\delta FS;k)$

By Corollary 8.4, $U_3(\delta FS;k)$ is the smallest *extension* (Definition 8.7) of

> (i) the upper level *justification structures* in which the *delegation* claim **channel_k_fail_safe_implementation.clm[2,3]** is predicated,
> (ii) of the *elementary justification substructures* of the claims of this *delegation* claim expansion.

Thus, $U_3(\delta FS;k)$ must satisfy the Propositions (9.105), (9.107), (9.108) and (9.110), or equivalently:

$$\forall k: \ U_2(FS;k) \subseteq U_3(\delta FS;k)$$
$$U_3(HAZ_k^3) \prec U_3(\delta FS;k)$$
$$U_3(FS;k) \prec U_3(\delta FS;k).$$

Following Theorem 8.8 in Section 8.9.2, the "smallest" *expansion structure* which satisfies these Propositions is, for every channel k:

$$(9.114) \qquad U_3(\delta FS;k) = < \{FS(t)\}\, \{h^2(t)\}\, \{h^3(t)\}\, \{i^k(t)\}\, \{s^k(t)\}\, \{o^k(t)\};$$
$$\{IN_k\}\, \{OUT_k\}\, \{REQ_k\}\, \{SOF_k\}\, \{HAZ^2_k\}\, \{HAZ^3_k\};$$
$$\{k_2\}\, \{k_3\}>$$

where $\{k_2\}$ includes the constants relevant to channel k and $\{HAZ^2_k\}$; and $\{k_3\}$ is the set of designated constants relevant to $\{SOF_k\}$ and $\{HAZ^3_k\}$.

9.6.7 Expansion of Channel Independency Delegation Claim[2,3]

The delegation claim (j,k)_**channel_independency[2,3]** in the level-2 argument (Section 9.5.17) requires independence at the design level for pairs of channels that belong to the sets $\mathcal{R}^1$, $I^2(x,r)$ and $\mathcal{D}^2(h')$ defined at the upper levels. Two channels are mutually independent at level-3 if any *undesired event* affecting one channel processing unit does not affect the other at the same time. This obligation requires an anticipation of common cause *hardware and software* failures and an expansion into level-3 claims on independency with regard to these two types of failures.

This delegation claim is predicated in the level-2 *justification substructure* $U_2(ind(j,k))$ (9.79). The level-3 *expansion structure*, say $U_3(\delta ind(j,k))$, must therefore be an extension of this substructure:

$$(9.115) \qquad \forall(i,j): \ U_3(\delta ind(j,k)) \supseteq U_2(ind(j,k)).$$

$U_3(\delta ind(j,k))$ valuates the *expansion (logical consequence)* of the delegation claim into claims on level-3 claims on independency; thus, $U_3(\delta ind(j,k))$ constituents include those of its *elementary justification substructures* and those are identified in the two following sections.

9.6.7.1 Hardware Independency

Let us deal first with the *random undesired events* of the kind defined in Section 9.6.3.1. Two channels j and k are mutually independent with respect to a given *random undesired* event $h^3(t)$ if this event cannot affect the processing units in both channels and cause a coincident failure in these two channels. Let us then consider the following *substructure*:

$$(9.116) \qquad U_3 \, (\mathrm{ind}(j,k)) = \, < \, \{h^3(t)\} \, \{s^j(t)\} \, \{s^k(t)\} \; ;$$
$$\{HAZ^3_j\} \, \{HAZ^3_k\} \; ;$$
$$\{k_3\}>$$

where $\{k_3\}$ includes the constants of channels j, k. These channels are mutually independent with respect to a postulated undesired random event $(h^3(t) = h')$ if the following property holds in $U_3(\mathrm{ind}(j,k))$:

$$(9.117) \qquad (\forall \, t : \; HAZ^3_j(h^{'}, s^j(t)) \wedge HAZ^3_k(h^{'}, s^k(t)) = 0).$$

Channels requiring independency are those processing monitored variables which are assumed independent at level-1 by the *primary functional diversity and dependability claims* of Sections 9.4.10.3 and 9.4.12. These monitored variables are defined by the sets $\mathcal{D}^1(h')$ and $\mathcal{R}^1$, Definition 9.11 and Definition 9.14; and the corresponding channels that must be independent are those defined at level 2 (Definition 9.21, and Definition 9.22, pages 170 and 173):

$$\mathcal{D}^2(h') = \{k \mid IN_k(m_x(t), i^k(t)) \; and \; (\exists h' : x \in \mathcal{D}^1(h', d) \,) \}$$
$$I^2 = \{k \mid IN_k (m_x(t), i^k (t)) \; and \; x \in I^1\},$$
$$\mathcal{D}^2(h'), \, I^2 \in \{k_2\}.$$

The sets $\mathcal{D}^2(h')$, $h' \in h^1(t)$, are specified at level-2 in the sets of constants $\{k_2\}$ of $U_2(\delta\mathcal{D}^1 div)$ (9.80), and the set I^2 in the set $\{k_2\}$ of $U_2(\delta\mathcal{R}^1 rel)$ (9.92). $U_2(\mathrm{ind}(j,k))$ is a *substructure* of these two expansion structures and also, by (9.115), for the same pairs (j,k) of $U_3(\delta \mathrm{ind} \, (j,k))$. In order to preserve the independency assumed at level-1, mutual design independency between these pairs (j,k) of channels must be claimed and hold in the *justification substructures* $U_3(\mathrm{ind} \, (j,k))$:

$$(9.118) \qquad \forall \, pair \, (j,k) \in (I^2 \cup (\forall h' \in h^1(t): \mathcal{D}^2(h'))):$$

$$\forall h' \in h^3(t): (j,k)\_\textbf{hardware\_} h'\_\textbf{independency.clm3:}$$
$$U_3(\mathrm{ind}(i,j)) \models (\forall \, t : \; HAZ^3_i(h^{'}, s^i(t)) \wedge HAZ^3_j(h^{'}, s^j(t)) = 0))$$

$$\text{with } U_3(\mathrm{ind}(j,k)) \text{ given by (9.116)}$$
$$\text{and } U_3(\mathrm{ind}(j,k)) \prec U_3(\delta \mathrm{ind}(j,k)).$$

The evidence provided by the valuation of (9.117) and (9.118) can be deterministic or stochastic. For events $h^3(t)$ that may not occur in some processing units, the relations (9.117) can obviously be deterministically valuated to true. For other events, the evidence must be based on limiting values of the probabilities of joint occurrences.

This evidence guarantees independency with respect to random *undesired events* that are potential causes of common failures. But common cause failures may also result from defects in the design. Additional evidence on design diversity is therefore needed.

9.6.7.2 Design Diversity

This delegation claim (j,k)_**channel_independency[2,3]** also requires at the design level a behaviour independent from the *undesired events* of the kind defined in Section 9.6.3.2; that is, independency from undesired events caused by design defects that may escape verifications and validations, may rest in hardware or software components common to two or more channels and may cause the simultaneous failure of these channels. Of course, the *acceptability* (correctness, completeness and accuracy) of the hardware and software implementation constitute the primary defence against these coincident failures. But the obtainable evidence to valuate the *acceptability* claims (9.98), (9.100), (9.102) may not be sufficient; *i.e.* not commensurate with the high levels of dependability expected from the design for high risk applications.

Design diversity[14] is then a classical second echelon of defence to improve dependability and prevent hardware or software design flaws from defeating safety functions [62]. Design diversity can be implemented in several ways; the basic objective is to design the diverse channels processing units so that they are in different states at the time they process inputs that must be protected against coincident design defects. Design diversity may resort to different techniques:

- Different processing hardware,
- Different operating systems,
- Different input data sampling sequences,

---

[14] Interestingly enough, the idea of applying design diversity to computers is far from being new. In his book [105], U. Voges reminds that, as early as 1834, in his paper [58] "Babbage's Calculating Engine; from the Edinburgh Review", D. Lardner wrote: "The most certain and effectual check upon errors which arise in the process of computation, is to cause the same computations to be made by separate and independent computers; and this check is rendered still more decisive if they make their computations by different methods". And the weakness of the method was also already recognised as Lardner continues: "It is, nevertheless, a remarkable fact, that several computers, working separately and independently, do frequently commit precisely the same error; so that falsehood in this case assumes that character of consistency, which is regarded as the exclusive attribute of truth. Instances of this are familiar to most persons who have the management of the computations of tables. We have reason to know, that, M. Prony experienced it on many occasions in the management of the great French tables, when he found three, and even a greater number of computers, working separately and independently, to return him the same numerical result, and *that result wrong*".

- Different coding, naming, data configurations, addressing schemes,
- Different programming languages,
- Different plausibility checks or exception mechanisms,
- Asynchronous execution trajectories,
- Different algorithms,
- Independent design teams,
- Manual intervention.

Design diversity is motivated by the dual concern that (i) hardware design defects and especially software defects are credible sources of common mode failures and (ii) that design often cannot be proven defect-free. Regulators, however, often insist [28] that design diversity must be seen as an addition to – and not as a substitute for – evidence that the possibility of residual causes of coincident defects is reduced to the barest possible minimum.

This viewpoint is sound. The independency between channels needed and claimed at the architecture level by the delegation claim (j,k)_**channel_independency[2,3]** (9.79), must be primarily ensured by the conjunction of the level-3 claims on hardware independency (9.117), on hardware-software correctness (9.100) and completeness (9.98). Design diversity is intended to bring additional evidence and not to be a substitute for a lack of evidence justifying these claims. Diverse unacceptable channels never make an acceptable one.

Thus, whatever means of design diversity is used, the acceptability claims (9.98), (9.100) on hardware-software completeness and correctness must hold. Let us now define the conditions for two channels to be mutually diverse *at design level*. Let us consider two channels j and k which are assumed to satisfy the claim (9.100) on correctness and are designed to produce the same output value vector o(t) at the same time t. Note that this last assumption is necessary; otherwise no diversity would be needed at *design level*. Hence, we have:

$$REQ_j\ (i^j(t),\ o(t))\ _{U_3(SOF_j)} \dashv SOF_j(i^j(t)\,s^j(t),\ o(t)\,s^j(t+1)),$$
$$REQ_k\ (i^k(t),\ o(t))\ _{U_3(SOF_k)} \dashv SOF_k(i^k(t)\,s^k(t),\ o(t)\,s^k(t+1)),$$

where the *substructures* $U_3(SOF_j)$, $U_3(SOF_k)$ are defined by (9.99). If we now trace back the reason why these two channels are required to be independent and immune to a coincident failure caused by a design defect of the kind discussed in Section 9.6.3.2, we find that this level-3 independency is required when the two channels are, at level-1, intended to be either redundant to support reliability claims, or functionally diverse to support safety claims (*cf.* the level-2 expansions, Figure 9.4, page 157).

If the channels j and k are functionally diverse, we have:

$$REQ_j(i^j(t)) \equiv REQ_k(i^k(t))$$

where – following the notation defined in Section 8.2.1 – $REQ_i(i^i(t))$ denotes the *range* $o^i(t)$ of $REQ_i(i^i(t))$, $i = j, k$.

And, if they are redundant, one has also:

$$i^j(t) \equiv i^k(t) \text{ and } REQ_j \equiv REQ_k.$$

In either case, a *necessary* (but not sufficient) condition for channels j and k processing units to have diverse behaviours – when processing at the same time input values that must independently produce identical output values – is to be in different internal states. Then, considering the substructure:

(9.119) $\qquad U_3(div(j,k)) = < (i = j,k: \{i^i(t)\} \{s^i(t)\}\{o^i(t)\});$
$$\{REQ_i\}\{SOF_i\};$$
$$\{k_3\}) >,$$

where $\{k_3\}$ includes the constants related to the implementation of $\{SOF_i\}$, $i = j,k$, the claim for design diversity between two channels j and k is:

(9.120) $\qquad$ **(j,k)_design_diversity.clm3:**
$\qquad U_3(div(j,k)) \vDash$
$$\qquad\qquad (\forall t, \forall i^j(t), \forall i^k(t) : REQ_j(i^j(t)) \equiv REQ_k (i^k(t)) \Rightarrow (s^j(t) \neq s^k(t))),$$

$\qquad\qquad$ with $U_3(div(j,k))$ defined by (9.119)
$\qquad\qquad$ and $U_3(div(j,k)) \prec U_3(\delta ind(j,k)),$

where $REQ_i(i^i(t))$ designates the *range* $o^i(t)$ of $REQ_i(i^i(t),o^i(t))$, $i = j,k$.

Channel processing units which may benefit from design diversity are the same as those for which the hardware independency is claimed in the previous section; they are defined at level 2 by the sets $\mathcal{D}^2(h')$, $h' \in h^1(t)$, specified in the sets $\{k_2\}$ of $U_2(\delta\mathcal{D}^1 div)$ (9.80), and the set I^2 specified in the set $\{k_2\}$ of $U_2(\delta\mathcal{R}^1 rel)$ (9.92). $U_2(ind(j,k))$ is a *substructure* of these two expansion structures and also, by (9.115), for the same pairs (j,k) of $U_3(\delta ind(j,k))$.

If (9.120) is satisfied, the states of the two processing units are different when processing input values $i^j(t)$ and $i^k(t)$ that are intended to produce redundant output values. Thus, two *exceptions* or two *non-acceptable* $SOF_i(i^i(t),s^i(t),o(t)s^i(t+1))$ relations activated in two channels at the same time t, would necessarily be different.

If these two different *exceptions* or *non acceptable* SOF relations result in output values that are not equal although intended to be redundant, then the voting relations (OUT) (Definition 9.18) will filter out the misbehaviour. A problem remains when these output values are nevertheless equal. If condition (9.120) is satisfied, production of an identical *non-acceptable* output value o(t) in the two diverse channels therefore requires that one of the two following conditions are satisfied:

1. either the two different *exceptions* or *non-acceptable* SOF relations are the results of concomitant but different *design defects*,
2. or a single *design defect* affects different states.

The first condition can be considered as violating the single failure criterion and in this sense can be considered as less likely to occur than the latter. The second condition can indeed be satisfied whenever a single *design defect* affects more than one single state and possibly even a large subset of states. These situations are not difficult to conceive; they can for instance result from incorrect conditions in *while, if* or *for* loop control statements that cause the software to execute wrong parts of a program, or from address mis-calculations when accessing access tables, arrays, or queues.

This is why the condition of claim (9.120) is a necessary but not sufficient condition for guaranteeing the diversity between two channels. Further considerations on this condition for diversity and its limitations are given below in Section 9.6.7.4.

9.6.7.3 *The Expansion Structure* $U_3(\delta in(j,k))$

By Corollary 8.4, $U_3(\delta in(j,k))$ is the smallest *extension* (Definition 8.7) of:

(iv)   the upper level *justification structures* in which the delegation claim (j,k)_**channel_independency[2,3]** in the level-2 argument (Section 9.5.17) is predicated,

(v)   and of the *justification elementary substructures* of the claims of the expansion of this delegation claim.

Thus, $U_3(\delta in(j,k))$ satisfies the Propositions (9.115), (9.118), (9.120):

$$U_3(\delta ind(j,k)) \supseteq U_2(ind(j,k))$$
$$U_3(in(j,k)) \prec U_3(\delta ind(j,k))$$
$$U_3(div(j,k)) \prec U_3(\delta ind(j,k)).$$

According to Theorem 8.8, Section 8.9.2, the "smallest" *expansion structures* which satisfy these Propositions are, for a pair of channels j,k:

$$(9.121) \qquad U_3(\delta ind(j,k)) = <\{h^3(t)\}(i = j,k:\ \{i^i(t)\}\{s^i(t)\}\{o^i(t)\})\ ;$$
$$(i = j,k:\ \{REQ_i\}\{HAZ^3_i\}\{\ SOF_i\})\ ;$$
$$\{k_2\}\{k_3\}>,$$

where $\{k_2\}$ includes the constants of $U_2(ind(j,k))$ (9.77) for channels j,k and $\{k_3\}$ is the set of designated constants relevant to $\{SOF_i\}$ and $\{HAZ^3_i\}$, i=j,k.

Moreover, $U_3(in(j,k))$ and $U_3(div(j,k))$ must be the *elementary substructures* of $U_3(\delta ind(j,k))$; thus $U_3(\delta ind(j,k))$ should satisfy the Tarski–Vaught condition (Theorem 8.1, page 90).

9.6.7.4 A Diversion on Design Diversity

The condition in claim (9.120) clarifies what is meant by diversity at the design level; it shows its distinct role, different from the objective of functional diversity at the process level. All types of diversity implementations mentioned in Section 9.6.7.2 have the sole objective of satisfying this condition (9.120) for different types of combinations of time, input and system state values. The condition also suggests possible implementations of diversity that would be guaranteed for specific combinations of input and internal states only. In principle, the necessary condition (9.120) could be restricted to certain subsets of input pairs $i^j(t)$, $i^k(t)$ values. Convincing evidence of such restricted types of diversity, focusing for instance on safety critical input signals, might be easier to establish.

The condition in claim (9.120) and its implications define the potentialities and the limitations of the defence that can be achieved by *design diversity*. The valuation to true of the condition (9.120) in the *justification substructures* $U_3(\mathrm{div}(j,k))$ is *necessary* but per se not *sufficient* to prevent a single design defect from being a common cause of failure in redundant channels. A single defect can simultaneously affect more than one single state in a channel, so that channels in different states can be affected by the same defect. The condition in claim (9.120) should in this case be generalized for subsets of states; the provision of evidence to valuate (9.120) would then require the identification of the state subsets that can be affected by the common cause design defect one wish to protect from. This identification can be envisaged for certain types of *exceptions* mentioned in Section 9.6.3.2, but is more than likely impossible for design defects that cannot be anticipated. This problem is at the root of the difficulties to obtain in practice convincing evidence of the clear benefits of *design diversity*.

Remark 9.10 Jointly with claim (9.120), the claim (9.117) must remain satisfied for two channels processing units implemented with design diversity, *i.e.*:

$$(\forall\, t:\ \mathrm{HAZ}^3_j(h^3(t),\, s^j(t)) \wedge \mathrm{HAZ}^3_k(h^3(t),\, s^k(t)) = 0\,)$$

so as not to be affected at the same time by a same given *random undesired event* $h^3(t)$. Thus, in the expansion of the *delegation* claim (j,k)_**channel_independency**[2,3], as shown in the expansion on Figure 9.6, the claim (9.120) cannot be claimed without (9.117). Note that, since the states of the processing units are now different, the evidence for (9.117) may be easier to obtain.

Taking a step back and looking at the three types of diversity that have so far been analysed and discussed throughout this chapter – functional, architecture and design – we come to the conclusion that diversity

- Is a property that cannot be dispensed with when justified stringent dependability requirements have to be met,
- Is in these cases a lesser evil that has to be carefully justified,
- Is all the more problematic to implement as the subsystems involved need to share resources or to communicate.

"Lesser evil" may appear quite strong a qualification; it is difficult, however, to think of a situation where it would not be better to do without diversity, should a single means exist that can be demonstrated reliable enough for the task. The absence of such a means, therefore, always needs careful consideration. It may be the result of reliability targets that are unjustifiably high or, alternately, of targets that are too high for a single level-1 system. The first alternative calls for a careful analysis of the reliability targets; the second one for considering defences in depth and a distribution of the responsibility of the level-0 dependability requirements over two or more independent diversified systems.

Resource sharing and communication raise great problems because of the necessity to guarantee independency between subsystems claimed to be diverse. A resource sharing protocol or a communication interface has, by definition, to remain unique and can therefore always be a potential single point of common cause failure. Diversity cannot be a property of protocols and interfaces, whatever they are. It would simply be non-sense to talk of diversity for the design of voting or priority schemes between redundant channels, operating systems, communication protocols, as it would be for standards, procedures and laws that regulate interactions and life in our society[15]. For all these systems, the design objective is the exact opposite of diversity: "*E pluribus unum*".

9.6.8 Expansion of Reliability and Safety Delegation Claims[2,3]

The *grey delegation* claims **channel_k_reliability.clm[2,3]** (9.90), and **channel_k_safety.clm[2,3]** (9.91) introduced in Section 9.5.13.3 are predicated in the *justification substructure* $U_2(REQ_k)$ (9.47) and $U_2(FS;k)$ (9.63) respectively.

Reliability and safety were explicitly claimed at level-1 on the assumption that some degrees ($\mathcal{R}^1$) of redundant processing and diversity ($\mathcal{D}^1$) are available at level-2 (*cf.* Section 9.4.12.2). These delegation claims must therefore be expanded into level-3 claims requiring not only reliability and safety at the design level - but also independency.

Let $U_3(\delta rel_k)$ and $U_3(\delta saf_k)$ be the *expansion structures* in which must hold the expansions (*logical consequences*) of (9.90) and (9.91), respectively. According to Proposition (iii) in Theorem 8.7, these structures have to be *extensions* (Definition 8.7) of the level-2 *justification substructures* in which the *delegation* claims are predicated:

$$(9.122) \qquad U_2(REQ_k) \subseteq U_3(\delta rel_k)$$
$$U_2(FS;k) \subseteq U_3(\delta saf_k).$$

9.6.8.1 The Relation $\widetilde{SOF}_k$

The level-2 *grey delegation claims* (9.90) and (9.91), introduced in Section 9.5.13 are based on the assumption that there is a non-zero probability that the implementations $\widetilde{REQ}_k$ of the REQ_k relations are flawed or affected by defects

---

that have not been anticipated, and are not acceptable in the sense of the delegation claim $\mathcal{A}^2\_$**channel_k_acceptable_implementation.clm[2,3]** (in argument 9.5.17). More precisely, these claims assume that there is a non-zero probability that $R\widetilde{E}Q_k$ does not satisfy the level-3 claims of Section 9.6.5 on channel k hardware and software completeness, correctness or accuracy or the claims on design robustness of Section 9.6.6.

These level-2 *grey delegation claims* require level-3 statistical evidence on the dependable behaviour of the implementation $R\widetilde{E}Q_k$ of the channel processor software and hardware, and must be expanded into appropriate level-3 claims.

Assuming that *undesired events* and relations HAZ^3_k, IN_k, REQ_k, SOF_k, OUT_k are validated by level-3 claims for channel k, we can define for each channel k the following probabilities:

$$(9.123) \qquad \mathit{fsw}_k(t) \; = \text{Prob} \; \{ \, \widetilde{o} \, (t) \neq SOF_k \, (i^k(t) \, s^k(t)) \mid \; S\widetilde{O}F_k \, \},$$

where $S\widetilde{O}F_k$ denotes a possibly imperfect implementation of the SOF_k relations and $\widetilde{o} \, (t)$ denotes the vector of output variable values delivered by this implementation.

This probability is defined for a given state $s^k(t)$ and is taken over the set of possible values of the input variables $i^k(t)$, $i^k(t)$ being considered as a *random variable vector* conditioned on the values of the monitored variables by the relation IN_k. Its probability distribution can be obtained in the following way:

$$(9.124) \qquad F_3(x, t) \; = \text{Prob} \; \{ \, i^k(t) = x \}, \qquad x \subseteq \mathit{range} \, (IN_k)$$

$$= \sum_y \; F_1 \, (y,t), \qquad y \subseteq \mathit{domain} \, (IN_k) \text{ and } y = IN_k^{-1}(x),$$

$$= \sum_x \; F_1 \, (\, IN_k^{-1}(x), t), \qquad x \subseteq \mathit{range} \, (IN_k),$$

where $F_1(y,t)$ is the level-1 probability density function of the random variable vector $m(t)$, defined at level-1 by (9.26), and IN_k^{-1} is the inverse of the relation IN_k.

Let also $h^3(t)$ be a *random variable* with values in *domain* (HAZ^3) and define the probability density function:

$$(9.125) \qquad H_3(x,t) = \text{Prob} \{ \, h^3(t) = x \}, \qquad x \subseteq \mathit{domain} \, (HAZ^3).$$

The probability of the implementation $S\widetilde{O}F_k$ being unsafe upon occurrence of some *undesired event* $(h^3(t),t)$ is then defined as:

$$(9.126) \qquad \mathit{usw}_k(t) = \text{Prob} \{ \, \widetilde{c} \, (t) \neq (\, h^3(t) \, (HAZ^3_k \circ SOF_k \circ OUT_k) \notin FS(t) \mid S\widetilde{O}F_k \, \}$$

$$= \text{Prob} \; \{ \mathit{range} \, (HAZ^3_k \circ SOF_k \circ OUT_k) \notin FS(t) \mid S\widetilde{O}F_k \, \}.$$

Probabilities (9.123), (9.126) model each two types of randomness. The randomness of input data and undesired event occurrences ($h^3(t)$,t) modelled by the probability distributions $F_1(x,t)$, $H_3(x,t)$ respectively, on the one hand and the possibility of an imperfect implementation $\widetilde{SOF}_k$ on the other.

The relations $\widetilde{SOF}_k$, analogue to $\widetilde{REQ}$ (Definition 9.13), are not known with certainty and to some degree are undefined. The domain of $\widetilde{SOF}_k$ is the same as *domain* (SOF_k) but the range is unknown as it includes values of the output variables that are not permissible. Should $\widetilde{SOF}_k$ be "perfect", then it could be replaced by the relation SOF_k in the conditional probabilities (9.123), (9.126) which would then be necessarily equal to 0.

Being undefined, $\widetilde{SOF}_k$ does not belong to the structure $U_3(SOF_k)$, but can be valuated by statistical test data. These $\widetilde{SOF}_k$ data on the behaviour of the processing hardware and software implementation are part of the evidence to be provided at level 3.

The vectors $i^k(t)$ and $h^3(t)$ in the Expressions (9.123), (9.126) are *random variables* and the probabilities are taken over their possible values, *i.e.* the *domains* of IN_k and HAZ^3_k respectively. The probability distribution of $i^k(t)$ is determined by the probability density function $F_1(x,t)$ (9.26) of the monitored variable m(t) and therefore correspond to the use and environment profiles specified by the *dependability requirements* **CLM0**; these data are included in the set of level-1 parameter values $\{k_1\}$ of the structure $U_1(REQ)$ (9.11) (*cf.* claim (9.32) in Section 9.4.12).

9.6.8.2 Design Reliability and Safety

If we approximate relations by functions, a claim on an upper limit Fsw_{kmax} for the expected number of failures of channel k processing hardware and software during an interval $[t_1, t_2]$ can be formulated as:

(9.127) **software_k_reliability.clm3**:

$$U_3(SOF_k) \models (\ \overline{Fsw}_k\ (t_1,\ t_2) \leq Fsw_{kmax}\ ),$$

$$\text{with } \overline{Fsw}_k(t_1,t_2) = (t_2 - t_1)^{-1} \int_{t_1}^{t_2} fsw_k(t)dt$$

$$U_3(SOF_k) \text{ defined by (9.99)}$$

$$U_3(SOF_k) \prec U_3(\delta rel_k).$$

Fsw_{kmax} belongs to the set of constants $\{k_3\}$ of $U_3(SOF_k)$ and must be calculated from the constant F_{kmax} specified in the set $\{k_2\}$ of the structure $U_2(REQ_k)$. The interval $[t_2 - t_1]$ is specified at level-1 as the interval over which the system is expected to deliver a correct service (Claim (9.32) page 142); this interval and the profile probability density $F_1(x,t)$ (9.26) also belong to this set $\{k_2\}$ of the *structure* $U_2(REQ_k)$ (*cf.* the level-2 claim (9.88) on reliability in Section 9.5.13.3), which is a *substructure* of $U_3(\delta\ rel_k)$; thus, they are also included in the constants of $U_3(\delta\ rel_k)$ and of its *elementary substructure* $U_3(SOF_k)$.

In practice, the *justification substructures* $U_3(SOF_k)$, together with the profile distributions $F_3(x,t)$, constitute the specification of the random tests of the $\widetilde{SOF}_k$ implementation, the results of which provide the statistical data which valuate the probabilities *$fsw_k(t)$* of a failure of the channel software.

Likewise, a claim on an upper bound of the expected number of channel k processing hardware or software unsafe failures on occurrence of a level-3 *undesired event* during a time interval $[t_2 - t_1)$ can be formulated as:

(9.128) software_k_safety.clm3:

$$U_3(FS;k) \vDash (\ \overline{U}sw_k\ (t_1,t_2) \leq Usw_{kmax})$$

$$\text{with } \overline{U}sw_k(t_1,t_2) = (t_2 - t_1)^{-1} \int_{t_1}^{t_2} usw_k(t)dt$$

$$U_3(FS;k) \text{ defined by } (9.109)$$

$$\text{and } U_3(FS;k) \prec U_3(\delta\ saf_k).$$

The $h^3(t)$ probability density function $H_3(x,t)$ is part of the evidence to be provided at level-3 and must be included in the set $\{k_3\}$ of parameter values of $U_3(FS;k)$. Usw_{kmax} also belongs to $\{k_3\}$ and is calculated and derived from the constant U_{kmax} of the set $\{k_2\}$ of $U_2(FS;k)$. The interval $[t_2 - t_1)$ is specified at level-1 as the interval over which the system is expected to deliver a correct service (Claim (9.32) page 142); this interval is included in the set of constants $\{k_2\}$ of the *structure* $U_2(FS;k)$ (*cf.* level-2 claim (9.89) on safety in Section 9.5.13.3), which is a *substructure* of $U_3(\delta saf_k)$; thus, it is also included in the constants $\{k_3\}$ of $U_3(\delta saf_k)$ and of its *elementary substructure* $U_3(FS;k)$.

In practice, the *justification substructures* $U_3(FS;k)$, together with the profile distributions $H_3(x, t)$, constitute the specifications of the random tests to be performed on the $\widetilde{SOF}_k$ implementation, the results of which provide the statistical data which valuate the probabilities *$usw_k(t)$* of an unsafe failure of the channel software.

Similar to levels 1 and 2, these probabilistic claims – although formulated in terms of relations SOF_k and HAZ^3_k – address the channel hardware and software processing implementation of these relations; the conditional probability Expressions (9.123), (9.126) express the specific contribution of this possibly not perfect implementation onto the system reliability and safety. The claims do not address the validity of the relations, a validity ensured by the acceptability and safeness claims of Sections 9.6.5 and 9.6.6.

These probabilistic claims require level-3 statistical evidence only, and therefore are *white.*

Remark 9.11   If, beyond the single failure criterion, the system is required to sustain the occurrence of concomitant *undesired events* at the environment, architecture and design levels, joint distributions of the random variables $h^1(t)$, $h^2(t)$ and $h^3(t)$ would have to be considered.

9.6.8.3 *The Expansion Structures* $U_3(\delta\,rel_k)$ *and* $U_3(\delta\,saf_k)$

We can now define the constituents of the *expansion structures* introduced in Section 9.6.8 for the expansion of the channel reliability and safety delegation claims.

According to Corollary 8.4 these *expansion structures* are the smallest *extensions* (Definition 8.7) of the upper level *justification structures* in which the delegation claim is predicated and of the *elementary substructures* of the claims of the expansion.   Thus, $U_3(\delta rel_k)$ and $U_3(\delta saf_k)$ must satisfy the following Propositions:

$$U_2(REQ_k) \subseteq U_3(\delta rel_k)$$
$$U_2(FS;k) \subseteq U_3(\delta saf_k)$$
$$U_3(SOF_k) \prec U_3(\delta\,rel_k)$$
$$U_3(FS;k) \prec U_3(\delta\,saf_k)$$

The smallest expansion structures which satisfy these Propositions (Theorem 8.8) are:

$$(9.129) \qquad U_3(\delta rel_k) = <\, \{m(t)\}\,\{c(t)\}\,\{i^k(t)\}\,\{s^k(t)\}\,\{o^k(t)\}\,\{c^k(t)\};$$
$$\{IN_k\}\,\{REQ_k\}\,\{SOF_k\}\,\{OUT_k\};$$
$$\{k_2\}\,\{k_3\}>$$

where $\{k_2\}$ is the set of constants of $U_2(REQ_k)$ and $\{k_3\}$ includes the constants of $U_3(SOF_k)$; and:

$$(9.130) \qquad U_3(\delta saf_k) = <\, \{FS(t)\}\,\{h^2(t)\}\,\{h^3(t)\}\,\{i^k(t)\}\,\{s^k(t)\}\,\{o^k(t)\};$$
$$\{IN_k\}\,\{REQ_k\}\,\{HAZ^2_k\}\,\{HAZ^3_k\}\,\{SOF_k\}\,\{OUT_k\};$$
$$\{k_2\}\,\{k_3\}>$$

where $\{k_2\}$ is the set of constants of $U_2(FS;k)$, and $\{k_3\}$ of $U_3(FS;k)$.

Moreover, $U_3(SOF_k)$ and $U_3(FS;k)$ must be *elementary substructures* of $U_3(\delta rel_k)$ and $U_3(\delta saf_k)$ which must thereby satisfy the Tarski–Vaught condition (Theorem 8.1, page 90).

9.6.9 Level-3 Expansions

The *expansion structures* which have been constructed at design level-3 are for every channel k: $U_3(\delta\mathcal{A}^2 REQ_k)$ (9.104*)*, $U_3(\delta FS;k)$ (9.114), $U_3(\delta ind(j,k))$ (9.121), $U_3(\delta rel_k)$ (9.129), and $U_3(\delta saf_k)$ (9.130).   These structures constitute the upward interface of level-3 to levels-1 and 2. Each *expansion structure* is the smallest *extension* (Definition 8.7) of upper level *justification structures* in which the delegation claim being extended is predicated, and of the *elementary substructures* of the claims of the expansion (Corollary 8.4).

Figure 9.9, page 242, gives an overview of these *expansion* structures and their interdependencies with their *elementary substructures* and their *substructures* of the level-2. Claims at level 3 are generated by the expansions of the *grey* and *black delegation claims* of level-2 specified in the level-2 argument (Section 9.5.17, page 184). These *expansions* constitute the first step of the argument at level-3; they are given in Figure 9.6 with the notation introduced in Section 9.2. This figure shows how each level-2 *delegation subclaim* is expanded in its proper expansion structure as a *logical consequence* of level-3 claims.

9.6.10 Level-3 Evidence and Delegation Claims

Level-3 comprises the innermost parts of the system where most of the functionality of the system is collected. This is why, except for the claims requiring maintenance and operator control evidence, all claims are *white* at this design level. The only level-3 claims requiring additional level-4 evidence are the channel **hw_sw_accuracy** (9.103) and **fail_safe_control** (9.110) claims for which operator interventions may be needed for maintaining accuracy and, as an ultimate recourse, for dealing with the consequences of *undesired events* that the system cannot control alone.

All other level-3 claims are *white claims* requiring level-3 evidence only. The purpose of this section is to define what evidence is required, how it fits in the argumentation and which basic techniques can be effective to obtain this evidence. The implementation and use of these techniques are of course not trivial; references to papers are given for guidance and details. The evidence is specific of the different types of channel software such as, for instance, the system software, the library modules and the application software; examples of evidence required by these different types are given for the SIP system described in Appendix A.

9.6.10.1 Evidence of Hardware and Software Acceptability
The evidence for a hardware and software acceptable implementation – *i.e.* for this implementation completeness, correctness and accuracy – is defined by the *well-formed formulae* (9.98), (9.100) and (9.103). This evidence must justify for each channel k that all *restricted relations*

$$(9.131) \qquad REQ_k\,(i^k(t),\, o^k(t))$$

hold true if the relations

$$(9.132) \qquad (IN_k(m(t),\, i^k(t)) \circ SOF_k(i^k(t)\,s^k(t),\, o^k(t)\,s^k(t+1)) \circ OUT_k(o^k(t),\, c(t))$$
$$\wedge NAT(m(t), c(t))$$

valuate to true within the level-3 *expansion structure* $U_3(\delta\mathcal{A}^2 REQ_k)$ (9.104). In other words, one must validate $U_3(\delta\mathcal{A}^2 REQ_k)$ as a model (*interpretation*) of the software implementation and verify that in this model each relation $REQ_k(i^k(t),o^k(t))$ is implied by the extended relation (9.132).

Different techniques can be used; in practice. They resort either to static analysis (review, model checking) or to dynamic analysis (testing, simulation). Whatever technique is used, the relations (9.131) (9.132) specify what must be verified. For instance, if *active reviews* [76] are performed, the relations (9.132) are those that must be blindly reconstructed by independent reviewers and subsequently compared to the relations (9.131) for conformity. Similarly, for dynamic analysis, the relations (9.132) are the specifications of the test cases or the simulator behaviour, while the relations (9.131) specify the oracle.

In general, the verification that the source code satisfies relations (9.131) is facilitated when, contrary to what is often current practice, the source programming language tends to be of a declarative style. Whatever programming language is used, the point we wish to make is that the formal reasoning, analysis and verification of software is easier (i) when software specifications are expressed in terms of functional relations and (ii) the programming style tends to be declarative and functional rather than purely procedural, or worse when graphic and data flow type [52] languages are used, as it is often the case in practice to both specify and produce industrial embedded code. To illustrate this point, consider in the present context, the following paradigm of a declarative program fragment:

$$(9.133)$$

$$\textbf{begin}$$
$$t := t_0; \quad s^k(t) := s_0; \quad i^k(t_0) := i_0;$$
$$(o^k(t_f) s^k(t_f)) := SOF_k(i^k(t) s^k(t));$$
$$\textbf{end}$$

where $o^k(t_f) s^k(t_f)$ are the output and state values upon termination of the program at time t_f. The program is deterministic if $SOF_k(i^k(t) s^k(t), o^k(t) s^k(t+1))$ is a *function*. This program description is more amenable to formal reasoning, analysis and verification against channel requirements specified in terms of functional relations. It says what the software has to compute without saying how.

This description, of course, does not capture the computational aspects of the function to be calculated, aspects that are essential to allow a compiler to produce the necessary procedural machine code. The function to be computed is merely specified as a set of value pairs, without consideration for the "behaviour" or the "calculation" of the function. But the intent of functional programming languages [45] – founded on the Church's lambda calculus – is precisely to embrace these computational aspects of functions, while aiming at a declarative style by using recursion, lambda expressions and rewriting rules such as substitution.

An equivalent but more imperative archetypical form of programming based on functional relations could be:

(9.134)

$$\textbf{begin}$$
$$t = t_0;\ s^k(t) = s_0;\ i^k(t_0) = i^k_{\ 0};$$
$$\textbf{while}\ (o^k(t)\ s^k(t+1)) \neq (o^k_{\ f}\ s^k_{\ f}))\ \textbf{do}\ \mathrm{SOF}_k(i^k(t)s^k(t))\ \textbf{end do;}$$
$$\textbf{end}$$

Here, the condition of the **while** statement explicitly states the complement of the predicate that must hold true in the state $s^k_{\ f}$ when the program specified within the pair of delimiters **do** and **end do** terminates. The state of the program is explicitly used together with a sequence of constructs (commands) to specify how to modify this state. Because of their generality, *functional relations* – of which *functions* are a special case – can thus describe the semantics of various programming syntaxes.

Evidence is also needed to ensure that the source code is correctly compiled into machine code, and the machine code correctly executed by the underlying hardware. Compilers, assemblers and interpreters must be in principle qualified to the same level of integrity as the source code being compiled. Faulty compilers may inject into the object code faults that can be causes of common failures in redundant channels. Decompilation is a method which uses a separate tool to re-transform the object code into a readable form that can be verified by comparison with the original source code. A process of this kind is described in [83]; a well-known application in nuclear engineering is the assessment of the software of the Primary Protection System for the UK Sizewell B Nuclear Power Station [84].

9.6.10.2 Design Robustness Evidence

Evidence of the robustness of the design with regard to *random undesired events* (Section 9.6.3.1) is obtained by the verification that, for every channel k, the structure $U_3(\delta\mathrm{FS};k)$ (9.114) is a model for the software implementation in which the relations of the *well-formed formula*

$$\mathrm{HAZ}^3_{\ k}\ (h^3(t),\ s^k(t)) \circ \mathrm{SOF}_k(i^k(t)s^k(t),\ o^k(t)s^k(t+1)) \circ \mathrm{OUT}_k(o^k(t),\ c(t)))$$
$$\in \mathrm{FS}(t))$$

hold true. These relations are the specifications of test cases and/or the scope a failure mode, effects and consequence analysis (FMECA) that can be applied to the source code. *Undesired events* are specified by $h^3(t)$; their detection and effects on the states of the software are specified by the relations $\mathrm{HAZ}^3_{\ k}$ and are validated by the claim (9.107).

The absence of *design defects* (Section 9.6.3.2) should be ensured by evidence on the acceptability of the software discussed in the previous section. This absence can be cross-verified by a Failure Mode and Effects Analysis (FMECA) on the source code; some potentialities and limitations of such an analysis are reported in [72] for dataflow programming languages and function block diagrams. The consequences of potential design defects are, however, more easily and systematically postulated in functional programs of the type illustrated above, since the conditions that are violated when a program module does not terminate or terminates incorrectly appear explicitly in the source code.

9.6.10.3 Hardware Independency and Design Diversity

The claim formula on hardware independency (9.118):

$$(\forall t : \text{HAZ}^3_i(h', s^i(t)) \wedge \text{HAZ}^3_j(h', s^j(t)) = 0)$$

and the claim formula (9.120) on design diversity

$$(\forall t, \forall i^j(t), \forall i^k(t) : \text{REQ}_j(i^j(t)) \equiv \text{REQ}_k (i^k(t)) \Rightarrow (s^j(t) \neq s^k(t))),$$

define the required independency between two channels i, j.

These properties should be established by static analysis of the design. In practice, ensuring or demonstrating that the software of diverse channels satisfy the above diversity condition for all relevant states, is likely to be unfeasible or practical. It would however be good practice to attempt to identify the states $s^j(t)$ which do not satisfy the condition, and target these states for specific tests or further verifications, for instance by means of a specific FMECA analysis. An example is given in [90]; the FMECA is performed as a means to identify software components that may, if faulty, lead to a common cause failure, and the purpose is to show that either these components are fault-free or that the postulated component failures lead to a safe state.

9.6.10.4 Design Probabilistic Evidence

The statistical evidence required on the dependability of a possibly faulty implementation $\widetilde{SOF}_k$ was discussed in Section 9.6.8.2. If statistical testing is used to obtain this evidence, the expressions (9.123) and (9.126) define the random tests to be executed; the probability distributions of the *undesired events* $h^3(t)$ to be taken into account are specified by the sets $\{k_3\}$ of parameter data of *the expansion susbtructure* $U_3(\delta rel_k)$ (9.129) and $U_3(\delta saf_k)$ (9.130).

B. Littlewood and D. Wright give prediction-based stopping rules for the random operational testing of software-based systems in [63] that are both effective and practical.

9.6.11 Level-3 Argument

Summing up, the arguments at level 3 have the following generic structure; references in parentheses point to the valuations of the level-3 *substructures* which define the required evidence:

(9.135) **argument(CLM0, level-3)**

    {begin

    expansion(CLM0, level 3); /*see Figure 9.6 */

    /* **white claim justifications by U$_3$ valuations:** */

    $\forall k: U_3(SOF_k)$ ⊨ **k_hw_sw_completeness.clm3**;/* (9.98)*/

$\forall k: U_3(SOF_k) \models k\_\textbf{hw\_sw\_correctness.clm3}$; /* (9.100)*/

$\forall k: U_3(\mathcal{A}^3 I/SOF_k) \models \mathcal{A}^3\_k\_\textbf{hw\_sw\_accuracy.clm3}$; /* (9.103)*/

$\forall k: U_3(HAZ_k^3) \models h^3(t)\_k\_\textbf{effects\_ validity.clm3}$; /* (9.107)*/
$\forall k: U_3(HAZ_k^3) \models h^3(t)\_k\_\textbf{detectability.clm3}$; /* (9.108)*/
$\forall k: U_3(HAZ_k^3) \models h^3(t)\_k\_\textbf{mitigation.clm3}$; /* (9.110)*/

$U_3(\text{ind}(i,j)) \models (i,j)\_\textbf{channel\_} h^3(t)\_\textbf{independency.clm3}$; /* (9.118)*/
$U_3(\text{div}(i,j)) \models (i,j)\_\textbf{design\_diversity.clm3}$; /* (9.120)*/

$U_3(SOF_k) \models \textbf{software\_k\_reliability.clm3}$; /* (9.127)*/
$U_3(FS;k) \models \textbf{software\_k\_safety.clm3}$; /* (9.128)*/

/* <u>grey claim evidence and delegation:</u> */

$\forall k: U3(FS;k) \wedge \textbf{channel\_k \_ fail\_safe \_control.clm[3,4]}$
$\qquad\qquad \models_{U_3(FS;k)} \textbf{channel\_k \_ fail\_safe \_control.clm3}$; /* (9.113)*/

<u>end</u> argument (CLM0, level-3).

9.6.12 Level-3 Documentation

The analysis of this Section 7.6 allows us to determine the documentation which is required to expand and justify the level-3 claims. This material should necessarily be part of the level-3 documentation of a dependability case supporting **CLM0** claims; as a matter of fact, to the extent that these claims address essential properties of the design, it should be part of the documentation of any dependable system design.

The level-3 documentation should at least include:

1. For each *delegation* claim from level-2, the description of the universe and relations of the corresponding expansion model (**L₃**-*interpretation* and *expansion structure)*; these *expansion structures* are for every channel k: $U_3(\delta\mathcal{A}^2REQ_k)$ (9.104*)*, $U_3(\delta FS;k)$ (9.114), $U_3(\delta\text{ind}(i,j))$ (9.121), $U_3(\delta\text{rel}_k)$ (9.129), and $U_3(\delta\text{saf}_k)$ ((9.130). Together, they include:

    -   The set of relevant processor and memory states $s^k(t)$,
    -   The functional software relations SOF_k,
    -   The set of *undesired events* $h^3(t)$ and the relations HAZ^3_k,
    -   The parameter sets $\{k_3\}$ which include the $h^3(t)$ probability distributions that valuate the level-3 dependability claims and for every channel k the reliability assessment time intervals $[t_2 - t_1)$;

2. The arguments (9.135) detailing the *expansion* of level-2 *delegation claims* into level-3 claims, the level-2 *justifications* and the level-3 to level-4 *delegations*.

3. The description of the *justification elementary substructures* on the *interpretation* of which the argument is based: $U_3(SOF_k)$ (9.99), $U_3(HAZ_k^{\,3})$ (9.106), $U_3(ind(i,j))$ (9.116), $U_3(div(i,j))$ (9.119), $U_3(FS;k)$ (9.109) and $U_3(\mathcal{A}^{\,3}I/SOF_k)$ (9.101).

4. The documented evidence which validates – by valuation of these *justification substructures* – the level-3 claims. This evidence is the set of mappings onto these justification structures which valuate these claims to true.

This documentation is essential to the design, production and validation of the software, and must be accessible to:

-   The computer system design management, the software and hardware engineers for approval with respect to feasibility, completeness and readability; to programmers for coding; to the team in charge of specifying test cases of the design; to independent reviewers of the design.
-   To safety experts and licensing authorities, for compliance with standards, guidance and the applicable directives for programming, coding and testing safety critical systems.

9.7 Structure of the Operation Control

Man–machine interactions and the human–system interface devices are described at this level. They constitute a highly complex and sensitive part of a safety critical system; their design necessitates much expertise and experience in the relevant application domain. Laprie's book [57] may be a useful entry point to the huge literature concerning man–machine interactions, in particular for a survey of the issues raised by the cooperation and the repartition of tasks between a safety critical system and its operators. Concerning hardware interfaces, a rather comprehensive description of the alarms, displays and control devices used today by personnel to interact with nuclear power plants can be found in [102].

Our aim is not to go into the detail of the anthropometric and ergonometric characteristics of these interfaces, but to illustrate two important points. First, this section shows that man–machine interactions are an important and necessary source of evidence for the justification of the dependability of a system although it is often neglected, seldom used and documented in safety cases. And secondly, we show how models at this level can be developed and integrated into a safety justification framework on the same principles on which the previous levels are based.

In practice, the operation control interface can be very complex. Its main structure only will be analysed here, detailed enough, however, so as to formulate and justify the essential requirements. Simplicity is anyway a valuable objective at

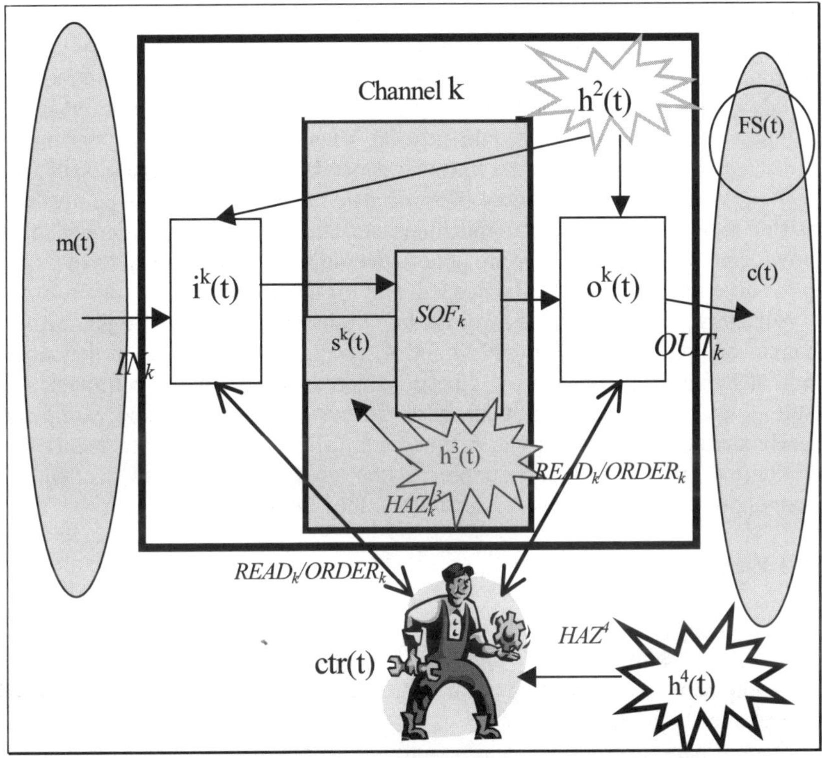

Figure 9.7. Control and operation level

this level. An important requirement of the operation control, indeed, is the economy of the control and the avoidance of unnecessary duplication. Each control or input or output device should be necessary, should avoid the use of superfluous resources and be the simplest effective control for the task concerned [102]. Our multilevel model will prove to be particularly apt at justifying these properties.

We found it convenient to consider the control of the system operation as being performed by a channel external to the system, see Figure 9.7, which will be referred to as the *control channel*. The *control channel* receives and sends information from and to the human operators. It also receives information from different components of the system and can re-act upon some of these components. But it cannot act directly on the environment.

This restriction is important. The scope activity of the external control channel must indeed be precisely delineated and documented. Dependability is at risk if the system and human control respective capabilities and responsibilities are not precisely defined.

First, the channel must be assumed to be unable to directly capture the state of the environment, modify or react upon the environment; it must be assumed that it can do so via the system input and output channels only. This restriction is an essential characteristic of the operation control channel: the operator has no other means of knowing the state of the system environment or acting upon this environment than the system itself. If he had access to other automatic means, the bounds of the system would not be defined, and the dependability that would ultimately be demonstrated would rely on other systems than the system itself. Our objective is to establish the intrinsic dependability of the system specified by the REQ relations – *i.e.* the grey box *substructure* in Figure 9.1 – controlled and assisted by a well-defined man–machine interface. It is when this dependability is known, and only then, that we are able to decide whether additional by-pass, back-ups or diverse systems might be needed, and for what purposes.

Secondly, 'control' at this level means 'human control', and 'operator' means 'human operator'. Any automatic control, diagnostic or assistance that operates without human intervention must be considered as part of the computer system itself. Otherwise, again, the bounds of the system being justified would not be clearly defined. Watchdogs and timer controllers such as those mentioned in Section 6.4 are a particular example. Although they may be independent of the channel processors and powered separately, they belong to the system.

9.7.1 Elements of the Operation Control Structure

Man–machine interface devices are the principal elements of the universes of the level-4 model *substructures*: operator control consoles, calibration and periodic test equipment, alarm and operator diagnostic systems, hardware and software maintenance systems.

System parameter values, alarms and other data that are "read" or "recorded" by the operation control channel are displayed by control room or console displays, recorders and alarm panels, by equipment rack back and front panel LEDs, and by the testing units. All this information is supposed to be delivered by the system to the operation control via the input and output registers of its different channel defined at level-2. For example, input register values are read by the testing unit for confirmation of the test data that are injected into the system for periodic testing; output register values include the results of these tests as well as those that inform operators of the state of actuators and mechanical or hydraulic equipment controlled by the system.

Data, preventive or corrective control commands and actuation signals are sent to the system by means of operator consoles, pushbuttons or switches used to modify the state of the system, the values of designated parameters (*e.g.* set points) and, when necessary, to manually activate actuators. This information is addressed to any processing unit, via the input register of the corresponding channel, or sent directly to an output register when its state has to be forced to safe values.

Therefore, the model assumes that the operation control channel has direct reading and writing access to the input and output registers $i^k(t)$ and $o^k(t)$ of the redundant channels only, and no access to the internal state $s^k(t)$ of each channel (see Figure 9.7). Such a man–machine interface structure may seem too restrictive. It is, however, consistent with the requirement that operators of safety critical systems are not allowed to alter the internal state, logic or software while the system is in operation. We will also see that this model, although simple, is sufficient to valuate all claims on operation control, that is all the *delegation claims* predicated by the upper levels. Because it is conceived as a feedback loop controller, the bounds of the system remain clearly defined and claims on properties such as negative feedback, reactivity, or stability can be naturally formulated and validated.

9.7.2 Operation Control Relations READ$_k$ and ORDER$_k$

The exchanges of data over the control channel are modelled by relations. Let ctr(t) be the vector which defines the sate of the control channel at time t.

A relation READ$_k$ specifies the capture by the control channel of the contents of the input and output registers of channel k; it is defined as follows:

Definition 9.26

- *domain* (READ$_k$) is a set of vector-valued time functions containing possible values of sub-vectors of $i^k(t)$ and $o^k(t)$;
- *range* (READ$_k$) is a set of vector-valued time functions containing possible values of the vector ctr(t);
- The pairs $< i^k(t), \text{ctr}(t)>$, $< o^k(t), \text{ctr}(t)>$ are in READ$_k$ if and only if the values of the vector crt(t) define the state of the control channel when $i^k(t)$ and $o^k(t)$ describe values taken by the channel k input and output variables, respectively.

A relation ORDER$_k$ specifies the effects of control commands on the channel k input and output register values:

Definition 9.27

- *domain* (ORDER$_k$) is a set of vector-valued time-functions containing possible values of ctr(t);
- *range* (ORDER$_k$) is a set of vector-valued time-functions containing possible values of sub-vectors $i^k(t)$ and $o^k(t)$;
- The pairs $<\text{ctr}(t), i^k(t)>$ and $<\text{ctr}(t),o^k(t)>$ are in ORDER$_k$ if and only if $i^k(t)$ and $o^k(t)$ describe the values taken by the channel k input and output variables respectively, when ctr(t) describes the state of the control channel.

The relations ORDER$_k$ specify the behaviour of the operators, users and maintenance teams when they follow operational, maintenance, calibration, testing and manual control procedures to intervene on channel k.

9.7.3 Control of Operation. Undesired Events (HAZ^4)

The control performed by the operation channel can be affected by *undesired events*. *Undesired events* $(h^4(t),t)$ that may occur at this level are human failures, consequences of deficiencies in control, maintenance and testing procedures. The consequences can be diverse: the sending of illegal or wrong control commands, failures to issue commands in time.

These *undesired events* affect the state of the control channel, *i.e.* the value of the state vector ctr(t).

We call HAZ^4 the relation between the state of the operation control and these *undesired events*:

Definition 9.28

- *domain*(HAZ^4) is the set of vector-valued time-functions containing all possible values of $h^4(t)$;
- *range* (HAZ^4) is the set of vector-valued time functions containing values of ctr(t);
- The pair $<h^4(t)$, ctr(t)$>$ is in HAZ^4 if and only if (ctr(t), t) is an event occurring whenever $(h^4(t),t)$ occurs.

9.7.4 Level-4 Expansion of Level-3 Delegation Claims

As for the upper levels, the level-4 claims result from the expansion of the upper level delegation claims. Level-4 expansions are interpreted and valuated by L_4 - *expansion structures*. A level-4 *expansion structure* $U_4(\delta\xi_4)$ (see *e.g.* (8.16)) – which valuates the logical sequence of the expansion of a delegation claim $\delta_{34}(\xi_4)$ – must be an *extension* (Definition 8.7, page 88) of the L_3 - *justification structures* in which this delegation claim is predicated (Proposition (iii) in Theorem 8.7, page 103).

The upper level delegation claims on level-4 are the channel reliability and safety delegation claims from level-2 (cf; argument (9.94) and the channel accuracy and fail-safe control delegation claims from level-3 (argument (9.135). Level-4 provides the ultimate recourse for these functions and for the *undesired events* that cannot be entirely ensured, controlled or mitigated by the upper levels. Delegation claims onto level-4 have to be expanded into *white* level-4 subclaims. There is no *black* claim or further claim delegation possible at level-4. All level-4 claims must be justified by level-4 evidence only.

The following sections develop the expansions of the upper level delegation claims mentioned here above and elicit the level-4 *white* claims. Again, as explained in Section 6.6.1, these expansions are essentially performed with regard to the three criteria of (i) model validity, (ii) control acceptability and (iii) robustness and dependability.

9.7.5 Expansion of Accuracy Delegation Claim[2,4]

The delegation claim from level-2 $\mathcal{A}^2$_k_**hw_sw_accuracy.clm[2,4]**, in argument (9.94), is predicated at level-2 in the justification structure $U_2(\mathcal{A}^2 I/O_k)$ (9.50). Its expansion is a logical consequence valuated in an *expansion structure* which, to be consistent with the notation of Section 8.8.4 (page 96), is called $U_4(\delta \mathcal{A}^2\, I/O_k)$. According to Proposition (iii) in Theorem 8.7, $U_2(\mathcal{A}^2\, I/O_k)$ must be a *substructure* (Definition 8.7) of this level-3 *expansion structure*:

$$(9.136) \qquad \forall k: U_2(\mathcal{A}^2\, I/O_k) \subseteq U_4(\delta \mathcal{A}^2 I/O_k).$$

The constituents of $U_4(\delta \mathcal{A}^2\, I/O_k)$ include those of its *elementary justification substructures* which are identified in the following sections.

The delegation claim on accuracy has to be expanded at level-4 into claims on the validity (completeness and feasibility) of the operation control, and on correct (accurate) maintenance ensured by this control.

9.7.5.1 Control Feasibility and Completeness

The data acquisition devices of the control channel must be able to observe and capture the input and output states of the system. The behaviour of these devices must be *specified* for every input and output state; thus one must have:

$$(9.137) \qquad (\forall k:\ domain\ (READ_k) \supseteq\ (range\ (IN_k)),$$

$$(9.138) \qquad (\forall k:\ domain\ (READ_k) \supseteq\ (domain\ (OUT_k)),$$

where *range* (IN_k) and *domain* (OUT_k) are given by Definition 9.15 and Definition 9.17.

The control channel must also be able to test and modify input and output states of the system. Its behaviour must be *specified* for every possible input and output state. Thus, one must have:

$$(9.139) \qquad (\forall k:\ range\ (ORDER_k) \supseteq range\ (IN_k)),$$

$$(9.140) \qquad (\forall k:\ range\ (ORDER_k) \supseteq domain\ (OUT_k)).$$

These formulae require that the behaviour of the control channel be *explicitly specified* for every input state and output state, but not that every state be actually readable or modifiable. Inclusion is used in these *sentences* and not identity because, as we shall see in Section 9.7.6.3 on operation control robustness, there exist – in *range* $(ORDER_k)$ – values produced by control commands that are feasible but might be illegal or erroneous; these values must be inhibited and thereby do not belong to *range* (IN_k) nor to *domain* (OUT_k).

The *range* $(READ_k)$ and *domain* $(ORDER_k)$ constitute the universe of the human operators' control and procedures. This set of controls and procedures must be *complete*.

That is, an operator response must be specified for every possible system input or output state that can be read:

$$(9.141) \qquad (\forall_k: domain\ (ORDER_k) \supseteq range\ (READ_k)).$$

The control operators must also be able to react for every condition that might occur in the system environment; thus, we must have:

$$(9.142) \qquad (\cup_k\ domain\ (REACT_k) \supseteq domain\ (NAT)).$$

But, since the control channel has direct access on the environment via the system itself, this completeness is ensured by (9.137) and by the level-2 claim on architecture completeness (9.41):

$$\{\cup_k: domain\ (IN_k)\} \supseteq domain\ (NAT),$$

The behaviour specified for the control channel must remain "feasible", *i.e.* compatible with the environment constraints specified by NAT. Again, since the control channel has no direct access on the environment except via the system itself, this compatibility is ensured by the level-2 claim on architecture feasibility (9.44), which guarantees that:

$$(\forall_k: range\ (OUT_k) \subseteq range\ (NAT)).$$

Therefore, the valuation of the formulae (9.137) to (9.141) provide the evidence for the completeness and feasibility of the control of channel k. These formulae are valuated in the *substructure*:

$$(9.143) \qquad U_4(ctr;k) = <\{i^k(t)\}\{o^k(t)\}\{ctr(t)\}\ ;$$
$$\{READ_k\}\{ORDER_k\}\ ;$$
$$\{k_4\}>$$

where $\{k_4\}$ includes the constants related to the control of channel k. Level-3 domains, relations and constants are absent in this structure, reflecting the fact that the level-4 control has no direct access to the software.

Hence, we have the claim:

$$(9.144) \qquad \textbf{k\_control\_feasibility\_and\_completeness.clm4:}$$

formulae (9.137) to (9.141) are satisfied in the
elementary justification substructure $U_4(ctr;k)$

with $U_4(ctr;k)$ defined by (9.143)

and $U_4(ctr;k) \prec U_4(\delta \mathcal{A}^2 I/O_k).$

9.7.5.2 Maintenance
It was mentioned in Sections 9.5.3 and 9.5.5 that the channel relations IN_k and OUT_k are relations and not functions because of the imperfections (imprecisions and drifts) in the measurements, in input and output devices and A/D and D/A converters.   Channel accuracy relies indeed on periodic re-calibrations, compensating corrections or even replacements of these devices during operation. These maintenance adjustments aim at ensuring that the claim (9.53) on channel accuracy remains satisfied in operation.   They are periodically performed by operator verifications and commands defined by $READ_k$ and $ORDER_k$ relations on channel k input and output devices. Hence, for the results that must be ensured by the maintenance, we have the level-4 claim which must hold true in the *justification structure* $U_4(\text{ctr};k)(9.143)$:

$$(9.145) \qquad (\forall t = nT_k, n = 1, 2,...:$$
$$(\forall m_i(t) \mid IN(m_i(t), k):$$
$$\| \ range \ (READ_k(I\tilde{N}_k(m_i(t)))) \ \| \leq \bar{u}_2 \ (m_i(t)))$$
$$(\forall c_i(t) \mid OUT(k,c_i(t)):$$
$$\| \ range \ (READ_k(domain(\widetilde{OUT}_k \ (o^k(t), c_i(t))))) \ \| \leq \bar{u}_2(c_i \ (t)))$$

$$\vdash_{U_4 \ (\text{ctr};k)} \textbf{k\_channel\_maintenance.clm4}$$

$$\text{with } U_4(\text{ctr};k) \text{ defined by (9.143)}$$
$$\text{and } U_4(\text{ctr};k) \prec U_4(\delta\mathcal{A}^2 I/O_k)$$

$$\text{with } \bar{u}_2 \ (m_i(t)), \bar{u}_2 \ (c_i \ (t)) > 0$$
$$\bar{u}_2 \ (m_i(t)), \bar{u}_2 \ (c_i \ (t)) \in \mathcal{A}^2 \subset \{k_4\},$$

where $I\tilde{N}_k$ and $\widetilde{OUT}_k$ describe the behaviour of the implementations of the corresponding relations and T_k is the periodicity of the adjustments for channel k. This periodicity is related to the regeneration time intervals used to evaluate channel k reliability and safety at level-2 (claims (9.90),(9.91)) and is included in the constants of the justification structures $U_2(REQ_k)$, (9.47) and $U_2(FS;k)$, (9.63) . The set $\mathcal{A}^2$ of upper bound values $\bar{u}_2$ is defined at level 2 by claim (9.53).

9.7.5.3 The Expansion Structure $U_4(\delta\mathcal{A}^2 I/O_k)$
The elements of the *expansion structure* of the delegation claim $\mathcal{A}^2\_\textbf{channel\_k\_hw\_sw\_accuracy.clm[2,4]}$ are now identified. By Corollary 8.4, this *expansion structure* is the smallest *extension* (Definition 8.7) of (i) the upper level *justification structures* in which the delegation claim is predicated and (ii) of the *justification elementary substructures* of the claims of this *delegation* claim expansion.

Thus, $U_4(\delta \mathcal{A}^2 I/O_k)$ must satisfy the Propositions (9.136), (9.144), (9.145):

$$\forall k: \quad U_2(\mathcal{A}^2 I/O_k) \subseteq U_4(\delta \mathcal{A}^2 I/O_k)$$
$$U_4(ctr;k) \prec U_4(\delta \mathcal{A}^2 I/O_k),$$

and following Theorem 8.8 in Section 8.9.2, the "smallest" *expansion structure* which satisfies these Propositions is, for every channel k:

$$(9.146) \qquad U_4(\delta \mathcal{A}^2 I/O_k) = <\{m(t)\}\,\{c(t)\}\,\{i^k(t)\}\,\{o^k(t)\}\,\{ctr(t)\}\ ;$$
$$\{IN_k\}\,\{OUT_k\}\,\{READ_k\}\,\{ORDER_k\};$$
$$\{k_2\}\,\{k_4\}>$$

where $\{k_2\}$ are the constants of $U_2(\mathcal{A}^2 I/O_k)$ (9.51) and $\{k_4\}$ those of $U_4(ctr;k)$ (9.143).

Moreover, $U_4(ctr;k)$ must be *elementary substructure* of $U_4(\delta \mathcal{A}^2 I/O_k)$; hence, $U_4(\delta \mathcal{A}^2 I/O_k)$ should also satisfy the Tarski–Vaught criterion (Theorem 8.1, page 90).

9.7.6 Expansion of Channel Fail-safe Control Delegation Claim[3,4]

For every channel k, the level-3 delegation claim **channel_k_fail_safe_control.clm[3, 4]** (in argument (9.135) is predicated in the level-3 *justification structure* $U_3(FS;k)$ defined by (9.109), and only in this structure:

$$U_3(FS;k) = <\{FS(t)\}\,\{h^3(t)\}\,\{i^k(t)\}\,\{s^k(t)\}\,\{o^k(t)\}\ ;$$
$$\{IN_k\}\,\{OUT_k\}\,\{SOF_k\}\,\{HAZ^3_k\}\ ;$$
$$\{k_3\}>.$$

Because of Proposition (iii) in Theorem 8.7, $U_3(FS;k)$ has to be a *substructure* (Definition 8.7) of the level-4 *expansion structure* of this delegation claim. In agreement with the notation introduced in Section 8.8.4, we call $U_4(\delta FS;k)$ these level-4 *expansion structures*; and we must have:

$$(9.147) \qquad \forall k: U_3(FS;k) \subseteq U_4(\delta FS;k).$$

$U_4(\delta FS;k)$ valuates the *expansion (logical consequence)* of the delegation claim into claims on level-4 fail-safe control, postulated undesired events validity, detection, and mitigation mechanisms. $U_4(\delta FS;k)$ other constituents than those of $U_3(FS;k)$ include those of its *elementary justification substructures* which are identified in the following sections.

9.7.6.1 Fail-safe Control

We have identified several cases where the system alone cannot control and mitigate the consequences of *undesired events* that may occur at levels 2 and 3. In all these situations, the operation control remains the last defence, and an overriding control command must be available to force the system output into a safe state within acceptable delays. More precisely, the following Proposition must hold for every channel k:

$$(9.148) \quad \forall\, t,\ \forall m(t),\ \forall s^k(t),\ \forall i^k(t) \mid IN_k\,(m(t),\ i^k(t)):$$

$$(o^k(t) \notin (REQ_k\,(i^k(t)) \wedge OUT_k\,(o^k(t)) \notin FS(t))$$
$$\Rightarrow (\exists\, crt(t+\Delta t),\ 0 \leq \Delta t \leq \tau : OUT_k((ORDER_k\,(crt(t+\Delta t))) \in FS(t+\Delta t)).$$

The value of the control channel state vector $crt(t+\Delta t)$ designates an overriding command such that *range* $(ORDER_k\,(crt(t+\Delta t))$ is a value that forces the output register contents of channel k to a safe state; τ is a maximum reaction time which is function of the specification REQ being violated. The two conditions:

$$o^k(t) \notin (REQ_k(i^k(t))\ and\ OUT_k(o^k(t)) \notin FS(t)$$

in (9.148) identify contents of the output register that are invalid and at the same time do not comply with the level-2 and 3 mitigation claims (9.64) and (9.110).

A typical overriding command is a command disarming the channel watchdog. More generally, the output value generated by the command and defined by *range* $(ORDER_k\,(crt(t+\Delta t)))$ must be safety-oriented or availability-oriented depending on the type of $c_i(t)$ variable being controlled and identified by the *voting relations* OUT, (Definition 9.18).

An important requirement for safety critical systems is also to ensure that the system, whatever its current state, cannot prevent an operator's overriding command from being effective. This is why (9.148) must hold true for all possible internal states $s^k(t)$ of the channel.

Hence, if we consider the *substructure*

$$(9.149) \qquad U_4(FS;k) = <\{m(t)\}\,\{FS(t)\}\,\{ctr(t)\}\ \{i^k(t)\}\ \{s^k(t)\}\,\{o^k(t)\}\ ;$$
$$\{IN_k\}\,\{OUT_k\}\,\{REQ_k\}\,\{READ_k\}\,\{ORDER_k\}\ ;$$
$$\{k_4\}>,$$

the following *white* claim must hold :

$$(9.150) \qquad \textbf{k\_channel\_control\_fail\_safeness.clm4:}$$
$$U_4(FS;k) \vdash$$
$$\left(o^k(t) \notin (REQ_k\,(i^k(t)) \wedge OUT_k\,(o^k(t)) \notin FS(t)\right)$$
$$\Rightarrow (\exists\, crt(t+\Delta t)\,,\ 0 \leq \Delta t \leq \tau : OUT_k((ORDER_k\,(crt(t+\Delta t))) \in FS(t+\Delta t)\right),$$

$$\text{with } U_4(FS;k) \text{ defined by (9.149)}$$
$$\text{and } U_4(FS;k) \prec U_4(\delta FS;k).$$

9.7.6.2 Level-4 Undesired Events Validity

Errors at the operation control level include human errors. Not all such errors can be anticipated or even detected. A syntactically illegal control command issued by an operator can be detected, but all misused or delayed legal commands cannot.

Nevertheless, as for the upper levels, all *undesired events* (UE's) of the kind defined in Section 9.7.3 that can be anticipated and postulated should be documented and described by the domain $\{h^4(t)\}$ of the set of relations HAZ^4 (Definition 9.28). HAZ^4 *range* must be a valid description of the effects of these *undesired events* on the variables ctr(t). More precisely, the *justification elementary substructure*:

$$(9.151) \qquad U_4(HAZ^4) = <\{ctr(t)\} \; \{h^4(t)\} \; ; \\ \{HAZ^4\} \; ; \\ \{k_4\}>$$

must be a *model*, as defined by (8.4), of the level-4 documented *undesired events* and of their effects on the control channel. This obligation lead to the validity claim for every channel k:

$(9.152) \qquad$ **$h^4(t)$_effects_ validity.clm4:**

$U_4(HAZ^4) \vDash$ **L_4-postulated control UE's and effects**

with *justification substructure* $U_4(HAZ^4)$ defined by (9.151)

and $U_4(HAZ^4) \prec U_4(\delta FS;k)$.

That is, $U_4(HAZ^4)$ defined by (9.151) must be a valid **L_4-*interpretation*** of the postulated errors and effects that may affect the control.

9.7.6.3 Robustness

All *undesired events* $h^4(t)$ should be detected and their consequences should be mitigated. *Detection* of these *undesired events* requires that the claim:

$(9.153) \qquad$ **$h^4(t)$_detectability.clm4:**

$U_4(HAZ^4) \vDash (\forall \; h^4(t) \mid (HAZ^4 \; (h^4(t), crt(t)):$
$\qquad\qquad (HAZ^4 \;)^{-1}$ is a surjective function *into* crt(t))

with $U_4(HAZ^4)$ defined by (9.151)

and $U_4(HAZ^4) \prec U_4(\delta FS;k)$

holds (valuates to true) in $U_4(HAZ^4)$. This claim guarantees that every crt(t) value in the *range* of HAZ^4 is related to a unique value of $h^4(t)$, *i.e.* $(HAZ^4 \;)^{-1}$ is a function, and all values of the domain of HAZ^4 are related to crt(t).

Concerning *mitigation*, there are two possibilities:

    (i)   The control command affected by an error is detected as being erroneous and is inhibited before execution;

    (ii)   The control command affected by an error is not detected as being erroneous; its effects only can possibly be mitigated.

The first option leaves the current output values unchanged and favours availability; the second option forces fail-safe output values and favours safety.

Inhibition of a control command, following a detected undesired *event* $h^4(t)$, requires that the values of the sub-vectors $i^k(t)$ and $o^k(t)$ are not modified by the control command when the value of $crt(t)$ is affected by $h^4(t)$, *i.e.* has a value in the range of $HAZ^4(h^4(t))$. The control must be designed in such a way that the $i^k(t)$ and $o^k(t)$ values are in this case outside the range of the $ORDER_k$ relation. Let us then consider the substructure:

$$(9.154) \qquad U_4(inhib;k) = < \{ctr(t)\}\,\{h^4(t)\}\ \{i^k(t)\}\,\{o^k(t)\}\ ;$$
$$\{ORDER_k\}\ \{HAZ^4\}\ \{IN_k\}\ \{OUT_k\}\ ;$$
$$\{k_4\}>$$

where $\{k_4\}$ includes the constants of the control channel and of $U_4(HAZ^4)$. The following claim must hold in this structure:

(9.155) $h^4(t)\_k\_$**inhibition.clm4:**

      $U_4(inhib;k) \models$

      $(\textit{range}\,(ORDER_k\,(HAZ^4\,(h^4(t))) \notin (\textit{range}\,(IN_k) \cup \textit{domain}\,(OUT_k)))$,

      with $U_4(inhib;k)$ defined by (9.154)

      and $U_4(inhib;k) \prec U_4(\delta FS;k)$.

Mitigation of the consequences of an $h^4(t)$ *undesired event* affecting a command which cannot be inhibited and which erroneously modifies the $i^k(t)$ or $o^k(t)$ values requires that the sole effect of this event be to force the output to a safe value.

Let us then consider the following substructure:

$$(9.156) \qquad U_4(mitig;k) = < \{ctr(t)\}\,\{h^4(t)\}\ \{i^k(t)\}\ \{s^k(t)\}\,\{o^k(t)\}\ FS(t)\ ;$$
$$\{ORDER_k\}\ \{HAZ^4\}\,\{IN_k\}\ \{OUT_k\}\ \{SOF_k\}\ ;$$
$$\{k_4\}>$$

where $\{k_4\}$ includes the constants of the control channel and of $U_4(HAZ^4)$.

The following claim must hold in this structure:

(9.157) $h^4(t)\_k\_$**mitigation.clm4** :

$U_4\,(\text{mitig};k) \vDash$

$$\big(\, range(\text{ORDER}_k(\text{HAZ}^4(h^4(t))\circ\text{SOF}_k(i^k(t)\ s^k(t),o^k(t)s^k(t+1))\circ\text{OUT}_k(o^k(t),c(t)))$$
$$\in \text{FS}(t)\,\big)$$

$U_4\,(\text{mitig};k) \vDash$

$$\big(range\ (\text{OUT}_k(\text{ORDER}_k\,(\text{HAZ}^4(h^4(t)))))) \in \text{FS}(t)\,\big)$$

with $U_4(\text{mitig};k)$ defined by (9.156)

and $U_4(\text{mitig};k) \prec U_4(\delta\text{FS};k)$.

In the first case, mitigation involves the channel processor and its software; in the second case, the channel output device.

Remark 9.12 This claim on mitigation must be considered as being complementary to the level-3 claim on software correctness (9.100), page 193. Remember that software correctness is strictly defined as the property of satisfying the channel specifications REQ_k, and leaves the software unspecified if the implementation of the predicates IN_k, OUT_k is faulty or if the values $\{i^k(t)\}$, $\{o^k(t)\}$ are out of range. Such faulty conditions are precisely those that this claim (9.157) must cover.

Because of their universes and relations, the following Propositions hold between structures for the *interpretation* of the language $\mathbf{L_4}$ used to describe the undesired events and their effects:

(9.158)
$$\forall k:\ U_4(\text{HAZ}^4_k) \subseteq U_4(\text{FS};k)$$
$$U_4(\text{HAZ}^4_k) \subseteq U_4(\text{inhib};k)$$
$$U_4(\text{HAZ}^4_k) \subseteq U_4(\text{mitig};k)$$
$$U_4(\text{inhib};k) \subseteq U_4(\text{mitig};k).$$

Since all these *justification structures* are *elementary substructures* of the *expansion structure* $U_4(\delta\text{FS};k)$, and because of Lemma 8.5, the following Proposition must also hold:

(9.159)
$$\forall k:\ U_4(\text{HAZ}^4_k) \prec U_4(\text{FS};k)$$
$$U_4(\text{HAZ}^4_k) \prec U_4(\text{inhib};k)$$
$$U_4(\text{HAZ}^4_k) \prec U_4(\text{mitig};k)$$
$$U_4(\text{inhib};k) \prec U_4(\text{mitig};k).$$

Again – as for the upper levels, *cf.* (9.9), (9.20) , (9.66) and (9.112) – these *elementary substructure* relations induce a logical ordering for the justifications that take place in these *substructures*. Quite naturally, the detectability and effects of the undesired effects (HAZ^4_k) should be validated first, then the inhibition and last the mitigation of their consequences by the control channel.

9.7.6.4 The Expansion Structure $U_4(\delta FS;k)$

The elements of the *expansion structure* of the delegation claim **channel_k_fail_safe_control .clm[3,4]** are now identified. By Corollary 8.4, this *expansion structure* is the smallest *extension* (Definition 8.7) of (i) the upper level *justification structures* in which the delegation claim is predicated and (ii) the *justification elementary substructures* of the claims of this *delegation* claim expansion. Since $U_4(\delta FS;k)$ must satisfy the Propositions (9.147), (9.155), (9.150), (9.157), we have:

$$\forall k: U_3(FS;k) \subseteq U_4(\delta FS;k)$$
$$U_4(FS;k) \subseteq U_4(\delta FS;k)$$
$$U_4(inhib;k) \prec U_4(\delta FS;k)$$
$$U_4(mitig;k) \prec U_4(\delta FS;k).$$

Hence, by following Theorem 8.8 in Section 8.9.2, the "smallest" expansion *structure* which satisfies these Propositions is, for every channel k:

$$(9.160)\ U_4(\delta FS;k) = \{m(t)\}\{crt(t)\}\{FS(t)\}\ \{h^3(t)\}\ \{h^4(t)\}\{i^k(t)\}\ \{s^k(t)\}\{o^k(t)\}\ ;$$
$$\{IN_k\}\{OUT_k\}\{REQ_k\}\{SOF_k\}$$
$$\{HAZ^3_k\}\ \}\{HAZ^4_k\}$$
$$\{READ_k\}\ \{ORDER_k\};$$
$$\{k_3\}\{k_4\}>$$

where $\{k_3\}$ are the constants of $U_3(FS;k)$ and $\{k_4\}$ the constants of $U_4(mitig;k)$.

Moreover, $U_4(inhib;k)$ and $U_4(mitig;k)$ must be *elementary substructures* of $U_4(\delta FS;k)$; $U_4(\delta FS;k)$ must therefore satisfy the Tarski–Vaught condition (Theorem 8.1, page 90).

9.7.7 Expansion of Channel Reliability and Safety Delegation Claims[2,4]

The *grey delegation* claims **channel_k_reliability.clm[2,4]** (9.90), and **channel_k_safety.clm[2,4]** (9.91) introduced in Section 9.5.13.3 are predicated in the *justification substructures* $U_2(REQ_k)$ and $U_2(FS;k)$ defined by (9.47), (9.63) respectively.

These delegation claims onto level-4 result from the necessity to ensure that the time intervals of the integrals (9.88), (9.89) that evaluate reliability and safety at level-2 are intervals between successive time instants at which the behaviour of the system can be considered as being independent of the past. The times at which periodic maintenance, re-calibrations and periodic functional tests of the specified channels are scheduled are examples of such regeneration times, and evidence of such regeneration points has to be provided by the operation control level.

Let $U_4(\delta\mathrm{rel}_k)$ and $U_4(\delta\mathrm{saf}_k)$ be the *expansion structures* in which must hold the expansions (*logical consequences*) of the delegation claims (9.90) and (9.91), respectively. According to Proposition (iii) in Theorem 8.7, these structures have to be *extensions* (Definition 8.7) of the level-2 *justification substructures* in which the *delegation* claims are predicated; thus we must have:

$$(9.161) \qquad U_2(REQ_k) \subseteq U_4(\delta\mathrm{rel}_k)$$
$$U_2(FS;k) \subseteq U_4(\delta\mathrm{saf}_k).$$

9.7.7.1 Periodic Functional Tests

It is obviously necessary – and required by most standards – to periodically test and verify the essential functions of a safety critical system while in operation. These recurrent tests re-confirm the correct functional behaviour. By their recurrence and coverage – complementary to the system self tests and diagnostics – they constitute regeneration points of the system behaviour and in this way contribute to the system reliability and availability.

Let test^k designate a test case, that is a vector of test data for channel k, and let T_k be the periodicity of these tests. The *range* of $ORDER_k(\mathrm{test}^k)$, Definition 9.27 page 219, specifies the contents $i^k(t)$ of the input register when this vector of test data is sent to channel k from an operator console.

The test verification of the functional correctness of the channels boils down to verifying that the level-2 claim (9.48) on channel correctness remains valid for the system in operations. *Ideally*, for these verification tests to be complete, the entire set of specifications REQ_k must be covered for each channel: a test case must satisfy the closed formula (9.48) for each possible content $i^k(t)$ of the input register that belongs to the *range* of $IN_k(m(t), i^k(t))$, for all possible value of the vector $m(t)$.

Let us then consider the structure U_4:

$$(9.162) \qquad U_4(\mathrm{test};k) = <\{m(t)\}\,\{c(t)\}\,\{\mathrm{ctr}(t)\}\,\{i^k(t)\}\,\{s^k(t)\}\,\{o^k(t)\};$$
$$\{NAT\}\,\{REQ\}$$
$$\{IN_k\}\,\{OUT_k\}\,\{REQ_k\}\,\{READ_k\}\,\{ORDER_k\};$$
$$\{k_4\}>.$$

The following *logical consequence* must hold true for the system implementation:

$$(9.163) \quad (\forall t = nT_k,\ n{=}1,\,2,\,...$$
$$\forall <m(t),i^k(t)> \in IN_k :$$
$$(\exists\ \mathrm{test}^k:$$
$$REQ(m(t),\,c(t))\ _{U_4(\mathrm{test};k)} \models$$
$$(IN_k(m(t), ORDER_k(\mathrm{test}^k)) \circ \widetilde{REQ}_k \circ OUT_k(READ_k(o^k(t)), c(t)) \wedge NAT(m(t), c(t))\,)\,)$$

with $U_4(\text{test};k)$ defined by (9.162)

$U_4(\text{test};k) \prec U_4(\delta\text{rel}_k)$

$U_4(\text{test};k) \prec U_4(\delta\text{saf}_k).$

$\widetilde{REQ}_k$ is the – possibly imperfect – implementation of the REQ_k relation (*cf.* Definition 9.13). Every valuation of U_4 (test;k) which valuates to true the *sentence* on the right hand side of (9.163) must also valuate to true the predicate $REQ(m(t), c(t))$.

This *sentence* (9.163) implies that the value $i^k(t)$ (= test^k) injected into the channel and the value $o_k(t)$ read at the output is a pair of values that satisfies the specification $REQ(m(t),c(t))$; thus, the Proposition (9.163) is the test criterion that the periodic functional test suite should *ideally* verify. Hence, the claim on channel k periodic functional testing:

(9.164)   **k_periodic_functional_test_coverage.clm4:**
            test criterion (9.163) holds true in the *elementary justification*
            substructure U_4 (test; k) defined by (9.162) and

$U_4(\text{test};k) \prec U_4(\delta\ \text{rel}_k)$

$U_4(\text{test};k) \prec U_4(\delta\ \text{saf}_k).$

The execution of these periodic functional tests is an essential contribution to the dependability of the system. This contribution is taken into account by the time intervals of the probability integrals for the evaluation of reliability and safety at levels 1, 2 and 3 (Sections 9.4.12.1, 9.5.13.2 and 9.6.8.2). Thus, the periodicity T_k of the claim (9.164) should be commensurate with the intervals chosen for the reliability and safety levels that are claimed for the system and each channel.

The test criterion (9.163) is an *ideal* testing objective. In practice, a channel k can seldom be tested for all input values $i^k(t)$ in the *range* of $IN_k(m(t), i^k(t))$, and for all possible values of the vector $m(t)$; remember the digression on testing of Section 9.5.16, page 184. For each channel k, the suite of test cases test^k has to be carefully selected, for example by partitioning the *domains* of the relations $REQ(m(t), c(t))$ and $IN_k(m(t), i^k(t))$ into *functional equivalence classes* (see *e.g.* [87]. A *functional equivalence class* of values $i^k(t)$ is a set of input values which receive a sufficiently equivalent channel treatment $REQ_k(i^k(t), o^k(t))$ so that a single or small number of test cases test^k is sufficiently representative of the class. For example, the *relation*:

$$\{< i^k(t), o^k(t) > \mid (i^k(t) \geq 0) \Rightarrow (o^k(t) = i^k(t))\ and\ (i^k(t) < 0\ ) \Rightarrow (o^k(t) = -\ i^k(t))\}$$

partitions the *domain* of $REQ_k(i^k(t), o^k(t))$ into two *functional equivalence classes*, one for the identity function, one for the negation function. In each class, one or a few test cases may be sufficient; for instance to test the mid-range of the interior, and/or the extremes of the class input domain, or any combination thereof. Similarly, each class has an output range and test cases may be selected to test the interior and/or the extremes of the range.

Within each class, special test cases may also be needed to verify special features or behaviours of the function. One must also ensure, if possible, that all error conditions and messages have been tested. In particular, test cases must verify the channel behaviour for out-of-range, missing or corrupted input values which correspond to the special $i^k(t)$ values within the range of the relation HAZ^2_k (*cf.* Remark 9.3, page 156).

Remark 9.13 Since the control channel has no direct access to monitored and controlled variables, these functional tests do not cover the behaviour IN_k and OUT_k of the input and output devices. An end-to-end testing of the channel requires special plant equipment – not part of the computer-based system – to test these devices and make the appropriate overlapping tests.

9.7.7.2 Verification of Self-tests

It is also good practice to periodically verify in operation that the system *undesired events'* detection and tolerance mechanisms remain active and keep working properly. Assurance of correct and continuous operation of these mechanisms implementation is also a contribution to the reliability and safety of the system. Hence adequate tests should periodically verify that the claims on *undesired events* mitigation at levels 2 and 3 remain satisfied in operation.

More precisely, a test should verify that the level-2 mitigation claim (9.64) remains satisfied for each anticipated *undesired event* $h^2(t)$ that may affect the channel I/O devices. There is, however, a difference between the tests that can be applied to the input and the output devices. As discussed in Section 9.6.6.3, the channel processor and its software have no control on the output devices being downstream. The effects of *undesired events* on these devices can be observed and tested by the values of the controlled variables $c(t)$ only. But, as explained in Section 9.7.1, the control channel has no direct access to these variables. For the *undesired events* affecting output registers and devices, operators must therefore have recourse to the defences and mitigation actions claimed in Section 9.7.6.1 on fail-safe control.

But, for each *undesired event* $h^2(t)$ that may affect the input devices, there must be a test case vector $test^k$ such that the following criterion formula hold true in the *structure* $U_4(test;k)$ defined in (9.162):

$$(9.165) \qquad (\forall\, h^2(t) \mid HAZ^2_k\, (h^2(t),\, i^k(t)) :$$
$$(\exists\, test^k :$$
$$range\,(HAZ^2_k(h^2(t),\, ORDER_k(test^k)))\circ\, \widetilde{REQ}_k\, \circ\, READ_k(o^k(t),\, crt\,(t))$$
$$\in FS(t)\,)\,).$$

Hence, we have the claim on channel k auto-test periodic testing:

(9.166)   **k_auto-test_periodic_test.clm4:**
         test criterion (9.165) holds true in $U_4(\text{test};k)$

         with $U_4(\text{test};k)$ defined in (9.162)

         and $U_4(\text{test};k) \prec U_4(\delta\text{rel}_k)$

            $U_4(\text{test};k) \prec U_4(\delta\text{saf}_k)$.

Concerning the level-3 *undesired events*, the control channel has no access to the channel internal states $s^k(t)$. The detection and mitigation mechanisms for level-3 $h^3(t)$ *undesired events* cannot therefore be verified by periodic tests. The verification of the continuous and correct operation of these mechanisms has to be dealt with by the software itself, an obligation which rests on the level-3 claims: $h^3(t)\_k\_$**mitigation.clm3** (9.110). This is why some standards and guidelines require that the correct activation and execution of processor auto-tests and self-diagnostics be verified in operation by the processor itself, cyclically or in response to a specific command from the operator. Again, testing the auto-tests contributes to the dependability of the system and the periodicity of these tests should be commensurate with the probability expressions given in the Sections 9.4.12.1, 9.5.13.2, and 9.6.8.2 for the evaluation of the claimed reliability and safety.

Remark 9.14 Complementarity of auto-tests and periodic functional tests. Some safety guides and standards require a demonstration that the coverages of the periodic tests and those of the system self-tests be somehow complementary, *i.e.* that the periodic test suites cover the level-3 defects that self-tests cannot cover. The models of levels-3 and 4 show that such a demonstration is not formally feasible, except if tests are exhaustive. Assuming that these defects could be anticipated at all, then specified and validated as members of the set of $h^3(t)$ of *undesired events*, if their effects cannot be detected by the design at level-3, they cannot be tested at level 4 either, since the specific periodic functional test specifications (9.163) that would be needed for testing the absence of these defects remain undefined. Exhaustive functional tests only could show this absence.

9.7.7.3 *The Expansion Structures* $U_4(\delta\text{rel}_k)$, *and* $U_4(\delta\text{saf}_k)$

According to Corollary 8.4, *expansion structures* are the smallest *extensions* (Definition 8.7) of the upper level *justification structures* in which the delegation claim is predicated and of the *elementary substructures* of the claims of the expansion.

Thus, $U_4(\delta rel_k)$ and $U_4(\delta saf_k)$ must satisfy the Propositions (9.161), (9.164), (9.166):

$$U_2(REQ_k) \subseteq U_4(\delta rel_k)$$
$$U_2(FS;k) \subseteq U_4(\delta saf_k).$$
$$U_4(test;k) \prec U_4(\delta rel_k)$$
$$U_4(test;k) \prec U_4(\delta saf_k).$$

The smallest expansion structures which satisfy these Propositions (Theorem 8.8) are:

(9.167) $U_4(\delta rel_k) = <\{m(t)\}\{c(t)\}\ \{ctr(t)\}\{i^k(t)\}\ \{s^k(t)\}\{o^k(t)\};$
$\{NAT\}\{REQ\}$
$\{IN_k\}\{OUT_k\}\{REQ_k\}\ \{READ_k\}\{ORDER_k\};$
$\{k_2\}\ \{k_4\}>$

where $\{k_2\}$ are the constants of $U_2(REQ_k)$ (9.47) and $\{k_4\}$ of $U_4(test;k)$. Similarly, we have:

(9.168) $U_4(\delta saf_k) = <\{m(t)\}\{c(t)\}\{ctr(t)\}\{i^k(t)\}\ \{s^k(t)\}\{o^k(t)\}\ \{FS(t)\}\ \{h^2(t)\};$
$\{NAT\}\{REQ\}$
$\{IN_k\}\{OUT_k\}\{REQ_k\}\{HAZ^2_k\}\{READ_k\}\{ORDER_k\};$
$\{k_2\}\{k_4\}>$

where $\{k_2\}$ are the constants of $U_2(FS;k)$ (9.63) and $\{k_4\}$ of $U_4(test;k)$.

Moreover, for $U_4(test;k)$ must be the *elementary substructure* of $U_4(\delta rel_k)$ and $U_4(\delta saf_k)$; hence these *substructures* must satisfy the Tarski–Vaught condition (Theorem 8.1, page 90).

9.7.8 Level-4 Expansion

Claims at level-4 are generated by the expansions of the *grey* and *black* claims of levels 2 and 3 that are specified in the arguments of those levels. These expansions constitute the first step of the argument at level-4 and are given in Figure 9.8 with the notation introduced in section 9.2. Again, as for the upper levels, these expansions are the links between, on the one hand, the evidence required by the upper levels and delegated to level-4, and on the other hand the properties of validity, robustness and fail-safeness that need to be demonstrated at level-4.

In particular, one remembers that evidence of regeneration points provided by periodic tests for the calculation of reliability, availability and safety was delegated from level 2 to level 4 in Section 9.5.13.3. One observes also that there are no direct delegation claims from level 1 to level 4. This is a consequence of the fact

---

expansion (CLM0, level 4)

  <u>begin</u>

$(\forall k: \mathcal{A}^2$**_channel_k_accuracy.clm[2,4]**

    $U_4(\delta\mathcal{A}^2 I/O_k)$ ⊣

        k_control_feasibility_and_completeness.clm4

        $\wedge$ **k_ periodic_functional_test_coverage.clm4**

        $\wedge$ **k_ channel_maintenance.clm4**$)$

  $(\forall k:$ **channel_ k_fail_safe_control.clm[3,4]**

    $U_4(\delta FS;k)$ ⊣

        k_control_feasibility_and_completeness.clm4

        $\wedge$ **k_channel_control_fail_safeness.clm4**

        $\wedge$ **h^4(t)_effects_ validity.clm4**

        $\wedge$ **h^4(t)_detectability.clm4**

        $\wedge$ **h^4(t)_inhibition.clm4**

        $\wedge$ **h^4(t)_mitigation.clm4**$)$

$(\forall k:$ **channel_k_reliability [2,4]** | **channel_k_safety [2,4]**

    $U_4(\delta rel_k) | U_4(\delta saf_k)$ ⊣

        k_control_feasibility_and_completeness.clm4

        $\wedge$ **k_ periodic_functional_test_coverage.clm4**$)$

        $\wedge$ **k_autotest_ periodic_test.clm4**$)$

<u>end</u> expansion (CLM0, level 4)

Figure 9.8. Level-4 expansions

that properties required at level-1 strictly depend *in the first instance* on the system, and not on operator assistance. The system has recourse to – and is assisted by – operation control only when it is needed, *e.g.* in case of failure.

Somehow, level-4 can be viewed as closing the system-environment control loop opened at level-1.

The *expansion structures* constructed at level-4 are for every channel k: $U_4(\delta\mathcal{A}^2 I/O_k)$ (9.146), $U_4(\delta FS;k)$ (9.160), $U_4(\delta rel_k)$, (9.167) and $U_4(\delta saf_k)$, (9.168). These structures constitute the upward interface of level-4 to levels-2 and 3. Each *expansion structure* is the smallest *extension* (Definition 8.7) of upper level *justification structures* in which the delegation claim being extended is predicated, and of the *elementary substructures* of the claims of the expansion (Corollary 8.4).

Figure 9.9, page 242, gives an overview of these *expansion* structures and of their interdependencies with their *elementary substructures* and their *substructures* at level-2.

9.7.9 Level-4 Evidence

Level-4 evidence is of different types, and comes from different sources. Ergonometric and anthropometric data must valuate the set of anticipated human errors $(h^4(t))$ and their effects (HAZ^4) on the control. Other potential sources of evidence are provided by the existence of:

- Periodic Functional Test Suites and Coverages Specifications,
- Alarms,
- Accident Conditions Monitoring Facilities,
- Periodic Re-calibration Procedures,
- Manual Controls and By-pass,
- Accident Procedures in Place.

The first three types of evidence are essentially modelled by READ functional relations; the last three types by ORDER functional relations which should lead the controlled variables $c(t)$ into safe fail-safe values belonging to FS(t).

Re-calibration contributes with evidence to channel accuracy, and periodic testing to dependability. Periodic inspections decrease the probability that a component fail undetected in the time interval between successive inspections. An optimal time schedule of periodic preventive operations is in fact determined by two factors: on the one hand a possible reduction in reliability during inspections when redundant equipment under test has to be put in stand-by and the reliability benefits of more frequent inspections on the other hand. An optimal frequency of inspections maximizing reliability results from a balance between these two conflicting factors [15].

Accident monitoring, manual controls and by-pass provide evidence for channel and design failures mitigation and also contribute to the fail-safe implementation of REQ_k relations.

Level 4 being the lowest, there is no black claim or delegation possible. If plausible level-4 evidence is not available to fully justify level-4 claims, the consequence would be that either the design or the *dependability case* needs revision, or that the *initial dependability requirements* are not justifiable.

9.7.10 Level-4 Argument

To sum up, the level 4 arguments have the generic structure below; in parentheses are the references to the level-4 *justification elementary substructures* the valuations of which define the required evidence.

(9.169)   **argument (CLM0, level-4)**

  begin

        expansion (CLM0, level 4); /*see Figure 9.8 */

      /* <u>white claim justifications by U_4 valuations:</u>*/

        k_control_feasibility_and_completeness.clm4; /*(9.144)*/
        k_periodic_functional_test_coverage.clm4; /*(9.164)*/
        k_auto-test_periodic_test.clm4; /* (9.166)*/
        k_channel_maintenance.clm4; /*(9.145)*/
        k_channel_control_fail_safeness.clm4; /* (9.150)*/
        $h^4(t)$_effects_ validity.clm4; /*(9.152)*/
        $h^4(t)$_detectability.clm4; /* (9.153)*/
        $h^4(t)$_k_inhibition.clm4; /* (9.155)*/
        $h^4(t)$_k_mitigation.clm4; /* (9.157)*/

  end argument (CLM0, level-4).

9.7.11 Level-4 Dependability Case Documentation

The analysis in Section 9.7 allows us to identify the documentation which is required to expand and justify the level-4 claims. This information should necessarily be part of the level-4 documentation of a dependability case supporting **CLM0** claims. Since these claims address essential properties of the operation control and man–machine interface, this material should also be part of the documentation of a dependable system design.

Should be included:

1. For each *delegation* claim from levels-2 and 3, the description of the universe and relations of the corresponding expansion model ($\mathbf{L_4}$-*interpretation* and *expansion structure*); these *expansion structures* are for every channel k: $U_4(\delta \mathcal{A}^2 I/O_k)$ (9.146), $U_4(\delta FS;k)$ (9.160), $U_4(\delta rel_k)$, (9.167) and $U_4(\delta saf_k)$, (9.168). Altogether, they include:
   - The set of states of the control channel ctr(t) and, for every channel k, the functional relations $READ_k$ and $REACT_k$;
   - The set of *undesired events* $h^4(t)$ and the relations HAZ^4;
   - The set of parameter values $\{k_4\}$ which includes the test periodicities;
2. The arguments 9.7.10 with the *expansion* of *delegation claims* of levels 2 and 3 into level 4 *white* claims (Figure 9.8).
3. The description of the *justification elementary substructures* on the *interpretation* of which the argument is based: $U_4(ctr;k)$ (9.143), $U_4(FS;k)$ (9.149), $U_4(HAZ^4)$ (9.151), $U_4(inhib;k)$ (9.154), $U_4(mitig;k)$ (9.156), $U_4(test;k)$ (9.162).

4. The documented level-4 evidence, expressed in terms of relations $READ_k$, $ORDER_k$, HAZ^4, which validates – by valuation of the *justification substructures* – the level-4 *white* claims; the level-4 evidence is the set of mappings onto the *justification structures* which valuate the claims to true.

This documentation should be available to:

- Plant safety and control room experts, in order to validate the model ($ctr(t)$, $h^4(t)$, HAZ^4, $READ_k$, $ORDER_k$), and to ensure that the documented control of operation is adequate in all eventualities;
- Designers of the operation control and of the computer system, and independent reviewers for approval with respect to feasibility and readability;
- Licensers and safety authorities, with respect to compliance with the plant and application safety requirements.

9.8 Guidance Provided by the System Substructure Tree

Figure 9.9, page 242, displays the oriented graph of the dependencies between the different *substructures* within and across the four levels of the justification. We proved in Section 8.11 that this graph is a tree, and we explained why it is an instructive model of the whole system design and justification. Its properties provide guidance on the way and the order to follow in the elaboration of the different submodels and in the construction of the arguments of the justification. All properties holding true in a *substructure* also hold true in the parent *structures* having this substructure as an *elementary substructure*. Consequently, as already suggested in Section 8.11, it is sensible to start by expanding and justifying at each level the claims that must be *interpreted* (valuated) in the "most elementary" *substructures*, *i.e.* in those *substructures* to which point a greatest number of oriented links in a graph as shown in Figure 9.9.

Thus, the justification should naturally start by formulating, expanding and justifying claims at level-1 and successively deal with claims at the next lower levels. Within each level, the "most elementary" *substructures* should be considered first, *i.e.* those that correspond to end nodes that support the largest oriented spanning subtree within the same level; these end nodes are *justification substructures*. We identified such "most elementary" *justification substructures* in different places in this chapter (*cf.* the relations (9.9), (9.20), (9.66), (9.112) and (9.158)).

At level-1 the end node to be considered first turns out to be (see Figure 9.9) the $U_1(NAT)$ *justification substructure* which valuates the environment constraints, followed by the $U_1(REQ)$ and $U_1(HAZ)$ *structures*, $U_1(FS)$ coming last. At level-2, the channel accuracy *justification substructures* $U_2(\mathcal{A}^2I/O_k)$ should be the first to be considered. At every level, the validity, detectability and effects of the undesired events (*justification structures* $U_i(HAZ^i)$) should be justified before their control (*substructure* $U_i(FS;i)$), or inhibition $U_4(inhib;k)$ and mitigation ($U_4(mitig;k)$).

At the same time, the graph also shows which justifications are mutually independent. For instance the *substructures* involved in the justification of claims on functional diversity are those which belong to the rather small subtree spanned at level-2 and 3 by the node $U_2(\delta\mathcal{D}^l div)$ at level-2. This independence implies a separation of concerns of particular interest for justifying analyses of system properties such as diversity. Because of the many system inner dependencies and the issues raised by the propagation of electric signals, hardware, software and timing errors, it is indeed often difficult to identify the extent to which separation of concerns and abstraction from details are justifiable for a given analysis.

9.9 Concluding Remarks: Model Inter-relations and Preservation Properties

This chapter, the longest of the book, is an attempt at proposing a realistic model of the dependability justification of an embedded computer system, and an illustration of the essential role played by models in the justification of dependability. Altogether we end up with 38 embedded *substructures* being models of the elements, relations and constants at the four levels that are needed to formulate and valuate essential dependability requirements and the claims into which they are expanded. Among these *substructures*, a limited number of *expansion substructures* at each level constitutes the interface with the upper levels.

The chapter gives a fair idea of both the complexity of the justification and the feasibility and advantages of the approach.

Interdependencies between *substructures* "glue together" the various arguments that must be made at the different levels, from the level-0 at which initial dependability requirements are expressed, through the design and implementation lower levels down to the operation and maintenance control lowest level. These interdependencies consist in claim *expansions* which specify, on the one hand, the evidence required at an upper level, and, on the other hand, the properties of validity, robustness against *undesired events* and dependability that need to be demonstrated at lower levels. Chapter 8 used concepts from model theory [6, 14, 55] to support the semantics of these *expansions*.

The flow of information across the different levels is materialised by *delegation claims* and parameter values. They specify downwards what to prove and under which constraints and assumptions, while evidence data valuate *justification substructures* at every level to justify grey and *black* claims made at the upper levels.

Information is passed from one level to the next lower ones in distinct specific ways: logical assertions for deterministic claims, statistical properties for probabilistic claims; constants as values assigned to free variables in the first case, statistical data assigned to probability distributions in the second case.

Relations appear to be an extremely useful tool to describe functions, properties and undesired events in universes as disparate as the environment, architecture, software and operation control of a system. They also provide a specification of the required testing, verification and validation evidence at these levels.

Evidence is nothing else but the valuation at each level of a model *substructure*. Valuation of a *substructure* is, in particular, the only possible justification of the property of completeness. This modelling approach in certain cases uncovers sources of evidence that are never explicitly considered and barely exploited in practice, like the distinct forms of channel diversity at architecture level.

Separation of concerns is another great advantage. Subtrees of *expansion* and *justification substructures* delineate the scope of a claim justification and determine the degree of detail of the analysis. The degree of detail and the complexity of the representations remain limited to what is necessary to formulate and expand the claims that are delegated onto each level and to valuate the evidence needed to support them. The ability of confining the scope of the representations to what is necessary and objectively dictated by specified evidence valuations is a powerful weapon against complexity which a claim oriented scheme only can offer. This ability proves particularly useful at the design level.

Propositions (9.9), (9.20), (9.66), (9.112) and (9.158) and others induce a logical sequencing amongst the corresponding justifications: because an **L**-well-formed formula is not satisfied in a *structure* unless it is satisfied in its **L**-*elementary substructure* (Definition 8.8), there is no sense in validating a *justification structure* if its *elementary substructure(s)* cannot be validated. This ordering figures as construction rule 11.1.6 in Chapter 11.

Several other more specific points have been clarified and illustrated, in ways that would not have been possible without the detailed analysis of a model of inter-related *structures*; in particular:

- The need for using functional *relations* both between values and between names of variables;
- The validation of the *system input-output specifications* against the *dependability requirements* as the acceptance process of only one REQ relation and *interpretation structure* by all protagonists involved (*cf.* Section 9.4.8, page 128). The acceptance process is akin to a process of verification, alleviating the difficulties of validation discussed at the end of the previous chapter (Section 8.14, page 113).
- The treatment of deterministic claims and probabilistic dependability claims in a same framework of arguments, probabilistic dependability claims being *black* claims at level 1, *grey* at level 2 and *white* at level 3.
- The absence of delegation claims from level 1 to level 4, only levels 2 and 3 having recourse to level 4; an absence which can be understood by the fact that a delegation claim from level-1 to level-4 would in some way "by-pass" the system and negate its existence, the environment calling for assistance upon the operation control without "trying" the system first.

- The precise definition of the concepts of redundancy, independency and diversity, and of the different roles they play in the justification of the initial dependability requirements.
- The different types of diversity that can be exploited at the environment, architecture, design and software levels, and the problems raised by the implementation of diversity;
- The demonstration of the absence of a possible argument for establishing complementarity between periodic tests and self tests;
- The justification of the design of the control of operation, the role of each control function and device being validated against upper level claims.

"As simple as possible but not more" (Albert Einstein) and "Everything simple is false but everything not is useless" (Paul Valery) are two antagonistic views between which this chapter can be viewed as an attempt at a compromise. The developments and arguments may seem rather detailed and even repetitive in some places. These details and repetitions have not been eluded. They reflect what a real *dependability case* is in practice, in its attempt to maintain a difficult balance between simplicity, pertinence, completeness and traceability. They also illustrate the systematic aspects of the approach followed.

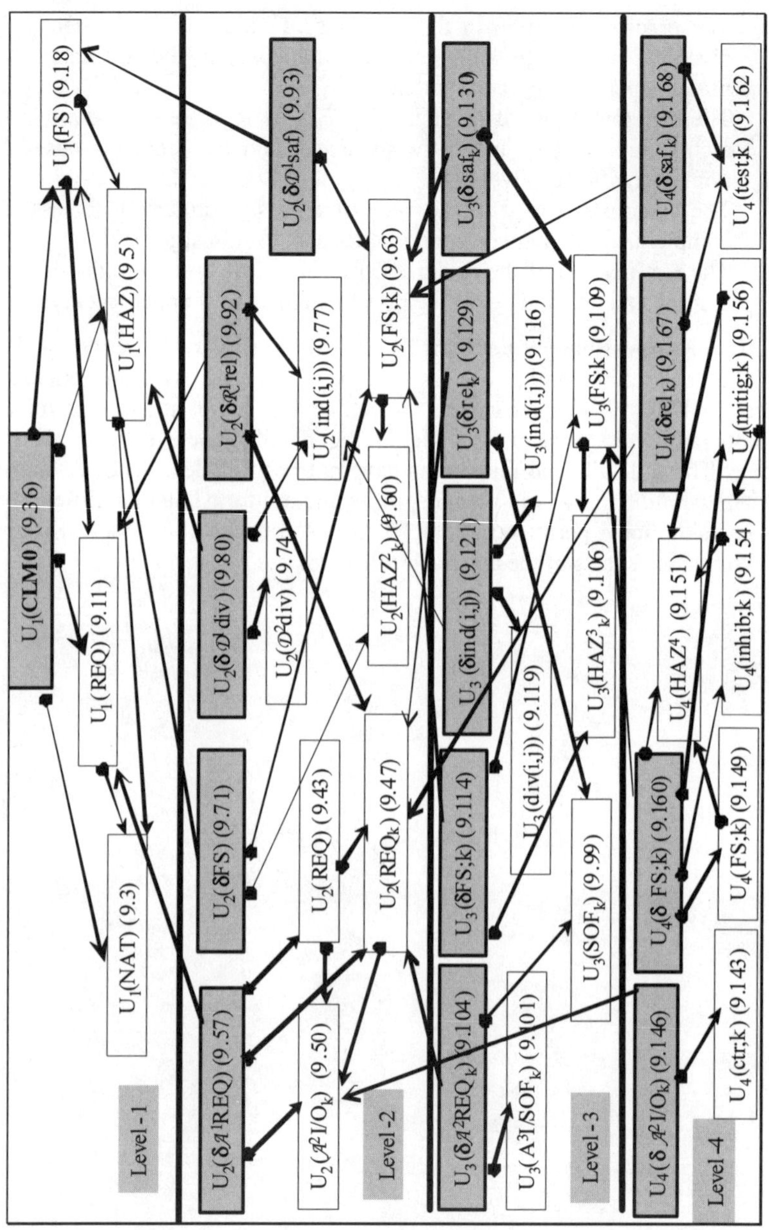

Figure 9.9 System *substructure* tree. *Expansion structure* nodes are *grey*; *white* nodes are *justification structures*. An oriented edge $U_i \bullet\!\!\longrightarrow U_j$ stands for "U_j is *elementary substructure* of U_i"; likewise, other oriented edges indicate *substructure* relations. References in parentheses point to the *substructures* definitions

PART IV

Methodological Implications

10

Pre-existing Systems and Components

It is rather natural to think that a system which has been in operation without failing, and for a long time, should receive credit for its dependability. Re-using a proven existing system, or a proven component should in principle be a safer option than to develop a brand new one, at least when, as it is the case for software, aging is not an issue. However, pertinent evidence of the dependability of a system, based on the experience and the feedback obtained from its past operation, is not always easy to collect. The previous chapter analysed the evidence needed to justify the specifications, the implementation and the control of operation of a system being developed. This chapter examines the evidence needed to justify the re-use of an a already existing system, previously and independently developed.

In general, operational history alone does not give sufficient evidence for the justification of the dependability of a system important to safety. It is often difficult to demonstrate that the past operation experience is in itself pertinent enough to guarantee the dependability requirements of a new application. Many standards clearly stipulate that the evaluation of the feedback of operational experience does not replace the evaluation of the design of the product itself and of its documentation (see *e.g.* [28, 48]). To allow true credit from operational experience, stringent requirements must be imposed on the documentation of the system, its history and the similitude between the conditions in which the system has been used and the new ones. An analysis needs to show that the system and the software perform the same function in the proposed new application as it performed during previous applications, and under similar conditions of use. Aspects to compare are, among others, the parameter input and output profiles, and performance and maintenance requirements.

Nevertheless, like other sectors, the computer and software industry finds the re-use of systems and components more and more economically attractive. There is an increased tendency to integrate within a system pre-developed hardware and run-time software such as operating systems, software libraries, programming platforms, user interfaces and various kinds of special purpose "intelligent" components, often referred to as Components-Off-the-Shelf (COTS). This pre-existing hardware and run-time software is originally designed for a wide range of applications, tested once and for all, and then embedded and used, sometimes as a "black box", in more complex systems.

These economical and marketing advantages exacerbate the difficulties raised by the usage of pre-existing components and systems in dependable applications. The pre-existing components may not have been developed according to the standards and levels of quality and reliability required by the application domain. And the documentation on their design, verifications and validation which is accessible to the user and to those in charge of the demonstration of dependability can often be scarce and deficient. Operational experience remains then the main or the only source of available evidence. Information on issues, research and recent developments of component-based computer and software systems can be found in [2, 38].

The following sections show how the multi-level dependability justification framework developed in the previous chapters helps to deal with these pragmatic problems.

10.1 Pre-existing Components

Typically, pre-developed Off-The-Shelf components (COTS), hardware or software are unit-tested once and for all, and for use in multiple products. They are tested again at integration time – but most of the time not individually: together with the hardware and customized software of the application in which they are embedded. These integrated tests offer no guarantee that all modes of interaction between the COTS component and the rest of the system have been verified.

In safety critical applications – in particular nuclear applications of the highest criticality level – the source code of COTS components may be required for inspection. But some regulatory guidelines, for instance those addressing the election electronic voting systems of some countries (see *e.g.* [68]), exempt COTS from these inspections. Unreviewed COTS components have been reported to have raised several problems, in election voting equipments and elsewhere.

The following sections show how the approach developed in the previous chapters helps to justify the dependability of systems making use of pre-existing components by giving means to specify (i) the evidence required from the component and (ii), what is new, the conditions of its use.

From now on, we shall generically refer to these operating systems, hardware/software programming platforms, COTS and various hardware/software components that provide a fixed set of services and functions to an application layer as "pre-existing components".

10.2 Composability and Re-use of Arguments

Before discussing the re-use of pre-existing components, we must pay attention to a more fundamental question. Basically, the justification of the "use" of a pre-existing component, assumed to be *a-priori* and independently validated, implies the "re-use" of *claims, arguments and evidence* addressing properties of the component. The question then is: What are the conditions under which this "re-use" is valid? Two types of conditions can be distinguished: conditions on the syntax of the claims and arguments being re-used and conditions on the semantics of the underlying models and assumptions. These conditions are re-examined in the following two sections, in the light of the developments in Chapters 6 and 8.

10.2.1 Syntactical Conditions

If an argument is re-used in a different justification context, the following syntactical definitions and properties must be maintained.

An *argument* is the tree of nested *subclaims* (*expansion*) and evidence components that *justifies* a *white, black or grey subclaim* at a given level; the argument comprehends the complete sequence of *inferences* (*expansions* and *delegations*) that link evidence components and nested *subclaims* (*cf.* Definition 6.3 and Figure 6.1 in Section 6.3).

A *subclaim* is always *justified* by a single *argument* (unicity property of Section 6.5.3). Only one instantiation and documentation of the argument is thus necessary, and this unique *argument* can support different *invocations* of a *subclaim* in the justification of a level-0 *dependability requirement*.

A *subclaim* is *invoked* by its predicate in one or several *expansions of delegation claims*, these *expansions* being always at the same level at which the *subclaim* is *justified* (*cf.* for instance Figure 8.3, page 108). This requires that *subclaims* be univocally identified by their level and by their predicate.

A *delegation claim* is a specification of an *expansion*, the level of the *expansion* and the set of values assigned by upper levels to the free variables of the *subclaim(s)* predicated in the *expansion* (Theorem 8.5, page 99).

A *delegation claim* can be made and replicated at different levels and with different values for the free variables of the *subclaim(s) of its expansion*, within the justification of a same level-0 *initial dependability requirement* or of different *initial dependability requirements*, but always above the level of *expansion*, and always with the same *expansion* valuated by the same *expansion structure*.

Thus, a *white, grey* or *black subclaim*, justified by a single *argument*, can be predicated in the *expansions* of different *delegation claims*.

10.2.2 Semantic Conditions

An *argument* obviously must be complete: all leaves of the tree must be evidence components. Furthermore, the following three conditions must be satisfied (*cf.* in particular Theorem 8.7, page 103):

- The *subclaim* must be justified at a given level i, valuated by a *justification structure* U_i that must satisfy the two conditions which follow;
- U_i must be an *elementary substructure* of every level - i *expansion structure* in which the *subclaim* is predicated;

- U_i must be a *substructure* of all lower level *expansion structures* in which are expanded the *delegation claims* predicated in the *justification* of the *subclaim*.

The three conditions impose constraints on the domains, relations and constants of the *structures* involved; these constraints result from the Definition 8.7 of *substructure* and the Definition 8.8 of *elementary substructure* in Section 8.8.2.

10.3 Guaranteed Services and Rely Conditions

A *subclaim* and its supporting *argument* is therefore not "context-free". We now have to determine if and how these rather strict syntactic and semantic conditions can be applied to the use of pre-existing components that are independently validated, with *interpretation structures* also independently developed. What do these conditions implicate regarding these *structures* which model both sides of the interface between the component and the embedding system?

The essential idea of our approach is the following. The basic justification problem is to establish that the behaviour of the pre-existing component conforms to the behaviour anticipated by the embedding system. Since the internal design of the pre-existing component is not part of the current development, the justification must be made possible by claiming properties of a precise and comprehensive interface between the system and the component, and by relying on a pre-established justification of these interface properties.

On the one hand, a pre-existing component can be reliably used as a black box, provided it offers guaranteed, *comprehensively and unambiguously defined services* to its environment (see Figure 10.1). In general, it is the design level-3 of the embedding system which uses services granted by pre-existing components, used at that level as channel input or output devices, processing units or software components[16].

On the other hand, a black box cannot, *a priori*, be expected to guarantee adequate services, whatever the environment is. The level-2 or 3 of the embedding system must be required to satisfy specific constraints and assumptions on which the pre-existing component must be allowed to rely. These conditions – we shall call them *rely conditions* – are the necessary and sufficient conditions that must be satisfied by the environment in which the pre-existing component operates to obtain the guaranteed services. A rely/guarantee concept is thus not much different from a client/server model, *cf.* for instance [96]; the same terminology was used in [53], but for the different purpose of documenting and reasoning about concurrent and shared-variable programs.

---

[16] Elements of the architecture, like a channel, could in principle also be implemented as a pre-existing component. This other possibility is briefly discussed in Section 10.5.

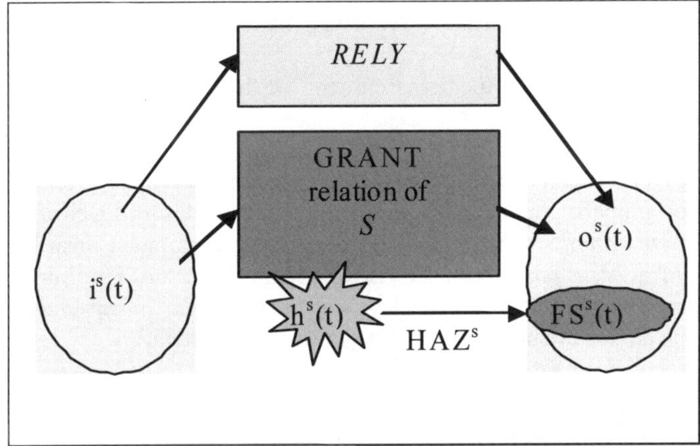

Figure 10.1. Embedded pre-existing component S

Hence, *rely and guarantee conditions* impose obligations on both the component and the system. And the justification of the use of a pre-existing component at the level 3 of an embedding system requires from these two "partners" three types of evidence:

- Evidence of the provision of guaranteed services by the component under the assumption that the *rely* conditions are satisfied;
- Level-3 evidence that the *rely conditions* are satisfied by the embedding system;
- Level-3 evidence that the *guaranteed services* satisfy the *delegation claims* from the embedding system upper levels.

A key point is that it should be possible to provide these types of evidence independently of one another. The pre-existing component supplier is responsible for providing evidence of the *guaranteed services* under the assumption that the rely conditions are satisfied. The user of the component, *i.e.* the embedding system, is responsible for providing the evidence that the *rely conditions* are satisfied and maintained in the embedding system and that the *guaranteed services* satisfy its claims.

The possibility for the supplier and the user to independently fulfill their proof obligations and provide the required evidence responds to the commercial constraints and the incentives which motivate the supply and use of pre-existing components. Each party, however, must be able to rely on the other's evidence and to justify that the conditions of use expected by the other are satisfied. This is possible only if a "compatible" interface model is used to formulate, interpret and valuate the evidence on each side. This model is the subject of the following sections.

10.4 The System-Component Interface Substructures

The embedding system is the "environment" of the pre-existing component. Thus, the model of the interface between a system and a pre-existing component has many common aspects with the model of the "environment-system interface" developed in Section 9.4.13. In particular, we will see in the following sections that, to be properly embedded at the architecture or design level of a system, the level-1 *justification structures* of the pre-existing component must be *elementary substructures* of *expansion structures* which are *extensions* (Definition 8.7) of the embedding system level-2 or level-3 *justification substructures* in which *delegation claims* are predicated by the embedding system.

10.4.1 The GRANT and RELY Relations

The "black-box" behaviour of a pre-existing component can be defined by a *relation* between its input and output variables, similar to the functional relations REQ defined in Section 9.4.7.

Let S denote a given pre-existing component or sub-system. Let $i^s(t)$ denote the vector of the values at time t of the input parameters and input variables of S; and let $o^s(t)$ denote the vector of the values at time t of the output variables of S. The $i^s(t)$ values are the contents of a shared register accessible by the embedding system for writing and by S for reading; the $o^s(t)$ values are the contents of a shared register written by S and read by the embedding system. A subset of the $o^s(t)$ values, denoted $FS^s(t)$, designates a safe subset of S; the role of that subset will become clearer in the following sections.

The input/output behaviour of S is described by a relation called GRANT.

Definition 10.1 GRANT is defined as follows:

- *Domain* (GRANT) is the set of vector-valued time-functions containing the possible values of $i^s(t)$;
- *Range* (GRANT) is the set of vector-valued time-functions containing the possible values of $o^s(t)$;
- $\{i^s(t), o^s(t)\}$ is in GRANT if and only if the pre-existing component lets the output variables take the values described by $o^s(t)$ when the values of the input variables are described by $i^s(t)$.

The GRANT relation is a part only of the specification of the S behaviour. It is a relation against which the implementation of S must be validated by the developer of S, under the assumption that the constraints in the environment in which S is embedded let the values and time variations of the input and output variables satisfy pre-determined conditions.

These conditions are specified by the relation RELY, also part of S specifications.

Definition 10.2 RELY is defined as follows:

- *Domain*(RELY) is a set of vector-valued time-functions containing all permissible values of $i^s(t)$;
- *Range* (RELY) is a set of vector-valued time-functions containing all permissible values of $o^s(t)$;
- $\{i^s(t),\ o^s(t)\}$ is in RELY if and only if constraints in the environment of S allow the output variables to take the values described by $o^s(t)$ when the values of the input parameters and variables are described by $i^s(t)$.

RELY plays here a role similar to the role played by the relation NAT (Section 9.4.3) in Chapter 9. Only values of $i^s(t)$ and $o^s(t)$ that belong to the *universe* of RELY are assumed by S to occur. The *domain* of RELY specifies the valid formats and ranges of values of the parameters and variables $i^s(t)$. The behaviour of S is undefined for $i^s(t)$ values that are not in this *domain*. The *range* of RELY specifies all valid format and values of the output variables $o^s(t)$. Any value not in this *range* is invalid and is considered as being either not produced by S or the result of an invalid behaviour of S. RELY specifies values for initialisation, calibration and resetting input and output registers, and other conditions like formatting and power conditions which must prevail to use and access S via its input-output registers. Only the values of the RELY *universe* (*domain* and *range*) are known to S. The behaviour of S is undefined for values not in this *universe*.

An important difference is that, contrary to the NAT relation defined in 9.4.3, RELY is a relation imposed by the pre-existing component S upon its environment, and not by the environment upon S.

Together with the GRANT relation, RELY constitutes the input-output specification of the behaviour of S and must be provided by the developer of S to the user of S. The user of S has the obligation of ensuring that the RELY relation is satisfied.

10.4.2 The HAZ^s Relation

Undesired events that may affect the internal behaviour of S must be anticipated by the developer of S and must be properly documented to the user. These UE's are part of the system-component interface specification. Let $h^s(t)$ be the vector of S entities that are sources of internal *undesired events* (Figure 10.1). The current value of an element indicates whether the corresponding entity has caused an event or not and which type of event (*cf.* in Section 9.4.5). These events directly affect the internal state of S, but this state is hidden from the embedding system. If S is properly isolated from the embedding system, the consequences of these *undesired events* should affect the behaviour of the embedding system by their impact on the output values $o^s(t)$ of S only.

This impact is modelled by a relation we call HAZ^s.

Definition 10.3 The relation HAZ^s is defined as follows:

- *Domain*(HAZ^s) is the set of vector-valued time-functions containing the possible values of $h^s(t)$;
- *Range*(HAZ^s) is the set of vector valued time functions containing values of $o^s(t)$;
- $<h^s(t), o^s(t)>$ is in HAZ^s if and only if $(o^s(t), t)$ is an event occurring whenever $(h^s(t), t)$ occurs.

The set of anticipated S *undesired events* $h^s(t)$ and the relation HAZ^s are also part of the specifications of S.

10.4.3 Proof Obligations for the Developer of S

It is the responsibility of the developer of S to justify the validity of the S specifications.

The input-output specifications of S must be complete. They must specify the behaviour of S for all cases and modes of execution that are anticipated in the environment, *i.e.* those defined by the RELY relation. The *input-output system specifications* must therefore satisfy:

$$domain \text{ (GRANT)} \supseteq domain \text{ (RELY)}.$$

The *input-output system specifications* must also be feasible with respect to the RELY constraints, *i.e.* the S behaviour described by GRANT must be allowed by RELY. Thus, we must also have:

$$domain \text{ (GRANT)} \subseteq domain \text{ (RELY)}.$$

The two sentences imply:

$$domain \text{ (GRANT)} \equiv domain \text{ (RELY)}.$$

Then, if we define the *justification structure* (Definition 8.10):

(10.1) $$U_s(\text{GRANT}) = <\{i^s(t)\}\{o^s(t)\ ; \{RELY\}\{GRANT\}\ ; \{k^s\}>,$$

the following primary claim must hold true:

(10.2) **GRANT_completeness_and_feasibility.clm1**:

$$U_s(\text{GRANT}) \vDash (domain \text{ (GRANT)} \equiv domain \text{ (RELY)})$$

with *justification structure* $U_s(\text{GRANT})$ defined by (10.1).

Feasibility is here defined with respect to the constraints assumed satisfied by the embedding system. It does not imply that the input-output system specifications are realizable. Realizability is a property of the lower levels (2 and 3) of the S implementation.

The input-output specifications of S must also be shown to comply with the functional requirements of S as they are documented in a language ($\mathbf{L_s}$) for the user of S. The $\mathbf{L_s}$-*interpretation* $U_s(GRANT)$ must be a *model* (8.4) of these functional requirements; hence the following claim must hold in $U_s(GRANT)$:

(10.3) **GRANT_correctness.clm1**:

$U_s(GRANT) \vDash S\_$**documented functional requirements**

with *justification structure* $U_s(GRANT)$ defined by (10.1).

As part of the level-1 justification of S, the set of anticipated *undesired events* described by the values of $h^s(t)$ must be shown to be as complete as possible. These *undesired events* must be unambiguously *detected, controlled* and lead to a *safe state*. The detectability, controllability and mitigation of these *undesired events* are properties of S which have to be justified by the developer. These properties imply that $h^s(t)$ and the subset of the $o^s(t)$ safe values, denoted $FS^s(t)$, are valid sets and that the RELY conditions are satisfied. The evidence to support this validity is provided by valuations of the *justification structure* (Definition 8.10):

(10.4) $U_s (HAZ^s)= <\{h^s(t)\}\ \{o^s(t)\}\ \{FS^s(t)\}\ ;\ \{RELY\}\{HAZ^s\}\ ;\ \{k^s\}>$

where $\{k^s\}$ includes constants relevant to HAZ^s.

Of primary importance to the user of S is the mitigation of these undesired events; hence, the necessity of a claim made by the developer that all S anticipated undesired events lead S to the well-defined subset $FS^s(t)$ of safe output sates:

(10.5) $h^s(t)\_$**mitigation.clm1**:

$U_s (HAZ^s) \vDash (\forall h^s(t):\ range\ h^s(t)\ (HAZ^s \circ GRANT) \subset FS^s(t))$

with *justification substructure* $U_s (HAZ^s)$ defined by (10.4).

At this point, it is worthwhile to observe that, *mutatis mutandis,* at the level-1 of the S justification, in comparison with the justification of a newly developed system (Section 9.4.10), the roles of the relations are inter-changed as shown in Table 10.1.

Table 10.1. Relation substitutions for a pre-existing component S dependability justification

Relation	substituted with
REQ	GRANT
NAT	RELY
HAZ^1	HAZ^s
$FS(t)$	$FS^s(t)$
$h^1(t)$	$h^s(t)$

Remark 10.1 Attention is restricted here to the most usual claims that are made on a pre-existing component S, and to the most essential level-1 evidence required from S to justify its use in an arbitrary embedding system. Obviously, depending on S functionality and the application supported by the embedding system, other specific claims on S may have to be made, *e.g.* on accuracy, reliability and availability; these claims must be treated in a similar way as above.

Remark 10.2 It is S developer's responsibility to demonstrate that the architecture and design of the S implementation satisfy the GRANT input/output specifications under the condition that the RELY relation is maintained by the embedding system. To this end, claims at S architecture level-2 and design level-3, similar to Sections 9.5 and 9.6 of Chapter 9, must be justified by the developer of S. More generally, as explained in Section 6.6.1 and illustrated by the *arguments* in Chapter 9, the basic properties of S that must be justified are properties of validity, implementation acceptability and dependability at the levels of architecture, design and operation control. The corresponding arguments must be part of a pre-existing *dependability case* and documentation of S, independently produced.

10.4.4 The $U_s(\delta\xi_s)$ Expansion Structures

According to Theorem 8.3 and Theorem 8.4, the *justification structures* introduced in the previous section must be *elementary substructures* of the *expansion structures* in which *grey or black claims* predicated in the embedding system are *extended* as *logical consequences* of claims made on S.

Let $U_s(\delta\xi_s)$ denote the *expansion structure* (*cf.* Definition 8.11) for a *delegation claim* $\delta_{ks}(\xi_s[B_1...B_k])$, with predicate ξ_s and assignment of values $[B_1...B_k]$ to free variables of S. The delegation claim is made on S by the embedding system to justify a *grey or black claim* at an upper level k of an argument.

Whatever the *delegation claim* $\delta_{ks}(\xi_s[B_1...B_k])$ is, $U_s(\delta\xi_s)$ must be an *extension* (Definition 8.8) of the S *elementary justification structures* necessary to justify the properties of S claimed with $\delta_{ks}(\xi_s[B_1...B_k])$ by the embedding system; in particular:

$$(10.6) \qquad U_s(GRANT) \prec U_s(\delta\xi_s) \qquad and/or \qquad U_s(HAZ^s) \prec U_s(\delta\xi_s).$$

The set of the *expansion structures* $U_s(\delta\xi_s)$ needed in a dependability case for valuating all delegation claims $\delta_{ks}(\xi_s[B_1...B_k])$ is the description of the *interface* of the pre-existing system with the embedding system.

$U_s(\delta\xi_s)$ is also the *structure* which must valuate to true the properties claimed with $\delta_{ks}(\xi_s[B_1...B_k])$ on S and in which the expansion of $\delta_{ks}(\xi_s[B_1...B_k])$ must hold; hence, $U_s(\delta\xi_s)$ is also an *abstraction* of the implementation of S.

Although they may in principle be either *grey* or *black*, there is no loss of generality if we consider the *delegation claims* made by the embedding system on properties of S as being all *black delegation claims* $\delta_{ks}(\beta_s[B_1...B_k])$; since S has to be considered as a "black box", to do so is even useful in order to keep claims requiring evidence from S separate from others in the arguments of the embedding system *dependability case.*

More specific instances of interface *expansion structures* will be discussed in Section 10.4.6.

10.4.5 Proof Obligations for the Embedding System

When a pre-existing component S is embedded in a system, the environment of S is instantiated by the architecture level-2 or the design level-3 of the embedding system. This embedding system architecture level is responsible for guaranteeing and providing evidence that the RELY conditions for using S are satisfied and maintained at the level where S is used.

Let us take an example to illustrate and simplify the exposition. Consider the level-1 and level-2 structures analysed in Chapter 9 as models of the embedding system, and let us assume that a pre-existing component S is to be used as a level-3 implementation of the relation REQ_z in some channel z. Note that the proof obligations would be very similar, *mutatis mutandis*, if S were to satisfy and implement one of the relations IN_z, OUT_z or SOF_z.

At the level-2 of this embedding system, there must necessarily be a *black claim $\mathcal{A}^2$_channel_z_acceptable_implementation.clm2*, similar to the *black claim* made in argument (9.5.17). This *claim* would be predicated at level-2 in a *justification structure* $U_2(REQ_z)$, which is similar to the *structure* (9.47) and is an *elementary substructure* of a level-2 *expansion structure* $U_2(\delta\mathcal{A}^1 REQ)$ (9.57). This *black claim* has to be expanded at level-3 (as in Section 9.6.5) into claims on *acceptability* (Definition 9.12), *i.e.* into claims on *completeness, correctness* and *accuracy.*

However, compared with argument (9.5.17), the dependability justification of a system using a pre-existing component presents two important differences that are examined in the two following sections.

10.4.5.1 Validity and Satisfiability of Conditions of Use
A first difference is that the level-2 *expansion* of the claim on channel implementation must also include a claim for evidence that the RELY conditions required for the use of the component are satisfied in the embedding system. The satisfiability of these conditions is a level-2 property. The level-2 claim for a

channel z acceptable implementation therefore becomes a *grey claim*; its level-2 *justification* in the level-2 argument (9.5.17) takes the form:

(10.7) $(S$_**conditions_validity and satisfiability.clm2**

$\wedge\ \mathcal{A}^2$_**channel_z _acceptable_S_implementation.clm[2,3]**$)$

$\vdash_{U_2(REQ_z)} \mathcal{A}^2$_**channel_z_acceptable_implementation.clm2.**

Let us first make the validity and satisfiability claim explicit. First of all, the user of S must have a correct *interpretation* of the RELY conditions for using S as they are imposed and documented by the designer of S. Let us consider the following *structure* at level-2 of the embedding system:

(10.8) $U_2(\text{RELY}) = <\ \{i^z(t)\}\{o^z(t)\}\ ;\ \{\text{RELY}\}\ ;\ \{k_2\}>,$

where the relation RELY is given by Definition 10.2 and the set $\{k_2\}$ designates the constants of the embedding system relevant to S environment. The *interpretation* provided by $U_2(\text{RELY})$ is obtained by assignment: an element of the universe $\{i^z(t)\}\{o^z(t)\}$, a member of the set RELY and a constant of $\{k_2\}$ are assigned to each variable symbol, predicate symbol, function symbol, operation symbol and parameter respectively, of the language $\mathbf{L_s}$ in which the RELY conditions for using S are documented by the designer.

Given this assignment, the valuations (Definition 8.2) of $U_2(\text{RELY})$ must *satisfy* (Definition 8.3) the S environment constraints. If these environment constraints valuate to true in $U_2(\text{RELY})$, $U_2(\text{RELY})$ *is a valid interpretation (model) for the constraints specified by the developer of S.*

Moreover, according to Lemma 8.7, the evidence provided by the valuations of the *interpretation* $U_2(\text{RELY})$ is valid if and only if (Theorem 8.4) $U_2(\text{RELY})$ is an *elementary justification substructure* of the $\mathbf{L_2}$-*expansion structure* $U_2(\delta\mathcal{A}^1\text{REQ})$ (9.57).

We therefore need the *claim*:

(10.9) S_**conditions_validity.clm2:**

$U_2(\text{RELY}) \vdash S$ **documented conditions of use**

with $U_2(\text{RELY}) \prec U_2(\delta\mathcal{A}^1\text{REQ}),$

the *expansion structure* $U_2(\delta\mathcal{A}^1\text{REQ})$ being defined by (9.57), and the *justification substructure* $U_2(\text{RELY})$ defined by (10.8).

The conditions for using S must not only be correctly interpreted; they must also be satisfied by the embedding system. In other words, all formulae with free variables in the universe of $U_2(\text{RELY})$ must hold in $U_2(\text{REQ}_z)$, and reciprocally. Hence, the claim:

(10.10)   **S_conditions_satisfiability.clm2:**

$$U_2(\mathrm{RELY}) \prec U_2(\mathrm{REQ_z})$$

with $U_2(\mathrm{REQ_z}) = < \{m(t)\}\,\{c(t)\}\,\{i^z(t)\}\,\{o^z(t)\}\ ;$
$\qquad\qquad\qquad \{\mathrm{REQ}\}\,\{\mathrm{IN_z}\}\,\{\mathrm{OUT_z}\}\,\{\mathrm{RELY}\}\,\{\mathrm{REQ_z}\}\ ;$
$\qquad\qquad\qquad \{k_2\} >$

where $U_2(\mathrm{REQ_z})$ is constructed by adding the relation RELY to the level-2 structure defined in (9.47) and $\{k_2\}$ includes the set of constants of channel z.

The upper level of Figure 10.2 displays the *structures* introduced so far in this section with their inter-relations.

10.4.5.2 S Acceptability

The second difference of the justification comes from the fact that the user of S must justify the *acceptability* of S as it is *specified* by the designer, and not of the S implementation itself, which is the designer's obligation.

More precisely, the *delegation claim*

$$\mathcal{A}^2\_\text{channel\_z\_acceptable\_implementation.clm[2,3]}$$

predicated at level-2 in a *justification structure* $U_2(\mathrm{REQ_z})$, must be expanded within a level-3 *interpretation structure* into claims on *S acceptability* (Definition 9.12), *i.e.* into claims on the *completeness*, *correctness* and *accuracy* of the S *specifications*, in reference to the embedding system requirements. The expansion is a logical consequence valuated in a level-3 *expansion structure* which we designate $U_s(\delta\mathcal{A}^2\,\mathrm{REQ_z})$ and will define later. According to Proposition (iii) of Theorem 8.7, $U_2(\mathrm{REQ_z})$ defined in (10.10), must be a *substructure* (Definition 8.7) of this level-3 *expansion structure*. Thus, the following Proposition must hold between the two embedding and embedded system *structures*:

(10.11) $U_2(\mathrm{REQ_z}) \subseteq U_s(\delta\mathcal{A}^2\,\mathrm{REQ_z}),$

Completeness of the S specifications implies that the universe of the relation GRANT includes the universe of $\mathrm{REQ_z}$:

(10.12) $domain\,(\mathrm{GRANT}) \supseteq domain\,(\mathrm{REQ_z})$

$\qquad\qquad\qquad range\,(\mathrm{GRANT}) \supseteq range\,(\mathrm{REQ_z})$

Moreover, the z channel specification $\mathrm{REQ_z}(i^z(t),o^z(t))$ constrained by a level-2 claim on architectural completeness such as (9.45) and satisfying a claim on channel correctness such as (9.48), must have a S implementation which satisfies a Proposition similar to (9.97) provided that the conditions for using S are satisfied.

In other words, every channel specification $REQ_z(i^z(t), o^z(t))$ must have a S implementation and one must have:

(10.13)
$$\forall\, i^z(t),\, o^z(t) \mid REQ_z\,(i^z(t),\, o^z(t))\ :$$
$$REQ_k\,(i^z(t),\, o^z(t))) \wedge RELY(i^z(t),\, o^z(t))$$
$$\vDash_{U_s(\delta\mathcal{A}^2 REQ_z)}\, (\exists\, i^s(t) = i^z(t), o^s(t) = o^z(t):\ GRANT(i^s(t),\, o^s(t))).$$

Hence, the following claim :

(10.14)
(z, S) specifications_completeness.clm3:
(10.12) and (10.13) are *satisfied* in the *structure* $U_s(\delta\mathcal{A}^2\, REQ_z)$

with $U_s(\delta\mathcal{A}^2 REQ_z)\ \succ\ U_s(GRANT)$
and $U_s(GRANT)$ defined by (10.1).

We observe that, in contrast to the claim (9.98) on software completeness in Chapter 9, this claim which addresses specifications and not an implementation is valuated in $U_s(\delta\mathcal{A}^2\, REQ_z)$ which is not a *justification structure* but an *expansion structure* – of which $U_s(GRANT)$ must be an *elementary justification substructure*.

The specifications of the pre-existing component S must not only be complete but also *correct* in relation to the channel z requirements specified by the relation REQ_z. The following claim must be *satisfied* (valuate to true) in the *expansion structure* $U_s(\delta\mathcal{A}^2\, REQ_z)$ for any monitored variable m(t) such that $IN_z\,(m(t),\, i^z(t))$:

(10.15)
(z,S) specifications_correctness.clm3:
$$\forall\, i^z(t) \mid IN_z\,(m(t),\, i^z(t)):$$
$$REQ_z\,(i^z(t),\, o^z(t))$$
$$U_s(\delta\mathcal{A}^2 REQ_z) \vDash (GRANT(i^z(t),\, o^z(t))\,) \wedge RELY(i^z(t),\, o^z(t)))$$

with $U_s(\delta\mathcal{A}^2\, REQ_z)\ \succ\ U_s(GRANT)$
and $U_s(GRANT)$ defined by (10.1).

which claims, according to (8.3), that a valuation of the *structure* $U_s(\delta\mathcal{A}^2\, REQ_z)$, valuating to true the right hand side closed well formed formula of (10.15), also *satisfies* the left hand side predicate. In other words, the z channel input-output requirement specification predicate must be the *logical consequence* of the GRANT relation whenever the relation RELY holds, and for values $i^z(t)$ such that the relation $IN_z\,(m(t),\, i^z(t))$ holds.

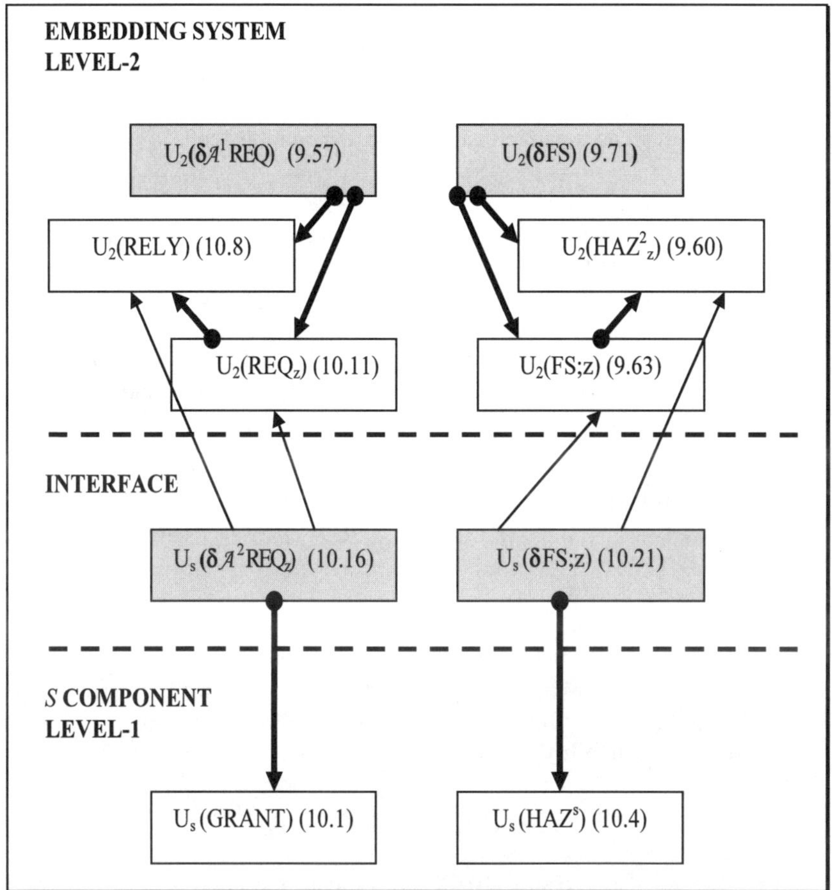

Figure 10.2 *Structures* of the interface between a pre-existing component S and its embedding system. Grey nodes are *expansion structures*; white nodes are *justification structures*. A bulleted oriented edge from U_i to U_j stands for "U_j is *elementay substructure* of U_i"; in the same way, a non-bulleted oriented edge stands for a *substructure* interdependency. References point to definitions given in Chapters 9 and 10

10.4.5.3 Justifying the Domain of Use

This last restriction is important. In principle and in practice, the input-output behaviour of S described by the GRANT relation usually covers more functionality than what is needed by the embedding system and described by the relation $REQ_z\,(i^z(t), o^z(t))$. This is why the Proposition (10.15) has to explicitly apply to the bounded variables $i^z(t)$ of the domain of $REQ_z\,(i^z(t), o^z(t))$ only. Otherwise it would imply that the REQ and the GRANT functionalities are identical.

It is thus noteworthy that the correctness of S specifications cannot be claimed and justified without making specific reference to the domain of use by the embedding system. This is a formal confirmation that the correct use of a pre-existing component cannot, in general, be justified independently of the conditions under which it is used by the embedding system. Much of the commercial success of COTS and pre-existing components is based on the ignorance of this constraint. Often, dependability cases dangerously ignore this point as well.

10.4.5.4 The Interface Expansion structure $U_s(\delta\mathcal{A}^2 REQ_z)$

As before, because the user of S must justify the *correctness* of the S specifications and not of the S implementation, evidence for the claim is obtained by valuation of an *expansion structure* of which $U_s(GRANT)$ must be an *elementary justification substructure.*

We can now define this *expansion structure* $U_s(\delta\mathcal{A}^2 REQ_z)$ as being the smallest *extension* (Definition 8.7) of the *substructures* $U_2(REQ_z)$ (9.47), and $U_s(GRANT)$ (10.1) that have been used in (10.11), (10.14) and (10.15):

$$(10.16) \qquad U_s(\delta\mathcal{A}^2 REQ_z) = <\ \{m(t)\}\,\{c(t)\}\,\{i^z(t)\}\ \{o^z(t)\}\ \{i^s(t)\}\ \{o^s(t)\}\ ;$$
$$\{REQ\}\,\{IN_z\}\,\{OUT_z\}\,\{REQ_z\}\ \{GRANT\}\,\{RELY\};$$
$$\{k_2\}\ \{k^s\}>$$

where $\{k_2\}$ is the set of constants of channel z, and $\{k^s\}$ the relevant constants of S.

Moreover, in order to be an *elementary substructure* of $U_s(\delta\mathcal{A}^2 REQ_z)$, $U_s(GRANT)$ must also satisfy the Tarski – Vaught condition (Theorem 8.1, page 90). Moreover, quite naturally, by definition of an *elementary substructure*, since claim (10.10) on satisfiability imposes that

$$U_2(RELY) \prec U_2(REQ_z),$$

it makes sense to justify claim (10.9) and validate the conditions of use before justifying the other claims.

The interface *expansion structure* $U_s(\delta\mathcal{A}^2 REQ_z)$ and its connections with the other *structures* introduced so far are displayed on Figure 10.2 where connections resulting from transitivity (Lemma 8.1 and Lemma 8.4) are also indicated.

Remark 10.3 The *logical consequence* (10.15) constrains the acceptability of the pre-existing component S, and bears a special meaning. If the predicate RELY is false, then any S behaviour is considered acceptable by (10.15); RELY is part of S specifications.

Remark 10.4   Propositions (10.13) and (10.15) imply that, for all pairs $<(i^z(t),\ o^z(t))>$ such that $IN_z\ (m(t),i^z(t))$ and $REQ_z\ (i^z(t),\ o^z(t))$, the relation $REQ_z\ (i^z(t),\ o^z(t))$ valued in the *interpretation structure* $U_2(REQ_z)$ is a *restricted* relation (Definition 8.7) of the relation $GRANT(i^z(t),o^z(t))$ in $U_s(GRANT)$. Note that the relations GRANT and REQ_z are not identical. In particular, as said earlier, GRANT may describe extra-functionality not included in REQ_z, and also includes the relations dealing with $h^s(t)$ *undesired events*. The claims, however, ensure that none of these "extra relations" invalidates REQ_z.

10.4.5.5 S Robustness

The *mitigation* of the consequences of the *undesired events* $h^s(t)$ that may affect the behaviour of *S* is the responsibility of the embedding system and can be treated in a similar way as *S acceptability*.

At level-2 of the embedding system, there is a *black claim* **channel_z_fail_safe_implementation.clm2,** which is similar to the claim made in argument (9.94).   This claim is predicated in a level-2 *justification structure* $U_2(FS;z)$ which is similar to the structure defined by (9.63) and is an *elementary substructure* of a level-2 *expansion structure* $U_2(\delta FS)$ (9.71):

$$(10.17) \qquad U_2(FS;z) = < \{FS(t)\} \, \{h^2(t)\} \, \{i^z(t)\} \, \{o^z(t)\}; \\ \{IN_z\} \, \{OUT_z\} \, \{REQ_z\} \, \{HAZ^2_z\} \, ; \\ \{k_2\}>$$

where $\{k_2\}$ includes the constants relevant to channel z and $\{HAZ^2_z\}$.

The level-2 *black claim* **channel_z_fail_safe_implementation.clm2** has to be expanded at level-3 (as in Section 9.6.6) into claims on validity and robustness. According to Proposition (iii) of Theorem 8.7, $U_2(FS;z)$ has to be a *substructure* (Definition 8.7) of the level-3 *expansion structure* of this *black claim* which we call $U_s(\delta FS; z)$; we must have:

$$(10.18) \qquad U_2(FS;z) \subseteq U_s(\delta FS; z).$$

These *structures* and their interdependencies are displayed on Figure 10.2, page 259. $U_s(\delta FS; z)$ valuates the *expansion (logical consequence)* of the *black claim* into level-3 claims on the validity, detection, and mitigation mechanisms of postulated undesired events. $U_s(\delta FS; z)$ constituents include those of its corresponding *elementary justification substructures*.

Claiming and providing evidence for the validity and detectability of *S* postulated undesired events $h^s(t)$ are obligations of the *S* developer and not of the embedding system. Hence, in contrast with the validity and satisfiability discussed in Section 10.4.5.1, these properties need not and could not be claimed by the embedding system.   But the fail-safe use of *S* by channel z is the user's responsibility; it is guaranteed under the condition that the following level-3 claim holds for the embedding system:

$$(10.19) \qquad \text{z\_FS}^s(t)\_\textbf{mitigation.clm3}:$$
$$U_3(\delta FS; z) \models (\forall o^s(t) \in FS^s(t) : OUT_z(o^s(t), c(t))) \in FS(t))$$
$$\text{with } U_s(\delta FS; z) \succ U_s(HAZ^s)$$
$$U_s(HAZ^s) \text{ defined by (10.4).}$$

Note that $FS^s(t)$ is part of the specification of S. This claim together with the validity of the set of *undesired events* $h^s(t)$ and of the relations $RELY(i^s(t),o^s(t))$ and $HAZ^s(h^s(t), o^s(t))$ - guaranteed by the developer of S - ensures that:

$$(10.20) \qquad (\forall h^s(t), i^s(t), o^s(t):$$
$$RELY\ (i^s(t), o^s(t)) \Rightarrow$$
$$range\ (HAZ^s(h^s(t), o^s(t)) \circ OUT_z(o^s(t), c(t))\ ) \in FS(t)).$$

An important issue, however, remains. The risk exists that *undesired events* at the architecture level ($h^2(t)$) or design level ($h^3(t)$) of the embedding system directly or indirectly affect the RELY conditions. The mitigation of this risk and the preservation of the RELY conditions remain the duty of the embedding system and are delicate aspects of the use of pre-existing components. We shall come back to this issue in Section 10.6.2.

10.4.5.6 The Interface Expansion Structure $U_s(\delta FS;z)$
We can now identify the constituents of the *expansion structure* $U_s(\delta FS;z)$. It is the smallest *extension* (Definition 8.7) of the *substructures* $U_2(FS;z)$ and $U_s(HAZ^s)$ that have been introduced by (10.17) and (10.19):

$$(10.21) \qquad U_s(\delta FS;z) = <\ \{FS(t)\}\ \{h^2(t)\}\ \{i^z(t)\}\ \{o^z(t)\}\ \{h^s(t)\}\ \{o^s(t)\}\ \{FS^s(t)\};$$
$$\{IN_z\}\ \{OUT_z\}\ \{REQ_z\}\ \{HAZ^2_z\}\ \{HAZ^s\};$$
$$\{k_2\}\ \{k^s\}>$$

where $\{k_2\}$ includes the constants relevant to channel z and $\{HAZ^2_z\}$, and $\{k^s\}$ the relevant constants of S. For $U_2(FS;z)$ and $U_s(HAZ^s)$ to be its *elementary substructures*, $U_s(\delta FS; z)$ must also satisfy the Tarski – Vaught condition (Theorem 8.1, page 90).

10.4.6 System-Component Interface Revisited

Expansion structures – such as $U_s(\delta \mathcal{A}^2 REQ_z)$ and $U_s(\delta FS; z)$ defined by (10.16) and (10.21) – valuate the *expansions* of the *delegation claims* that the embedding system makes on properties of S; the set of these *structures* constitutes the interface between the embedding system and the pre-existing component S. These interface *structures* and their interdependencies are displayed on Figure 10.2. $U_s(\delta \mathcal{A}^2 REQ_z)$ and $U_s(\delta FS; z)$ are particular instantiations of the *expansion structures* $U_s(\delta \xi_s)$ already introduced in Section 10.4.4 for valuating delegation claims $\delta_{ks}(\xi_s[B_1...B_k])$.

Figure 10.2 clearly brings out that these interface structures are upwards related to the structures of the embedding system by mere *substructure* relations, and not by "stronger" more constraining *elementary substructure* dependencies.

Other *claims* on the accuracy (defined by the $\mathcal{A}^2$ parameters of the delegation claim on acceptable implementation) or on the dependability of the pre-existing component should be dealt with in a similar way. If the pre-existing component is an input/output device or a software component, a *claim* sentence equivalent to (10.15) in which REQ_z is substituted with a predicate ξ_z ($\xi_z = IN_z$, OUT_z or $SOFT_z$) must be formulated and valuated by evidence in a corresponding *expansion structure* $U_s(\delta\xi_z)$. In all cases, the proof obligations (10.4.3) for the developer of the pre-existing component remain independent from the embedding system.

Quite naturally, Section 10.4.3 shows that the proof obligations resting on the *developer* of a pre-existing component are rather similar to those of the level-1 of the justification of a newly developed system. Table 10.1 indicates the similitudes of the *structures* and *relations* for both cases.

Regarding the *user*, we observed in Section 10.4.5 three essential differences between the justification obligations of a system which uses a pre-existing component at some lower level and a system which does not:

- Claims and evidence on the *acceptability* (completeness, correctness, accuracy) of a pre-existing component must be complemented with claims on and evidence of the validity and satisfiability of the conditions of use for the pre-existing component;
- The user of a pre-existing component S must provide evidence of the *acceptability* of the S specifications and not of the S implementation, and this evidence is obtained by the valuation of an *expansion structure,* the *elementary justification substructures* of which are supplied by the designer of S;
- The user of a pre-existing component S must specify the domain of use for which the correctness of the use is justified.

It is remarkable that, while all properties claimed by the system and by the pre-existing component are justified by the valuation of *justification structures*, the properties of their interface are valuated in *expansion structures*.

Expansion structures – such as $U_s(\delta\mathcal{A}^2REQ_z)$ and $U_s(\delta FS; z)$ defined by (10.16) and (10.21) – provide a general model of an interface between a pre-existing component and a system using it. The model is applicable to the valuation and justification of the various types of component-based software or hardware interfaces, protocols and services currently met in practice: client-server protocols, middleware layers, wrappers, connectors (see for instance [2, 38]) for more detail on design, encapsulation, compatibility and validation issues raised by the deployment of component based systems).

10.5 Documentation

The proof obligations of Sections 10.4.3 and 10.4.5 allow us to specify the documentation required to justify the use of a pre-existing component S.

The documentation to be provided by <u>the supplier of S</u> should at least include:

- The description of the **L_s**-*interpretation structures* $U_s(GRANT)$ (10.1) and $U_s(HAZ^s)$ (10.4) since the elements of these structures are needed by the user of S; this description must include:

- the S input-output variables,
- the documented functional relation of S (GRANT),
- the documented conditions for using S (RELY),
- the set of *undesired events* $h^s(t)$, the relation HAZ_s and the set of fail-safe states $FS^s(t)$
- the set of designated S parameters $\{k_s\}$;

- The evidence which validates – by valuation of the *structures* $U_s(GRANT)$ and $U_s(HAZ^s)$ – the functional relation GRANT under the assumption of the RELY conditions, the anticipated S *undesired events* $h^s(t)$ and the relation HAZ_s; this evidence consists in mappings of documentation data onto the $\mathbf{L_s}$-*interpretations* $U_s(GRANT)$ and $U_s(HAZ^s)$.

The justification documentation required from the user of S must include:

- Evidence of a valid *interpretation* of the documented conditions of use (claim (10.9)) and of their satisfiability (claim (10.10)); this evidence consists of mappings of documentation data on the interpretation *structures* $U2(RELY)$ and $U_2(REQ_z)$.
- The *delegation claims* from the upper levels of the embedding system on the pre-existing component S, and the *expansions* of these *delegation claims* into *claims* on properties of the *specifications* of S such as completeness, correctness, and mitigation.
- The description of the interface between the embedding system and the pre-existing system S; *i.e.* the *expansion structures* – such as $U_s(\delta \mathcal{A}^2 REQ_z)$ (10.16) and $U_s(\delta FS;z)$ (10.21) – needed to valuate the *expansions* of all *black claims* made onto S. The *expansion structures* of the interface must include the *substructures* of the embedding system, *e.g.* $U_2(REQ_z)$ (10.11) and $U_2(FS;z)$ (10.17), in which the black claims are predicated, as well as the *elementary justification substructures* of S, *e.g.* $U_s(GRANT)$ and $U_s(HAZ^s)$.
- The evidence which valuates to true the *expansions* of the embedding system black *claims* onto properties of S specifications. This evidence consists of mappings onto the interface *expansion structures* which valuate to true the *expansions* into *claims* on S.
- The domain for which the correctness of the use of S is to be justified, *i.e.* the input domain of a relation REQ_z $(i^z(t), o^z(t))$.

The documentation to be provided by the supplier of S must be made available to the user and, together with the justification documentation required from the user, be part of the *dependability case* documentation of the embedding system provided to licensors and safety authorities.

Remark 10.5 A pre-existing component may also be implemented as a part of the architecture at level-2 (*e.g.* a whole channel or any subsystem) instead of being a single component of this architecture at level-3. In this case, the proof obligations, the interface *expansion structures*, and the evidence that must be documented and provided, *ceteris paribus,* belong to the level-2 of the embedding system. The evidence and documentation required from the pre-existing component remain unchanged.

10.6 Concluding Remarks and Justification Issues

We have identified the conditions under which the use of a pre-existing component can be justified while its behaviour is specified by input-output specifications that have been independently validated, and its detailed internal behaviour remains hidden to the user.

Guaranteed services and conditions for use are modelled by GRANT, RELY and HAZ functional relations between monitored and controlled variables, with given sets of anticipated *undesired events* and fail-safe states. Independent justifications can show that these relations hold in two sets of *interpretation structures*, which can be independently developed to describe respectively the design of the pre-existing component and the conditions of its use in the embedding system. One of the important conditions is to make explicit the domain of the embedding system for which the correctness of the use of *the pre-existing system* is claimed and has to be justified.

It is worth noting that such encapsulation conditions are another application of David Parnas' information-hiding principle [73]: they are akin to the conditions under which the interfaces between the modules of a program satisfy this principle. As a general principle, it would be useful to identify typical sets of guaranteed services that may be claimed from different families of pre-existing software components (operating systems, application library modules, black box COTS, intelligent sensors), together with conditions of use that must be satisfied by the embedding system. The justification of the re-use of pre-existing components would be made easier in dependable applications.

10.6.1 Completeness of S Specifications

One should not, however, be overconfident. The GRANT/RELY/HAZ justification does not elude all difficulties. Two essential problems are discussed hereafter.

First, the difficulty of justifying the completeness of the specifications provided by the GRANT, HAZ and RELY relations should not be underestimated. Some aspects of these specifications are trickier than they may appear at first sight because the relations may have to document information on the behaviour of the pre-existing component implementation, which may be very detailed and difficult to anticipate. The valuation of the GRANT relation primarily provides evidence on the correctness and completeness of the input-output functions of S. But this evidence may not be sufficient. One may also have to justify claims on dependability properties of the pre-existing component such as its real-time performances, its maintainability, reliability or the diversity of its design in relation to the embedding system.

Diversity and reliability are in fact of particular importance when a given pre-existing component is used in multiple copies to contribute to various distributed, redundant or diverse functions a plant. A hardware or software defect of the component could simultaneously affect systems assumed to be independent in the plant safety analysis. An input-output relation such as GRANT may have to be complemented by other relations to model these non-functional properties.

Likewise, the RELY relation must specify *all* the conditions that must prevail in the embedding system so that the lower levels of the pre-existing component hardware-software implementation can operate, be controlled and maintained in the exact same environment that was anticipated by its developer. As shown in the previous section, it is on the basis of the documentation provided by the RELY relation that delegation claims on the corresponding lower levels of the embedding system are formulated and justified so as to guarantee the conditions of use and adequation between embedded and embedding implementations.

Many incidents, as the one here below, attributed to software are caused by a lack of attention paid to the conditions of use and not by incorrect functional specifications:

Example 10.1 These incidents[17] occurred in a nuclear plant in 2001 when new overload protection devices controlled by software were installed. The devices were intended to protect the controlled subsystems from damaging heat in case of electrical current overloads. Pre-installation tests had revealed that, under certain conditions, the protective device could trip on overload even if the applied current did not exceed the tripping threshold. The protective device behaved as expected when tested with stationary currents at normal operational levels but the malfunction occurred when heavy transient short circuit currents were applied periodically with a frequency that caused interference with the sampling frequency used by the device. The sampling frequency was a function of the time constant selected for the overload protection. In order to avoid these testing inconveniences, the supplier modified the software and a fixed and increased sampling frequency was used instead of the previous one which was a function of the selected time constant. The modification was regarded as minor modification of the sampling frequency and the supplier ensured that no other protective function had been modified. When the plant was restarted with the modified protective devices, one of the devices tripped on overload while the actual load was however far below the threshold. When analysing the software it was found that modifications implemented into the computation algorithm had introduced a previously non-existing and unanticipated sensitivity to unsymmetrical loads and in particular to changes in load symmetry.

Another example of conditions of use not adequately documented is the software defect involved in the accident of the ARIANE 5 launcher flight 501. One of the

---

[17] Event N°629, rated level one on the seven level scale (INES) of the International Atomic Energy Agency (IAEA).

reasons contributing to the accident was probably the lack of systematic and clear documentation of the assumptions and limitations of the floating point calculations and exception handling routines of a re-used predeveloped alignment   control software component which had successfully operated in previous ARIANE 4 flight launches [60].

More generally, a GRANT/RELY/HAZ relation based justification of the use of a pre-existing component is a justification, which necessarily implies the re-use of *sub-claims* with their *arguments* (Definition 6.3). A *subclaim* and its supporting *argument* is not entirely "context-free". It can only be re-used at a specific level, within a same *model* (same assumptions), and must remain supported by valid evidence at levels below. The Sections 10.4.3 and 10.4.5 give a generic example of how this can be done; they identify the usual proof obligations that rest on the developer and the user of *S*, the *interpretation structures* at the interface between *S* and the embedding system and the usual claims that must be valuated by these *structures*.

Practical aspects and conditions for the re-use of arguments are recalled as construction rules in Section 11.1.15.   The application of a claim-based justification to systems using pre-existing software was originally suggested in the CEMSIS project [85].

10.6.2 Robustness of *S* Operation

The other problem mentioned at the beginning of the previous section was already alluded to in Section 10.4.5.6. *Undesired events* at the architecture ($h^2(t)$) or design ($h^3(t)$) level of the embedding system may directly or indirectly affect the RELY conditions. The control of these events so as to preserve the RELY conditions or mitigate their effects remains the duty of the embedding system and is another delicate aspect of the use of pre-existing components.

The problem lies in the very nature of the conditions modelled by the RELY relation and in their absolute and unrepealable statute. By definition, they are the minimal and sufficient conditions, which must be maintained in an environment that the designer of *S* takes for granted, and to which he anticipates no exception. They are minimally restrictive in the sense that they allow *S* to be of general enough purpose and use. They are sufficient in the sense that they allow the design of *S* to be feasible and predictable. In the absence of such conditions, the designer could not guarantee for *S* a correct (claim (10.15)), fail-safe (claim (10.19)) and predictable (10.20) behaviour.

At the same time, a related problem rests with the user of *S* who has no means for anticipating and mitigating an erratic behaviour of *S*, should the RELY conditions be violated, since *S* output is in this case undefined. Besides, it may not be easy or even possible to guarantee the maintenance of the RELY relation for all possible conditions and undesired events that may arise in the embedding system. Timing or transitory faults, communication failures or Byzantine faults in distributed systems, and human errors are instance of events the consequences and side-effects of which may be difficult to evaluate and circumscribe.   More seriously, it may simply be unfeasible to justify the use of the pre-existing component, the RELY relation being not satisfied under the assumption that certain specific *undesired events* occur at the architecture or design level of the embedding system.

10.7 Criticality Degrees and Integrity Levels

Quite naturally, requirements on the dependability of a system and on the rigour of its justification should remain proportional with the importance to safety of the application supported by the system. This is a viewpoint adopted by several standards (see *e.g.* [46]). The issue at stake is the ability to balance in a project the efforts and resources of the justification with the criticality of the functions and the systems involved. It is also a principle that was followed by the CEMSIS project for the qualification strategy and the taxonomy of pre-existing products (see [85]).

However, relaxations that might be safely applied to the rigour of the design, the verification or the validation process of systems of lesser criticality are not easy to define and to justify on scientific grounds. The claim-based approach followed in this book, by focusing on the justification of specific dependability properties, makes things easier. Initial dependability requirements as those discussed in Chapter 5 are intended to be an exact definition of the dependability expected from the system, no more, no less. These requirements, and only those should be necessary to justify. Thus, the tuning and adjustment of the requirements on design process and justification in proportion to the expected dependability somehow is a natural consequence of the selection of specific initial requirements. Different requirements on the degree of *plausibility* of the interpretation models (see 7.2) should in principle also offer some flexibility, but the consequences of relaxing requirements on model plausibility in function of the impact on safety are far from easy to anticipate.

11

Construction Methods

Building the dependability case of an industrial computer based system requires self-discipline and method in order to control the complexity of the argumentation, the collection of disparate evidence and the huge volume of documentation involved.

11.1 Dependability Case Construction Rules

The approach proposed in this book is based on principles which call for certain rules, practices and conventions to be followed. They concern the identification of the initial dependability requirements, the order in which sub-claims and evidence components have to be preferably dealt with, the labelling of claims and evidence components, the construction of argument levels (expansions, justifications, delegations), the use of conjunctions. Some of the key principles, rules and caveat's are recapitulated and discussed in this section, intended to be an enchiridion and a concise and handy compendium for the practitioner.

11.1.1 Initial Dependability Requirements Address Product, Not Process Quality Assurance

The level-0 initial **CLM0** functional *initial dependability requirements* specify the system functions. The level-0 **CLM0** non-functional *initial dependability requirements* specify the dependability properties of the system behaviour required by the application such as safety, availability, timeliness, testability, maintainability.

These non-functional initial requirements should not be confused – as unfortunately too often happens with guides and standards – with the qualities required from the processes and procedures followed to design, verify and validate the system. These required process qualities should be stated in separate quality assurance plans; not because they are of lesser importance but because they are means of justification and not part of what has to be justified; they are qualities of the processes by which evidence is to be provided and not qualities of the system that must be justified. In other words, the design and V&V qualities achieved by observing appropriate assurance plans and standards are qualities required from the sources of evidence that have to support the claims into which the *dependability requirements* are expanded at the various levels. (*cf.* Section 9.3.1)

11.1.2 A Unique Interpretation Model for the Specification of Initial Dependability Requirements and for the Validation of Preliminary Specifications

In practice, it is efficient to have both sets, the initial **CLM0** *dependability requirements* and the *system input-output preliminary specifications*, interpreted and valuated within a unique environment-system interface structure model (Section 9.4.8); in particular by a same set of relations, such as NAT, REQ and HAZ which also define the boundaries of the system (*cf.* Remark 9.1, page 128).

Then, the validation of the system preliminary specifications amounts to an acceptance of this one and only one set of model relations by all experts involved. The validation is carried out by an *expansion* into *white primary claims* (Section 9.4.9) requesting evidence of the validity, completeness, feasibility, adequate coverage of the robustness and functional diversity of these specifications. The possibility of using a single "consubstantiating" model, complete and simple enough to be understood by all parties involved in the validation is a great asset for obtaining the evidence for these *primary claims.*

11.1.3 Expansions

Expansion is the first construction that must be made at each level j, j= 1...4. Every delegation claim from an upper level i, i < j, requesting evidence from level j must be expanded into a *conjunction* of level j subclaims.

Constructing a valid and complete *expansion* for a delegation claim is one of the most delicate tasks of the justification, a task requiring much engineering judgment. The following principles are helpful. As explained in Section 6.6.1, in a valid and complete *expansion*, subclaims are always of one of three main kinds:

1. *Validity subclaims* on the validity of the *interpretation* model of the level at which the *expansion* takes place; all these subclaims are by definition *white*;

2. *Dependability subclaims* for the existence of system dependability properties (robustness, redundancy, independency, diversity, safety, reliability,...) at the level at which the *expansion* takes place; or

3. *Implementation acceptability subclaims* on the *acceptability* (*cf.* Definition 9.12, page 137) of the implementation of these dependability properties at the lower levels.

Subclaims of the first kind are *white*: they address evidence that is required by the upper levels and must be provided at the current level. Subclaims of the second kind also address evidence required from upper levels; but this evidence may be available at the current level and/or at levels below; these subclaims may be *white, grey* or *black*. The third kind requires evidence from lower levels; these sub claims are *black*. Examples of the three kinds are given by the *expansions* constructed in the generic models of Chapter 9. Remembering these three kinds of subclaims is useful to ensure the completeness of the *expansions*.

11.1.4 Claims First

Claims and their *expansions* into subclaims at each level should preferably be identified before evidence components are sought. The previous rule 11.1.3 on *expansions* indicates that the nature of the claims at each level is dictated by the dependability required from the application or from the current level; thereby these claims specify what an acceptable design and implementation of the lower levels should be. Thus, the claims at each level determine the evidence that has to be provided by the current and/or the lower levels' design and implementation.

On the other hand, an essential objective of the design of a dependable system is to achieve a design and an implementation which *maximise the available evidence* by making the system as verifiable and testable as possible. Thus, the design and implementation should be determined by the evidence required while the dependability properties that are claimed are determined by the application.

The best way to resolve this somewhat conflicting situation is to develop the dependability justification and the design hand in hand. If, for practical reasons, this is not feasible and the dependability case has to be made after the design is complete, this 'claims first' rule should in any case prevail so as not to compromise the dependability which has to be justified by making it contingent to evidence which "happens" to be available.

Let us also recall the claim refutation stratagem of Section 6.6.3. Before embarking in a perhaps costly argument and search for evidence, it may pay off to check that no counter-example refutes the existence or the universality of a proposition for which evidence is claimed. Of course, the validity of the counter-example must be ensured.

11.1.5 Expansion and Justification Interpretation Structures

At each level, claim assertions are formulated in languages that have to be interpreted and valuated by evidence in *interpretation structures*. No precise description of claims and evidence is possible without well-defined *interpretation structures* (*cf.* Chapter 8). System properties implying completeness cannot be justified without reference to well-defined models such as *structures*. Relations, operations and elements of the universe of a *structure* have to be mapped onto the predicate, function and constant symbols of the language in which the claim is asserted. Evidence provides data and values which valuate the *structure* elements and the claim assertion to true.

The expansion of a *delegation* claim is a *conjunction* of *white, grey or black claims* which is valuated in an *expansion structure*. The justification of a claim is valuated in a *justification structure*. The *justification structures* of the claims of an expansion must be *elementary substructures* of the *expansion structure* (*cf.* Definition 8.8, page 89 and Theorem 8.7, page 103). Thus, every formula expressed in the language of the *justification structure* and true in this *structure* must also hold in the *expansion structure* and conversely.

As a consequence, the *expansion* and the *justification structures* of the claims of the expansion are not mutually independent. They have to be somehow conceived and constructed together. As explained in Section 9.4.13 and at every level of the analysis of Chapter 9, a three-step inductive procedure is recommended at each level (environment interface, architecture, design, operation) to resolve this issue:

1. Start by identifying the main elements of the computer system domain at the corresponding level and the relations between these elements;
2. Then, for each *delegation claim* onto this level, identify a set of subclaims of the expansion and define the *elementary structures* for their interpretation;
3. Construct the *expansion structure* which is the smallest *extension* (*cf.* Definition 8.7, page 88 and Theorem 8.8, page 106) of these *elementary structures* for verifying (interpreting and valuating) the logical consequence of the *expansion*;
4. Repeat the three steps above until no modification is found necessary.

This strategy was followed in the analysis of Chapter 9, at every level and for every delegation claim.

11.1.6 Ordering for Claim Elicitation and Justification

Because a well-formed formula expressed in a language **L** is not satisfied in a *interpretation structure* unless it is satisfied in its **L**-*elementary substructure* (Definition 8.8, page 89), there is no sense in validating a *justification structure* if its *elementary substructure(s)* cannot be validated. Therefore, it is recommended to follow a sequential ordering for the elicitation and formulation of the claims and of their *justification structures*.

At level-1, for instance, the *elementary substructure* relationship between different *justification structures* which is defined by Proposition (9.20), page 135:

$$U_1(NAT) \prec U_1(HAZ) \prec U_1(FS)$$

$$U_1(NAT) \prec U_1(REQ) \prec U_1(FS)$$

induces a logical sequencing amongst the corresponding justifications: one should first address the elicitation and justification of claims on the validity of the NAT environment constraints interpreted by the NAT *substructure*, then the validity of the undesired events (HAZ *substructure*), of the input-output specifications (REQ *substructure*) and the treatment of postulated initiating events (FS *substructure*), in that order.

Likewise, at the lower levels, a logical ordering in the justifications is defined by the relations between substructures defined by the Propositions (9.49), (9.55), (9.66), (9.112), (9.159).

11.1.7 Primary Claims

The detailed *expansion* of a *dependability requirement* **CLM0** into the level-1 *primary claims* may considerably differ from one application dependability case to another. However, these claims always belong to a restricted number of types; this was already mentioned in Section 6.6.1 and rule 11.1.2 – and it is also illustrated by the *expansions* of Sections 9.4.9 to 9.4.12. Hence, a few principles can guide the elaboration of *expansions* and help the justification of their validity and completeness.

Dependability requirements are expanded into claims called *primary claims* that are *deterministic* and possibly also, if necessary, *probabilistic;* these claims always have the following properties:

- All *primary claims* are *white* or *black*, never *grey*. (*cf.* Section 6.6.4; the level-1 argument in Section 9.4.15 is an example).
- *Deterministic white primary claims* address properties of validity, completeness, feasibility, robustness and functional diversity (*cf.* Section 6.6.1,). These different types of *primary claims* can be expressed as formulae that are *interpreted* and *valuated* in *elementary substructures* of a level-0 *expansion structure,* an example of which is the *structure* (9.36), page 145; the completeness of the *expansion* into primary claims of a *dependability requirement* is a property of this *expansion structure*, and, as such, can be valuated and justified.
- *Deterministic black primary claims* address properties of the implementation (Section 9.4.11): acceptability (Definition 9.12), fail-safeness and diversity.
- *Probabilistic primary claims* are all *black*, and address reliability, availability, maintainability *etc.* of the implementation.
- All *black primary claims* delegate evidence to levels 2 and/or 3, *never* to level 4, which is a consequence of the fact that properties required at level-1 strictly depend *in the first place* on the system performances, and not on operator assistance (Section 9.7.8).

11.1.8 Level-1 Expansion

The *expansion* of the *dependability requirements* **CLM0** into *primary claims* is an initial and decisive step of the justification which, despite the guidance given in Sections 11.1.5, 11.1.6 and 11.1.7, requires much knowledge, skill and creativity. The *dependability requirements* are given and formulated in some language related to the application – referred to as L_1 in Chapter 9. Usually, no well-defined model (*structure*) is provided for the *interpretation* and valuation of this language often a natural language. And yet, the *dependability requirements* must be logical consequences of *primary claims*, and these logical consequences have to be *interpreted* and valuated in *expansion structures* that are *extensions* of the primary claim *justification structures*. The languages used to formulate the *white* and *black primary* claims should be *reductions* (*cf.* Definition 8.5, page 87) of the language L_1. Thus, the formulation of an initial *dependability requirement* and its *interpretation* cannot be dissociated from the elicitation and formulation of the *primary* claims of its *expansion*. These initial steps of the design of the justification have somehow to be carried out hand in hand.

To identify the *primary claims*, the initial guidance available is the obligation for these claims to cover the properties mentioned in rule 11.1.7: validity of the *expansion structure*, correctness and dependability of the implementation. Another guidance is available and results from the constraint for the *primary claims* to be interpreted and valuated in *structures* (models) that are *elementary substructures* of the *expansion structure*. But, as already explained in Sections 11.1.5 and 11.1.6, the *expansion and elementary substructures* are inter-dependent and somehow have to be jointly conceived.

Hence, the "prestructured understanding" provided by this rather limited guidance still leaves open many options so that innovative thinking and engineering judgment is needed. The selection of precise *primary claims* and the definition of the expansion and *interpretation structures* can only result from a reflexive interplay between this prestructured understanding and an in-depth acquaintance with the different level-1 elements (domains, relations) and the engineering aspects of the specific computer system interface.

11.1.9 Assumptions Are Not Evidence

There is a temptation and danger to confuse *delegation claims* with assumptions. Assumptions are parts of the models that are used at each level of the justification to formulate and to interpret the claims and the evidence components at that level (Section 5.6). Assumptions postulate at each level of the justification the existence of system components and the conditions of the environment in which these components are to operate: a given range of temperatures, pressures or seismic risks; a given communication protocol, programming language or hardware component. Assumptions should be precisely and explicitly stated for each of the four levels of the justification (*cf.* in particular Section 9.4.4 for the environment-system interface).

Delegation claims assert properties of these components and their behaviour such as correctness, completeness, reliability, *etc.* These properties cannot be assumed true. Evidence is what must make these claim assertions hold true within a model. In practice, evidence components of a *dependability case* are the documented results of tests, measurements, simulations, software static analyses or formal proofs; also data on the feedback of operational experience, the verified features of a hardware component, of an actuators, a sensing device; confirmed assurance of periodic testing or re-calibration procedures, alarms, manual controls and by-pass, *etc.*

11.1.10 Evidence Must Be Explicitly Claimed

Evidence should not remain implicit. It must always refer to and support a claim explicitly formulated. This is particularly important when two or more pieces of plausible evidence are intended to re-enforce confidence in a claimed property, on the assumption that they are independent and diverse. Then a claim is required to explicitly assert this re-enforcement by means of diversity; and specific evidence justifying the diversity being claimed as well as an adequate level of independence must support this claim on diversity.

11.1.11 Claim Identification and Documentation

As mentioned in Section 5.7, requirements, *subclaims* and evidence components should have a unique identifier, consisting of a name, a type (*clm* or *evd*) and a level expansion number; as for instance the notation we use:

<initial dependability requirement name>.*clm*0
<subclaim name>.*clm* <level=1,2,3,4>
<evidence component name>.*evd* <level=1,2,3,4>.

Because claims and evidence components can be re-used in various parts of a dependability case, their names should be unique. While retaining some concision, they should also be as self-explanatory as possible. Note that a grey or black delegation claim is by construction always referenced at two distinct levels of the justification: the level at which it is formulated and the level at which it is justified.

It may be tempting, in particular when using graph-based tools [7] for documenting dependability cases, to make use of a Dewey levelled notation w.x.y.z for the numbering of claims. This notation gives the possibility of tracing back the parents (w,x,y) of a claim z with the number of dots indicating the level in the expansion. It may, however, lead to confusion when a subclaim has to be reused in a same dependability case, since the parents are then no longer necessarily the same. Thus, the label should preferably identify the name of a claim and its expansion level only.

11.1.12 Fixed Levels of Expansion and Evidence

The four levels of expansion (environment interface, architecture, design and operation) should form the main structure of the dependability justification, and only those. From experience, this structure with its associated models, interpretations and properties appear to be both necessary and sufficient for consistently organizing the different types of evidence required.

So far, no counter-example making it impossible to adhere to a four level structure has been found. Difficulties that were reported were often due to missing or misused *delegation claims*. Confusion between levels of justification and levels of abstraction (*cf.* Section 9.6.1) can also mislead people to believe that four levels might not always be enough to deal with the complexity of a dependability case.

Exempting the justification from the four-level structure would make it difficult or impossible to re-use subclaims in different arguments or dependability cases. Besides, a property of the structure is that a claim is always supported by evidence of the same level and/or by delegation claims on lower level evidence, but never on the same level or levels above. Breach of this rule would introduce the risk of creating loops in claim-evidence arguments.

The four-level justification structure does not exempt, of course, a level and its model from being organized with an appropriate more detailed inner structure.

11.1.13 Strict Conjunctions

As explained in Section 6.4, a claim at any level i is justified by an *expansion* which is a *conjunction* of evidence components from the corresponding level i and/or *delegation claims requesting evidence* from lower levels. The *conjunction* reflects the fact that the set of *delegation claims* and evidence components must be both necessary and sufficient to imply the parent claim being justified. The expansion should therefore not be an accumulation – as large as possible – of possibly useful evidence components and *delegation claims*, rather arbitrarily chosen, for instance simply because evidence is available. The necessity and sufficiency is the primary goal and should appear clearly in the expansion (Section 6.4).

As stated in earlier chapters, constructing a valid and complete claim *expansion* is one of the delicate tasks of the justification, and a task which requires much engineering judgment. Some guidance on how to expand delegation claims is given in Section 6.6.1.

11.1.14 Delegation to the Adjacent Lower Evidence Level

When constructing a dependability case, in case of doubt, it is advisable to systematically delegate claims to the immediately lower level. As explained in Section 6.6.1, when expanding a claim at a given level, one is not necessarily aware of the exact evidence needed from the lower levels. Systematic delegation to the next lower level forces the designer to consider expansions at every level. Some of the delegations may then just be "short circuits", but the risk of missing potential useful sources of evidence is reduced. Besides, these "short circuit" delegations may be removed after all levels have been expanded.

11.1.15 Re-use of Arguments

The *argument* of a *white, grey* or *black claim* made at a given level is the tree of nested *subclaims* and *evidence* components that justifies a *claim*, together with the inferences that relates these *evidence* components and the nested *subclaims* to the *claim* (Definition 6.3, page 45, and Figure 6.1, page 44).

A *white, grey* or *black claim* at some level j is always supported by a single *argument* (Section 6.5.3, page 51), but can be predicated in the *expansions* of different *delegation claims* made at upper levels $i < j$ and with different values for the free variables (see Figure 6.4 page 51, for instance).

Section 10.2 reviewed the syntactic and semantic conditions under which a *claim* and its *argument* can in principle be re-used. The *argument* obviously must be complete: all leaves of the tree must be evidence components. Furthermore, a *claim* and its supporting *argument* is not "context-free". The following three conditions must be satisfied (Theorem 8.7 on *structures*, page 103):

- The *claim* must be *justified* at a given level *i* with evidence valuated in a *justification structure* U_i that must satisfy the two conditions which follow;
- U_i must be an *elementary substructure* of every level-*i expansion structure* in which the claim is predicated;
- U_i must be a *substructure* of all lower level *j* (*j* > *i*) *expansion structures* in which the *delegation claims* predicated in the level *i justification* of the *claim* are expanded.

The three conditions impose constraints on the domains, relations and constants of the *structures* involved; these constraints are dictated by the Definition 8.7 of *substructure* and the Definition 8.8 of *elementary substructure*, page 89.

Under these three conditions, a level-i *white, grey* or *black claim* can be predicated in the *expansions* of any *delegation claim* which is made at any level above level-i and which specifies the claim predicate, its level, and the set of values assigned by upper levels to the free variables (Theorem 8.5, page 99).

One can distinguish three different situations in which a *white, black* or *grey claim* and its *argument* can be re-used under these three conditions:

- Re-use of a level-i *claim* and its *argument* within the *justification* of a level-0 *dependability requirement*.

    This is the simplest case. As there is a unique *justification tree* of *interpretation substructures*, the *argument* is univocally identified by the name and the level of the *claim* and valuated by its *justification structure*. Under the three conditions above only one instantiation and documentation of the *argument* is necessary, and the *claim* can be predicated in the *expansions* of an arbitrary number of *delegation claims* made at the upper levels *j* (*j*<*i*) with various values for the free variables.

    An example would be a *claim* on the existence of a property required from every instance of a program executed in redundant identical channels.

- Re-use in the justifications of different level-0 *dependability requirements* of a given system with a unique set of architecture, design and operation level *interpretation structures*.

    The different justifications may have in common sub-trees of *interpretation substructures* valuated by the same evidence components. A single instantiation and documentation of the *claim* and its supporting *argument* may in this case be invoked at the same level of the different justification trees. The *justification substructure* in which the *claim* holds true must be an *elementary substructure* of every *expansion structure* in which the *claim* is predicated.

    Examples are *claims* on properties of devices or procedures to mitigate level-2, 3 or 4 undesired events which may affect different level-0 dependability requirements.

- Reuse in justifications of level-0 *dependability requirements* of another system, independently designed and modelled by sets of environment, architecture, design or operation level *interpretation structures* which may be different from those in which the *claim* has been originally justified.

Again, in this case, the *claim* and its *argument* must be used at the same level in the alien system; and, as above, the original *justification substructure* in which the *claim* holds true must be an *elementary substructure* of every *expansion structure* in which the *claim* is predicated in the alien system; it must also be a *substructure* of the lower level *expansion structures* of every delegation claim predicated in the justification of the *claim*. Examples are *claims* on properties required from an application programming platform, an operating system, a run-time library or any system utility.

These general conditions of re-use and the documentation required in this situation may be hard to establish if the imported and importing systems have been designed and validated independently of one another. In practice, the situation is very similar to the use of validated pre-existing components and a well-defined interface is needed to wrap the imported system in the importing system, similar to the GRANT/RELY relations. The main conditions such an interface must satisfy are recalled in the section here below.

11.1.16 Pre-existing Components

An interface between an embedding system and a pre-existing component (S) is detailed in Chapter 10. It is based on three relations: RELY, GRANT and HAZ[s]. The proof obligations of completeness, correctness and undesired event mitigation that rest on the designer of S are given in Section 10.4.3; those of the validity and satisfiability of the conditions of use and on the acceptability of the S specifications that rest on the user of S are given in Section 10.4.5. The documentation required is detailed in Section 10.5.

A set of *expansion structures* – such as $U_s(\delta A^2 REQ_z)$ and $U_s(\delta FS; z)$ defined by (10.16) and (10.21) – constitutes the model of the interface between the embedding system and S. These *structures* valuate the *expansions* of the *delegation claims* that the embedding system makes on properties of S; these *structures* and their interdependencies are displayed on Figure 10.2.

11.2 FFP (Functions/Failures/Properties) Method

To secure completeness and consistency in the construction of *arguments*, a systematic method should be followed. The method given here, called FFP for reasons that will soon become clear, is based on the precepts that have been identified in this book.

As argued in Section 11.1.4, the method encourages and allows a prior identification of dependability claims, and does not let the safety justification to be driven by the evidence available. The method proceeds level by level, starting from the justification level-1. *Delegation claims* are expanded and *subclaims* elicitated at each level by considering first all functions and properties that must be ensured, and then all the *undesired events* and safe states that can be anticipated and must be dealt with.

Maximum evidence is sought first at each level; that is, *justifications* and *expansions* receive priority. Then, in a second step, *delegation* claims are generated to specify the evidence that is missing at the current level and required from lower levels not yet fully explored. This two-step ordering is based on a classical precept (see *e.g.*[20]) for acquiring the understanding of complex and composite systems: to deal with the simplest and immediate things first, the simplest being the part that does not require any other assumption than itself to be perfectly known. A process of *stacking delegation claims* and their *expansions* is therefore central to the approach.

In practice, designing is always an *iterative* process and designing a dependability case is no exception. When dealing with the functions, properties, *undesired events*, safe states and evidence at a given level, one may realise that delegation claims at upper levels need to be added or modified, or that additional evidence must be supplied at these upper levels. Consequently, it is recommended that the construction of arguments proceeds iteratively in a way similar to the following three main steps (**A**, **B** and **C**):

A. INITIALISATION

A1. Dependability analysis: an analysis of the application must first identify three sets at the environment-system interface level:

- the set of initial functional and non-functional *dependability requirements*,
- the set of postulated initiating events,
- the set of fail-safe states.

These sets must be carefully defined and validated as they, alone, define the dependability required by the application (Section 5.2).

A2. Define a set of *input-output system preliminary requirements* (Sections 5.3, 9.4.6) specifying the relations that the system – viewed as a black box – is expected to maintain in the environment in order to satisfy the dependability requirements, to deal with the consequences of undesired effects and to satisfy the environment constraints.

B. ITERATION

Proceeding **downwards level by level**, and preferably in the order environment-system interface level-1, architecture level-2, design level-3, control of operation level-4, **for each level do**:

B1. Identify the entities of the *structures* at that level; consider the universe, the relations, modes, events, effects *(cf.* Sections **Error! Reference source not found.**, 9.1 and, depending on the level, the preliminaries of Sections 9.4, 9.5, 9.6 and 9.7 respectively).

B2. Expand every **deterministic** *initial dependability requirement* (Sections 5.4, 9.4.9) into primary claims (level-1) or every deterministic *delegation* claim predicate from the upper levels into *conjunctions* of *white, grey or black* sub-claim predicates of the current level.

Every *expansion* must aim at consistency and completeness; to this end, take into consideration, in the following order:
- *All modes of behaviour of the system* (initialisation phase, cycle execution mode, testing mode, ...);
- *All* subclaims of the current level needed on *validity, feasibility, completeness, robustness and acceptability of implementation* (as for example in Sections 9.4.10, 9.4.11.1, 9.5.10, 9.6.5, 9.7.5);
- *All functions* that must be implemented at the current level and/or can be claimed from (delegated to) lower levels; make the appropriate functional *claims* and/or *delegations*;
- *All undesired events* that must be prevented, discarded and /or tolerated by current level and/or anticipated at lower levels; make the appropriate functional and non-functional *claims* and/or *delegations* (as for instance in Sections 9.4.11.2, 9.5.11, 9.6.6, 9.7.6);
- *All possible abnormal states* (exceptions, error states,...).

In considering each of the above items, start if possible with the *expansions* of those *claims* that are expected to be the simplest and easiest to expand, *e.g.* those likely to generate the most obvious and/or the smallest number of *subclaims*.

B3. Refine and express these *claims* to the greatest degree of detail within the *justification structures* of the current level. Possibly revisit step **B1** and refine these *justification structures*.

B4. Construct the *expansion structures* which valuated the *expansions* of the *delegation claims* from the upper levels and constitute the interface towards these upper levels; every *expansion structure* is the smallest *extension* (Definition 8.7, page 88) of the *justification structures* of the upper level claims which predicate the *delegation claim*; and of the *elementary justification substructures* constructed in **B3** of the *claims* in the expansion.

B5. Identify and justify, in the set of *claims* identified in steps **B2** and **B3**, as many *white claims* as possible by providing *justifications*, *i.e.* **conjunctions** of *evidence components* that valuate to true these *white claims* in their *justification structures*.

B6. Identify and justify, in the set of *claims* identified in steps **B2** and **B3**, the *grey* and *black subclaims* and construct *justifications*, *i.e.* **conjunctions** of *evidence* components of the current level and of *delegation subclaims* on lower level *evidence* that valuate to true these *grey* and *black subclaims* in their *justification structures*. Always take first into consideration *evidence components* of the current level. For *delegation sub-claims,* consider all functions and properties that should be claimed on lower levels. In particular, consider the *undesired events* that cannot be dealt with at the current level and need to be dealt with by the lower levels. Formulate these *delegation subclaims* to the greatest degree of detail within the *structures* of the current level. Possibly revisit and refine these *structures.*

B7. Repeat steps **B2 to B6** for **probabilistic** *dependability requirements* or **probabilistic** *delegation claims* from upper levels **expanded** into *conjunctions* of *white, grey or black* sub-claim predicates of the current level; consider all relevant *non functional properties* (reliability, availability, safety…) as well as additional functions that may be required by these properties and must then be claimed on the lower levels; make the appropriate additional *delegation subclaims.*

B8. Verify that all *initial dependability requirements* and *delegation claims* from upper levels are adequately dealt with at the current level; if necessary, identify appropriate additional *claims* at step **B2**.

C. END OF ITERATION

Revisit all levels and **Iterate** steps **B** until no modification is needed.

Remark 11.1 Probabilistic claims are dealt with separately from deterministic claims because their expansions can be constructed independently, as illustrated by the *substructures* of Chapter 9.

Remark 11.2 The expansion construction step **B2** recommends we start expanding in each category, whenever possible, the "simplest" claims first. As suggested earlier, this ordering facilitates (*cf.* the old principle of [20]) the acquisition of the understanding of composite systems, the simplest claims being meant to be those that do not require any other assumption than themselves to be perfectly known. To a certain extent, it is a similar precept that guides the sequence of steps **B5** and **B6**: the **B5** justification of *white* claims with evidence at the same level precedes the **B6** generation of *delegation grey* and *black* claims on lower levels. The reason is that *white claim justification* takes place within a single level and requires less additional assumptions than the generation of *delegation subclaims* which must be formulated for lower levels, possibly not yet completely explored, modelled and determined at the time they are generated.

Postface

The profusion of guides and standards on computer and software quality assurance brings to mind R. Descartes' "Discours de la Méthode", published in 1637 [20]. He writes "La multitude des lois fournit souvent des excuses aux vices"[18], and then identifies four precepts that, he claims, should be sufficient if they are systematically observed.

His first precept is to never hold something for true before all evidence has been acquired without pre-conceived judgment (no precipitation, no prejudice). His second precept is to divide every difficulty in parts, as many as possible required to resolve them better. The third is to order thoughts, addressing the simplest and the easiest to understand objects first, and reaching by degrees the understanding of the most composite ones; assuming an ordering even between objects that do not naturally precede each other. The last precept is to carry out all along the way, for the divisions of difficulties into simple parts and for the deductions of the composite from the simple, verifications that are so general and so complete that one can be assured that nothing has been forgotten.

It was only recently that these precepts qualified by Descartes as being "easy and simple to follow" came to our attention; it was quite reassuring to realize *a posteriori* that the approach recommended here in many ways conforms to these fundamental precepts.

To epitomize, the main technical aspects of this approach for the justification of the dependability of computer systems are:

-   The use of models at four progressively refined levels with relations preserving validated properties across the different levels. This level decomposition serves well the representation of the system, the search and organization of evidence, and the specification of the required dependability case documentation;

---

[18] "The multitude of laws often provides excuses for vices".

- Concepts and mechanisms for the ordered construction of arguments: an inductive expansion of claims into more refined claims, conjunctive inferences, and delegation of evidence onto lower levels.

Some of the major technical benefits that can be drawn are:

- A better structuring of arguments to limit and to back up subjective expert judgment;
- More natural and transparent reifications of non-functional dependability properties into functions;
- Possibilities of assessing the relative weight of a particular component of evidence within the whole justification;
- Modular re-use of arguments for the integration of sub-systems and pre-existing software.

Basically, the approach attempts to give precedence to – and to focus on – the dependability properties of the system behaviour. By the same token it is more cost effective in terms of efforts and resources spent on the justification: two objectives which meet priorities of regulators and licensees. Meeting their priorities should also help making regulator-licensee negotiations more supple and efficient.

Not all aspects of this justification framework have been thoroughly explored yet. The approach assumes that a set of initial dependability requirements is predefined. While it may be relatively easy to identify tangible and classical risks and justify proper protection, a more difficult problem may be to anticipate less tangible threats and rare events and identify adequate protection measures. Those threats and events must be anticipated at each of the four levels of the justification. While the method does not deal explicitly with this issue, it gives at least a first decomposition of the problem.

Valid evidence, *i.e.* evidence that is correct, consistent and complete, has been defined as data that valuate to true a claim formula within a model which must be an *elementary substructure* of the model in which the claim is formulated. This approach clarifies the constraints on the relations, constants, values of free variables, descriptions and documentation of the respective claim and evidence models.

However, the production of convincing evidence and the relations between the different levels would need further investigations. The required evidence is dictated by the prior dependability analysis, the initial dependability requirements, the subclaims and the arguments. And not the other way round, as one is often tempted to proceed, arguments being biased by pre-existing available evidence. More research should be carried out by projects intimately associated with the development of real systems and dependability cases. They would allow a realistic assessment of claim-based justifications with the freedom to determine what best evidence is required and feasible, instead of assessing dependability against pre-existing designs, safety cases and predetermined evidence.

Demonstrability, or rather justifiability as we defined it, is a related key issue. Analysis, testing, and simulation produce evidence. Despite their respectable history, these methods need to improve and in many ways offer place for progress.

It remains unclear, however, how they could ever by themselves achieve the levels of assurance and conviction that an objective demonstration of dependability require in practice. Justifiability should be recognized as a fundamental quality of a safety critical system. Whereas it should be "built-in", this quality is most often ignored. How to specify requirements for validation, how to make traceable designs, how to write analysable and testable software are great challenges that deserve a lot more attention in research and industry. Not enough initiatives have been undertaken to pursue efforts like those of the precursor European projects PDCS [88] and DeVa [22], the goal of which at the end of the nineties was to design "predictable computing systems" and to "design for validation". When one reflects on today state of the art and the world of safety critical hardware and software design, Karl Marx's observation comes to mind: "Up until now, the philosophers contended themselves with interpreting the world in different ways; the point is now to change it"[19].

And yet, evidence and justifiability are only parts of the problem. When taken as the absolute and only goal, they may lead to pessimistic scenarios similar to those that courtrooms now seem to experience. A recent article [41] in the Scientific American discussed the possible effects that progress in forensic science, popularised by the media and television programs, may have on the behaviour of jurors. Prosecutors, judges and police officers have noted that forensics television programs may lead jurors to have in court ever higher expectations from the technical quality and the amount of forensics evidence made available. As the techniques to produce evidence improve, so increases the level of claims. Requesting ever more technical evidence may give a basis for evading responsibilities and unresolvable tensions in arriving at a verdict. Disoriented and indecisive quests for evidence are far from unusual, especially when risks at stake are high; professional assessors, licensors and regulators are no exception. It is hoped that the rational and top-down structure that has been proposed here may at least help stakeholders to circumscribe the evidence actually needed and to strike the right balance between risk and claims.

Nevertheless, as we said in the introduction, the problem of the safety justification of computer-based systems remains extremely difficult. With a singular strength, the analysis requires an obstinate desire for understanding and strenuous work to make concepts precise, establish their interdependencies and test their validity. This work here, needless to say, has not the last word but hopefully throws some light on the issues and indicates ways to attack them. At the very least, it could perhaps reward the attentive and persevering reader, as a sort of "structuring utopia", opening perspectives for decomposing the problem, structuring the arguments and for developing adequate application-specific techniques and instruments.

One last point should perhaps not be forgotten. The approach presented here has a sort of paradoxical and somewhat magical coherency which makes it perhaps

---

[19] Karl Marx (1845) *Theses on Feuerbach.*

unique. It has two antagonistic advantages: the ability to cover the huge and disparate universe embracing the computer system environment, architecture, design and operation while, at the same time, the possibility remains to analyse in each domain and in detail the specific evidence required by the issues at stake.

Appendix A

The SIP System

This appendix briefly describes some pragmatic aspects of the early steps in the justification of a real subsystem of a nuclear protection system; it presents a real context in which some important *initial dependability requirements*, as they are defined in Section 5.2, page 24, have to be elicitated and expanded into *primary claims* and *claims* at the architecture and design levels, under real environment and technological constraints.

A.1 Description

The SIP is the **P**rocess **I**nstrumentation **S**ystem of a Pressure Water Nuclear Power Reactor, *i.e.* the input processing level of the Protection Logic System (PLS) of the reactor. It consists of four independent data acquisition and processing channels. These channels process thermal-hydraulic parameters transmitted by the process instrumentation (temperatures, pressures, differential pressures for flow rates, and levels). They provide each of the two redundant trains of the PLS with binary signals that are the results of threshold comparisons. They also send signals to the a non-classified control system, the control room, the alarm system and some actuating devices. Neutron flux signals delivered by the neutron detection units are processed in this plant by a separate input processing system of the PLS, *i.e.* the Nuclear Instrumentation System (NIS).

Each of the four independent channels consists of four programmable processing units (PU). Each processing unit essentially consists of (see Figure A1):

- Sensing devices and one or more analogue-digital converters (ADC) for 16 input analogue signals each;
- One CPU (16 bits) card with a memory extension card performing the following functions:
    - Acquisition, validation, if necessary linearization of the input measurement data;

    - Processing of the measurement data (filters, derivatives, lead-lag,…);

> - Comparison with alarm thresholds;
>
> - Fibre optic transmission link to the non classified control and information systems;

- One or two fail-safe binary output cards providing the PLS with 16 logic signals each (the threshold comparator outputs);
- One digital-analogue converter (DAC) card for 16 analogue outputs to 1-E classified indicators and/or recorders;
- An independent power supply.

The processing of the input functional parameters is distributed among four functional groups (see Figure A2). To ensure functional diversity, most – but not all – accidental conditions are detected by at least two diverse parameters belonging to two different functional groups. This requirement corresponds to the level-1 *primary claim* (9.21), page 136, on functional diversity which requires that postulated initiating events be detectable by a minimum number of monitored variables.

Each functional group of parameters is processed in three or four redundant channels by an independent processing unit. This implementation has to comply with a *delegation claim* from level-1 to level-2 such as (9.25), page 139, on functional diversity implementation which requires separation at architectural and design levels so as to preserve the benefits of functional diversity.

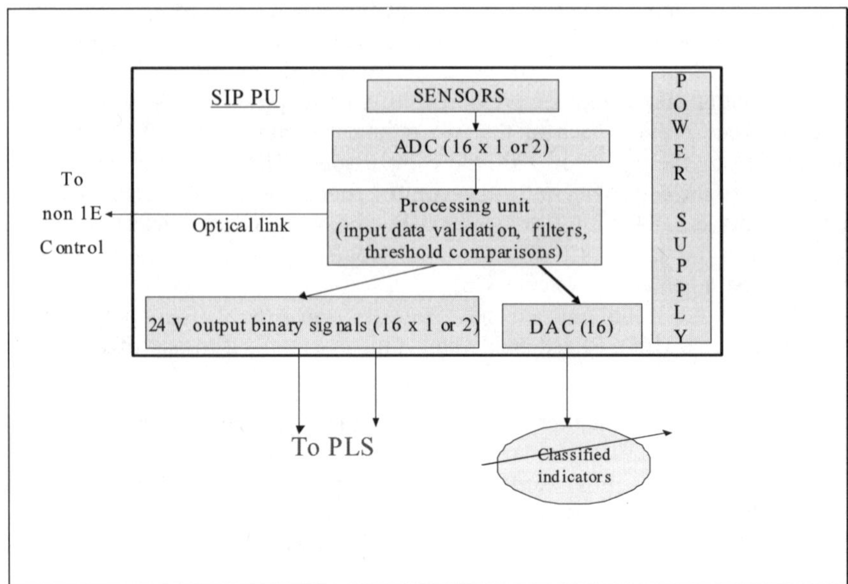

Figure A.1. SIP processing unit

A.2 Plant, Technology and Safety Replacement Constraints

The new SIP had to be implemented with the existing cabling system. The sensing devices, the interfaces to the PLS, to the control room and to the other systems remained unchanged.

The new software-based SIP is not backed up by a diverse system in a diverse technology. This decision was in part based on the argument that there is little to be gained by a replacement that would need to be backed up – even partially – by an equipment of the same obsolete technology.

All CPU's of the processing units are identical and support the same operating system and the same library of application modules. Only the application software – which makes use of these modules - may vary among processing units.

The plant accident analyses must remain valid. These analyses – based on the old hardware technology - postulated meeting the single failure criterion and thus an undisputed availability of every functional parameter processed by at least three independent measurement chains.

These analyses showed that pairs of "primary" and "back-up" functional parameters "cover" most design basis accidents hypothesized in the plant safety analysis, but not all. Besides, a same functional parameter may be primary for one accident, and back-up for another. The distribution of paired parameters over distinct functional groups – *i.e.* the functional diversity – was thereby not trivial to achieve and not necessarily complete.

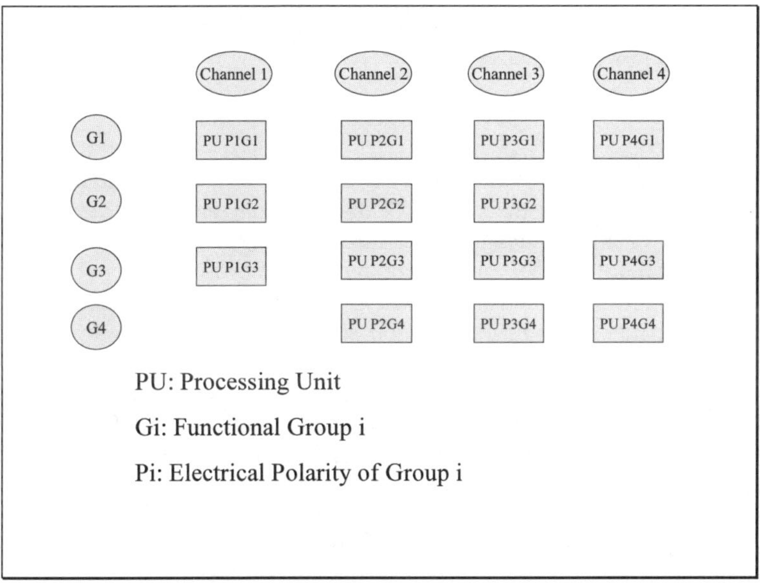

Figure A.2. Group architecture

A.3 Dependability Requirements (CLM0) and Primary Claims

Amongst many functional and non-functional requirements, the SIP is in particular required to operate continuously for 25 years. The MTBF of each processing unit must be at least 5 years. These requirements must be expanded into specific *primary dependability claims*, as in Section 9.4.12, page 139.

A safety analysis made at plant level evaluated that the reliability of a single SIP measurement chain generating an output signal to the PLS (shutdown or safety engineering feature), should have a probability P_M of failure per demand (pfd) less than or equal to:

$$P_M = (5\mathrm{x}10^{-3}/\text{demand}).$$

This value applies to the whole chain, from the data acquisition card to the SIP output relays; the sensing devices remaining unchanged are excluded from this calculation. This value estimation, however, is based on failure rates of electronic hardware components only and does not take into account possible design defects.

Thus, assuming that hardware failures occurring in different SIP channels are independent, the pfd of a SIP function (non generation of a shutdown or safety feature actuation signal) would be at most:

$$3P_M{}^2 = (7.5\mathrm{x}10^{-5}/\text{demand}).$$

Under the same assumptions of independence, the maximum probability of failures causing the non-generation of TWO diverse paired signals (at least if these signals activate loss of power relays), would be:

$$[3P_M{}^2]^2 = 5.67.5\mathrm{x}10^{-9}/\text{demand}.$$

The bounding probability value P_M is a basis to formulate *reliability primary claims* such as (9.32) on the reliability of individual measurement chains; for software, these claims are detailed in the next section.

No single failure of a SIP processing unit should cause the loss of a protection function actuation signal. To this end, in addition to redundant channels per function, comparators produce protection signals by loss of power. This single failure requirement has to be expanded into *validity primary claims* like (9.6), page 131, in order to ensure a definition of a complete set of anticipated undesired events; besides, *robustness primary claims* have to ensure detectability, controllability and mitigation of these events as claims (9.16), (9.17), and (9.19), pages 134 and *ff.*

The global accuracy of signal processing, including A/D and D/A conversions, must be at least 0.5%, accuracy that should be maintainable without re-calibrations for periods of at least 18 months. For the signals to the PLS leading to the plant shutdown, the SIP response time must be less than 100ms. These requirements are to be expanded into *implementation primary claims*, as defined in Section 9.4.11.

A.4 Primary Claims on Software CCF's

As it is often typical of nuclear modernization projects, the replacement of the SIP was primarily motivated by the aging of its hardware components and by the obsolescence of their technology; it was not motivated by a change of its conceptual design, nor by its functionality or its expected dependability. Hence, a natural claim on the new SIP was that its dependability, in particular its reliability and availability, be at least as good as that of the old one.

The single failure criterion, the full availability of every signal to the protection system, the deficit of functional diversity, and the absence of back-up and of software diversity raised severe demands on the reliability of software. In principle, these requirements demand that the software be without defect.

However, the plant safety analysis and accident analysis had taken into account the common mode failures of the existing hardware system. It was thus accepted to substitute the non-justifiable requirement of software zero defect with the requirement that the software common cause failure (CCF) level be not superior to the existing hardware CCF.

By following a method similar to those detailed in [70], the hardware CCF probability was estimated – taking into account failures resulting from design, calibration and other factors – to be equal to the value of P_M weighted by a global common cause factor equal to 0.11, hence of the order of 10^{-3}:

$$0.11\ P_M = (5 \times 10^{-4}/\text{demand}).$$

This order of magnitude was required to be also satisfied by software. It could then be concluded from this analysis that:

- The reliability requirement for software likely to fail concomitantly in more than one channel, but not in paired parameter processing chains, should be of the order of 10^{-3} - 10^{-4} pfd.
- The reliability requirement for software likely to fail concomitantly in paired parameter processing chains should be of the order of 10^{-6} - 10^{-8} pfd.

Claiming that these software reliability levels are satisfied by the implementation is the purpose of a *black primary reliability claim* (9.32), page 142, which should be a logical consequence of a *delegation claim* such as **(9.33)**. These levels are obviously extremely high; in particular, for achieving and demonstrating the second requirement, statistical testing is of no use in practice [63, 98].

To be justifiable, these claims have to be *expanded* into specific lower level *claims* modulated and adapted to the functionalities and required qualities of the different types of software of the SIP. These types, their functionalities and their specific implementation requirements are briefly described in the following section.

A.5 Software Architecture and Design. Claims and Evidence

The SIP has three types of executable software: an operating system, a library of modules and the application software. Distinct *claims* have to address these types in function of their functionalities and roles in the system.

A.5.1 Operating System

The operating system (OS) was developed in house by the supplier of the SIP. It is a rather simple system of less than 100 modules. A previous version of had been used in previous nuclear safety critical applications for many years. The current version included limited modifications. The operational experience was thus substantial, so that the risk of a residual defect was lower than for the application software.

However, this operating system is the same in all processing units of the four channels, and thus in the processing units which process diversified paired parameters. Hence, the consequences of a failure of the operating system could lead to the loss of a protection signal; the level of reliability required for the operating system software is thus of the highest order mentioned in the previous section.

It was necessary to ensure that the operating system be "fail-safe", fail-safeness meaning here that no defect could result in a loss of a safety signal. This fail-safeness had to be addressed by level-2 and level-3 claims requiring that no operating system software defect could be a cause of simultaneous failure in two or more parallel channels, or worse, in paired signal processing units.

Static and dynamic (tests) analyses of the code had been realised by the supplier. But, additional evidence was necessary. It was provided by an analysis of the possible failure modes of the operating system and consequences on the application. To reduce its complexity, the FMECA analysis was restricted to software defects that are potential causes of simultaneous failures in different channels. Such potential CCF's had to be anticipated and shown impossible by analysis; for those cases where analysis was too complex or not feasible or not sufficiently convincing, additional specific tests had to be made. A similar failure and consequence analysis was also performed for the software of the Tihange 1 nuclear instrumentation and is in part reported in [37, 90].

In order to further reduce the possibility of simultaneous failure, evidence was also required that the application code be as diversified as possible across the different channels so as to maintain in so far as possible the operating system in different states at the same instant (see A.5.3 below).

A.5.2 Library Modules

These logic, arithmetic and numerical modules – programmed in high level language (C) by the same supplier – enjoyed the same operational experience as the operating system software. The library consists of approximately 25 000 lines of code, including commentaries. The code is not recursive and modules are reentrant, care being taken that no internal variable can modify the state of a module upon termination.

In comparison with the OS, an additional element of evidence could be claimed. Most of the time, at a same instant, processing units dealing with diversified parameters of different functional groups concurrently execute different modules and with different input data. This diversification by design considerably reduces the risk of a simultaneous failure affecting diversified parameters caused by a software defect of the library.

The modules had been verified and tested independently from one another. They were to be tested again by the integrated tests applied to the application software. The combined evidence of operational feedback, unit and integrated tests, and the unlikely occurrence of common cause failures was claimed, argued and justified to be sufficient.

A.5.3 Application Software

The application software is executed under the control of the Operating System and basically consists of three tables: a stack of addresses describing the sequence of execution of the modules, a table of parameters values and a table of the variable current values. As it is specific to an application, this software does not benefit from operational experience. However, alike the library modules, and even more so, the application software can be made different from one channel to the other. Input parameters that are processed in less than 4-redundancy can be distributed differently over 3 channels. Besides, parameters of different functional groups are processed by different processing units in each channel. These differences – again a kind of diversification by design – can be exploited to minimise the possibility of simultaneous failures in parallel channels and in processing units of different functional groups.

In addition to evidence of correctness provided by verifications, unit and integrated tests, evidence was required that the design and coding of the application be:

- Defensive: the software must be shown to anticipate internal Undesired Events (Definition 9.4), and to privilege safe output states and values; in particular, the potential failure modes of the operating system anticipated by the analysis mentioned in A.5.1 have to be taken into account.
- Testable: the software of a signal measurement and processing chain must be kept as independent as feasible from that of other chains; interdependences had to be identified. Besides, execution cycles must be – in so far as possible – memory-less; indispensable memorisations of data between cycles had to be identified.
- Self-corrective and disclosing: errors that cannot be trapped and mitigated by the system itself (auto-tests) need to be detectable by periodic tests of the SIP functionality; these periodic tests need to be designed and shown complementary to the system auto-tests; tests of the auto-tests need to be periodically performed as well.

Evidence that the application software enjoys these three sets of qualities was the object of level-3 *design* and level-4 *operational claims*. Among these claims, one of the most difficult properties to justify was the complementarity between auto-tests and functional periodic tests. This complementarity, albeit essential, is often forgotten and not specified as an *initial dependability requirement*. Designs which enjoy this essential dependability property are scarce!

Appendix B

Automated Material Handling System

A typical automated nuclear material storage, retrieval and handling system is used as another practical example to illustrate a comprehensive *expansion* of a level-0 *initial dependability requirement*.

B.1. Description

Typically, nuclear material or waste is contained in cans placed on pallets and stored on fixed racks in storage facilities. Loading and unloading pallets onto or from the racks can be done by manually operated or automated material handling machines. Manual operation may require workers to be in proximity of the areas where the material is stored, exposing them to levels of radiation which might be potentially significant. Hence, the automation of storage operations brings with it many advantages. Dependability issues that can directly benefit from automation are workers' health and safety, the risk of potential accidents, the accuracy of inventories of material and of radiation levels, the prevention of criticality when storing fissile materials and the prevention of human operator and administrative errors.

In this case study, we consider the upgrading of a manually operated facility to a fully automated computerized system. A central computer has to be installed to control the inventory, transport, storage, retrieval and loading/unloading of the pallets of cans on the storage racks. The storage facility essentially consists of long racks with a few shelves of cabinets, each cabinet having the capacity of holding the load of one pallet of cans. An automatic guided vehicle (AGV) is guided by rail along the racks. The AGV deposits or picks up pallets of cans, one at a time, to cabinet locations specified by the central computer. Figure B.1 is a schematic plan view of the layout of this automated storage and retrieval facility.

The AGV is equipped with a microcomputer controller which communicates through a secure radio-link with the central computer system. The controller real-time software responds to load, unload and movement requests from the computer system which keeps an up-to-date record of pallets locations so as to give safe access to a properly selected cabinet and provide an accurate inventory of material cans. Once positioned in front of the specified rack location, the vehicle is locked and further moves are prevented; the pallet support is then lifted to the specified rack level and locked against further vertical displacements. A transversal or rotating telescopic fork (or a shuttle assembly in some systems) then horizontally moves to load or discharge the pallet of cans into or from the specified cabinet. Figure B.2 gives a schematic elevation view of an automated forklift vehicle.

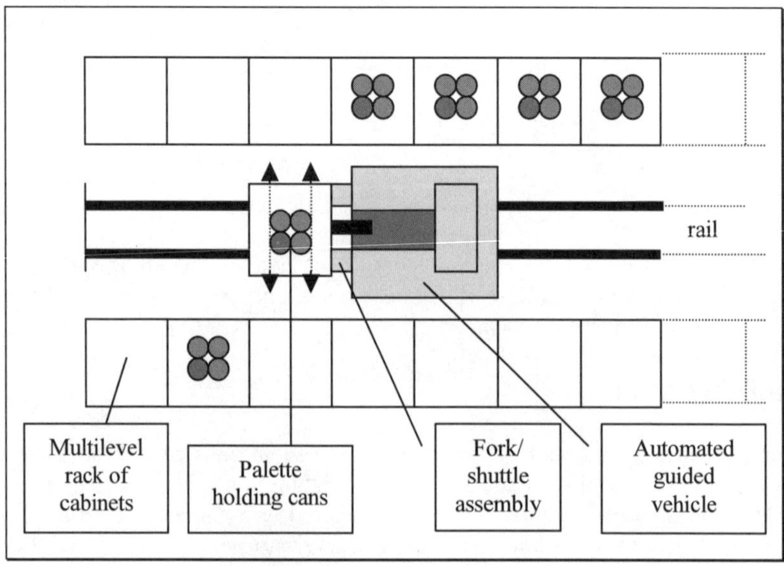

Figure B.1. Layout of automated material storage and retrieval system

The two following sections briefly describe some items of the safety case level-0 documentation of the system with references to corresponding concepts developed in Chapter 9.

B.2 Dependability Requirements (CLM0)

The safe behaviour of the AGV depends on several factors and its design must satisfy precise safety regulations. In particular, the Safety Standard of the International Atomic Agency (IAEA) on the storage of radioactive waste [47] specifies that the design of the handling system has to be considered in the safety assessment of the storage facility to ensure that handling failures cannot result in unacceptable consequences. Driving the vehicle or lifting pallets beyond position or speed limits, loading a cabinet already occupied, moving the pallet support while the vehicle is driving and many similar hazardous operations must be prevented so as to avoid material mishandling and damage. Dedicated transfer routes for loads should be designed so as to minimize the consequences of collisions. Equipment should therefore be provided with interlocks to prevent incorrect placement of containers, accidental release of loads or the application of incorrect forces in lifting and handling operations.

In this example we will restrict our attention to one of these important safety requirements, namely the assurance that proper locks are in place when a pallet of cans is being loaded into or discharged from a cabinet. It is indeed essential to avoid AGV or lift displacements when the pallet fork or shuttle support is being inserted or extracted from a cabinet while transferring radioactive material.

So, at level-0, among the **CLM0** deterministic dependability requirements that have to be justified, we select in this case study the dependability requirement:

(B1)   **no_displacmt.clm0:**
      horizontal vehicle and vertical lift displacements are prevented when pallet is being inserted or extracted from a cabinet.

A four level expansion of this dependability requirement is detailed in Table B.1.

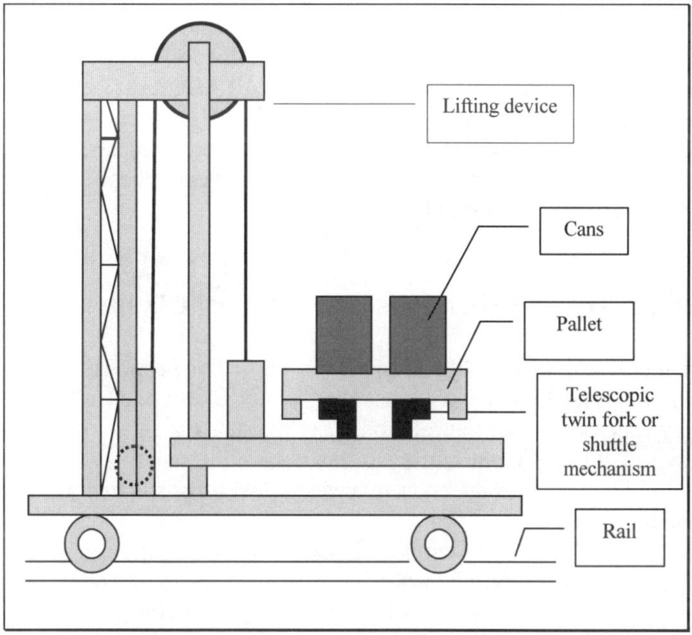

Figure B.2. Schematic view of an automatic guided vehicle

Of course, many other dependability requirements apply to a facility of this type; they address the real-time performances, the inspectability, repairability and maintainability over a determined life-time, the recoverability and return to a stable and safe state in the event of a malfunction or breakdown, the independence between process control and protection instrumentation, *etc.*

B.3 Constraints and Postulated Initiating Events

An essential part of the level-0 documentation (Section 9.3.4) is the description of the environment constraints the system has to comply with (modelled by the relation NAT), and the postulated initiating events (PIE's) and their effects (model relation HAZ[1]).

Let us observe that a typical NAT constraint on nuclear fissile material storage is the need to be aware of and to check against mass limits during material transfers. Mass limits may be determined by considerations on mass criticality or by limitations on the consequences of radiation releases.

The use of real-time software can greatly improve the safety of systems handling radioactive nuclear material; the software can keep track of the type and movement of containers and maintain an updated inventory [44].

This constraint, however, is of no direct concern in our particular case study example: it is a constraint for the functional dependability requirements addressing the inventory and storage system. Here, we are specifically concerned with the dependability of the computer implementation of the handling system. A typical constraint of this implementation is the modular design and extensive use of off-the-shelf components (COTS) and pre-developed software.

Among the *postulated initiating events* (HAZ[1]) that must be anticipated are those that can affect all electrical and mechanical components of the AGV, including lift and fork controls and motors, gear reducers, motor drives, communication, power systems, and all other components; these equipments must satisfy heavy-duty industrial standards.

B.4 Preliminary Specifications and Primary Claims

The other initial step of a justification (see in particular Sections 5.3 and 9.4.6) is the preliminary specification of "what the system is expected to do"; that is, a preliminary specification of the relation REQ that the system – viewed as a black box – is expected to maintain in the environment while satisfying the environment constraints and coping with the consequences of level-0 postulated initiating events (PIE's) and their effects.

To satisfy the **CLM0** requirement **no_displacmt.clm0**, the control logic of the micro-computer controller on board of the AGV has to control a set of interlocks that permit movements of the vehicle and the lift under the condition that these movements are safe. The controller has access to monitored and control variables to detect when the vehicle is clamped to the cabinet, the fork is engaged and if, *e.g.* following an operator error, drive or lift motors are started at the same time.

In terms of a relation REQ (*cf.* Section 9.4.7, page 128) between these variables, an input-output system specification can be expressed as:

(B2)   <no_ displacmt >:
        IF (AGV – cabinet clamped) **OR** (fork/shuttle engaged)
        AND ((drive mode on) **OR** (lift mode on))
        THEN <DISABLE VEHICLE DRIVE AND LIFTING DEVICE MOTORS
        WITHIN 0.1 SEC>.

The initial dependability requirement (B1) **no_displacmt.clm0** can now be expanded into *primary claims:*

-   A *white validity primary claim* requires evidence that this specification is valid (*cf.* Tables B.1 and B.2);
-   Two *black implementation primary claims* require evidence that the implementation of this specification is correct and failsafe (Table B.1).

The two *black claims* are *expanded* into sets of *delegation claims* on the architecture, design and operation levels (Table B.1); the evidence components are detailed in Table B.2. The expansions include various claims to justify the use of a COTS software platform. Figure B.3 shows the *argument* (Definition 6.3, page 45) justification tree at level-1.

A similar case study was developed in the research project CEMSIS (Cost Effective Modernization of Systems Important to Safety) [85]. The material handling device was based on a rotational mechanism. The *initial dependability requirement* studied in this case was that the rotation takes place only when conditions are safe. Figure B.4 is just intended to give an overview of the total expansion of this requirement; *claims* are in circles, evidence components in boxes.

The main structure of this expansion is – *mutatis mutandis* – quite similar to the expansion of Table B.2. This similarity suggests the possible existence of generic expansion structures for families of applications, and for systems of similar functionality with similar dependability issues. These structures would have in common, not only expansions at the level of the *primary claims* as suggested earlier (Section 11.1.7, page 273), but also common lower level claim-evidence expansions. Generic *justification structures* may help the design and validation of specific dependable applications and the reuse of claims and evidence.

The structures of Tables B.1, B.2 and Figure B.3 give, if not a measure, at least an idea of the complexity – in terms of inferences, claims and types of evidence involved – of the justification of a single initial dependability requirement, after all rather intuitive and simple in its initial formulation. Without the help of such structures, one may retrospectively wonder how convincing argumentations could at all be constructed in such disparate universes.

A look at these structures also prompts a last remark. Correctness of the software, despite its critical contribution to the dependability of the system, turns out to be only one property amongst others that needs to be claimed at level-3 of these argument expansions. If a dearth of evidence to support this property can possibly be compensated by other types of evidence, the expansion suggests justifiable ways to do it.

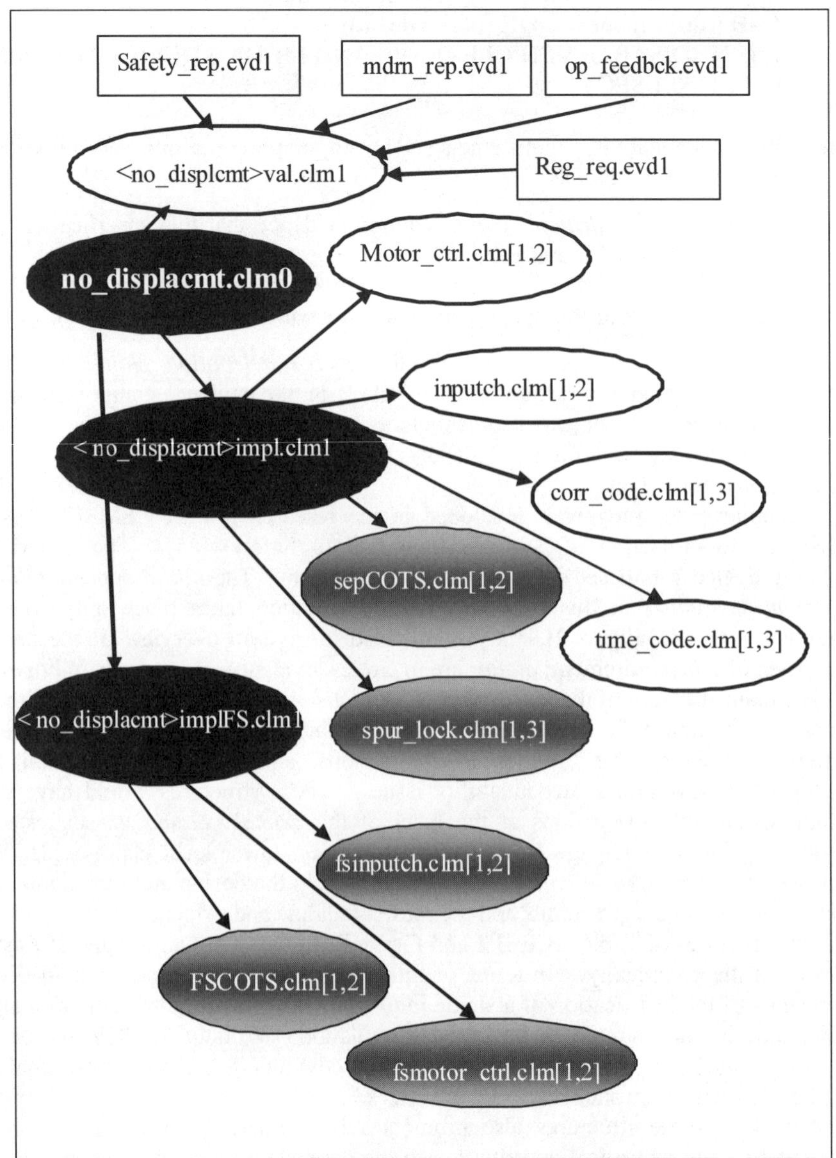

Figure B.3. Level-1 expansion of argument of claim **no_ displacmt.clm0**

Table B.1. Expansion of Dependability Requirement **no_ displacmt.clm0**

Expansion of Dependability Claim **no_ displacmt.clm0**: Horizontal vehicle and vertical lift displacements are prevented when pallet is being inserted or extracted from a cabinet.			
	Primary claims identifiers and descriptions	**Evidence Components**	**Delegations**
Environment-System Primary Claims	**<no_ displacmt >val.clm1:** The system specification <no_ displacmt > is valid; <no_ displacmt >: **IF** (vehicle – cabinet clamped) **OR** (fork/shuttle engaged) **AND** ((drive mode on) **OR** (lift mode on)) **THEN** <DISABLE VEHICLE DRIVE AND LIFTING DEVICE MOTORS WITHIN 0.1 SEC>;	AGV_spec.evd1 op_feedbck.evd1 safety_rep.evd1 mdrn_rep.evd1 reg_req.evd1	
	<no_ displacmt >impl.clm1: Implementation of the specification <no_ displacmt> is correct.		inputch.clm[1,2] motor_ctrl.clm[1,2] sepcots.clm[1,2] corr_code.clm[1,3] time_code.clm[1,3] spur_lock.clm[1,3]
	<no_ displacmt implfs.clm1: Implementation of the specification <no_ displacmt> is fail-safe; no single failure inhibit prevention.		fsinputch.clm[1,2] fsmotor_ctrlclm[1,2] fscots.clm[1,2]

Table B.1. (continued)

	Subclaim identifiers and description	Evidence components	Delegations
Architecture Subclaims	**inputch.clm[1,2]:** complete/adequate set of control unit sensor input channels. Transfer port status signals (docked/undocked, sealed/unsealed, fork/shuttle positions, drive mode) are correctly captured by corresponding sensing channels.	sens_spec.evd2	
	motor_ctrl.clm[1,2]: Adequate motor control and communication functional interface.	motorctrl_spec.evd2	
	fsinputch.clm[1,2]: sensor input channels are fail-safe.	sens_spec.evd2	valid_IO_chk.clm[2,3] autotsts_cov.clm[2,3] oper_fma.clm[2,4]
	fsmotor_ctrl.clm[1,2]: failsafe motor drive locking mechanism and control.	motorctrl_spec.evd2	fs_code.clm[2,3] autotsts_cov.clm[2,3] oper_fma.clm[2,4]
	fscots.clm[1,2]: adequate and failsafe COTS software platform.	COTS_exp.evd2 TSO_cert.evd2	*COTS platform justif. (sub-safety case);* oper_fma.clm[2,4]
	sepcots.clm[1,2]: no interference from non-used COTS functions.	COTS_exp.evd2; TSO_cert.evd2	*COTS platform justif. (sub-safety case);* segr_code.clm[2,3]

Table B.1. (continued)

	Subclaim identifiers and description	Evidence components	Delegations
Design Subclaims	**corr_code.clm[1,3]:** execution traces of the object code satisfy the specification <no_ **displacmt** >	code-ana.evd3 code_utst.evd3 code_itst.evd3 prog_exp.evd3	
	segr_code.clm[2,3]: protection of executable code against non-used code.	code-ana.evd3 code_utst.evd3 code_itst.evd3 COTS_exp.evd3;	
	time_code.clm[1,3]: (maximum execution + actuation time) is less than 0.5 sec.	code_itst.evd3	
	validIO_chk.clm[2,3]: adequate validity checks of input and output variables autotsts_cov.clm[2,3]	code_utst.evd3	oper_fma.clm[3,4]
	autotsts_cov.clm[2,3]: Complementary coverage of code by auto-tests and periodic tests.	autotst_spc.evd3	oper_fma.clm[3,4] pertsts_prcd.clm[3,4]
	fs_code.clm[2,3] : fail-safeness of application code wrt: - invalid input/output - errors trapped by auto-tests - other defects.	code_itst.evd3 code_utst.evd3 code_itst.evd3	oper_fma.clm[3,4]
	spur_lock.clm[1,3]: spurious locks are prevented.	code_itst.evd3	serv_prcd.clm[3,4]

Table B.1. (continued)

<table>
<tr><td rowspan="4" style="writing-mode:vertical-lr">Operation – Control Subclaims</td><td>**Subclaim identifiers and description**</td><td>**Evidence components**</td><td>**Delegations**</td></tr>
<tr><td>**oper_fma.clm[2,4] and oper_fma.clm[3,4]:** adequate anticipation of failure modes of vehicle, fork/shuttle, sensors, actuators, power supplies, motor lock.</td><td>io_rex.evd4
hw_rex.evd4
protype_rep.evd4
opr_rex.evd4</td><td>**nil**</td></tr>
<tr><td>**pertsts_prcd.clm[3,4]:** adequate periodic tests procedures.</td><td>hw_rex.evd4
io_rex.evd4</td><td>**nil**</td></tr>
<tr><td>**serv_prcd.clm[3,4]:** Adequate and robust (*e.g.* towards operators' errors) in-service procedures for periodic testing, maintenance of AGV and instrumentation equipment.</td><td>hw_rex.evd4
io_rex.evd4
protype_rep.evd4
opr_rex.evd4</td><td>**nil**</td></tr>
</table>

Table B.2. Evidence components for Dependability Requirement **no_displacmt.clm0**

	Identifier	Description	(Sub)claims supported
Evidence Components			
Environment-system Interface Evidence	AGV_spec.evd1	specification of AGV operation and interlocks	**no_displacmt_val.clm1**
	op_feedbck.evd1	operational feedback reports from AGV incidents	**no_displacmt_val.clm1**
	eng_exp.evd1	assesst. of competence and past experience of plant engineers	**no_displacmt_val.clm1**
	safety_rep.evd1	plant safety analysis report	**no_displacmt_val.clm1**
	mdrn_rep.evd1	upgrade specifications and motivation report	**no_displacmt_val.clm1**
	reg_req.evd1	regulatory requirements	**no_displacmt_val.clm1**
Architecture Evidence	sens_spec.evd2	specifications of AGV sensors	**inputch.clm[1,2]; fsinputch.clm[1,2]**
	motorctrl_spec.evd2	specifications of AGV communication /control interface	**motor_ctrl.clm[1,2], fsmotor_ctrl.clm[1,2]**
	COTS_exp.evd2	field experience report of the COTS programming platform	**fsCOTS.clm[1,2]; segrCOTS.clm[1,2]**
	TSO_cert.evd2	TSO certification of COTS programming platform	**fsCOTS.clm[1,2]; segrCOTS.clm[1,2]**

Table B.2. (continued)

	Identifier	Description	(Sub)claims supported
Desigh Evidence	code-ana.evd3	report of code static analysis.	**corr_code.clm[1,3], segr_code.clm[2,3];**
	code_utst.evd3	reports of code unit tests	**validIO_chk.clm[2,3] corr_code.clm[1,3]; segr_code.clm[2,3]; fs_code.clm[2,3];**
	code_itst.evd3	integrated on-site test suite and tests reports	**corr_code.clm[1,3] segr_code.clm[2,3]; time_code.clm[1,3]; fs_code.clm[2,3] spur_lock.clm[1,3];;**
	autotst_spc.evd3	auto-tests COTS platform specifications	**autotsts_cov.clm[2,3]**
	prog_exp.evd3	assesst. of competence and past experience of programmers and of suppliers of instrumentation	**corr_code.clm[1,3]**
	COTS_exp.evd3	Field experience report of the COTS programming platform	**segr_code.clm[2,3]**
Operation Control Evidence	io_rex.evd4	operational data on sensors, motors and actuator failure modes	**oper_fma.clm[2,4]; oper_fma.clm[3,4]; pertsts_prcd[3,4]; serv_prcd.clm[3,4];**
	hw_rex.evd4	operational data on AGV and computer failure modes	**oper_fma.clm[3,4]; pertsts_prcd.clm[3,4]; serv_prcd.clm[3,4]**
	opr_rex.evd4	reports of probation period, operator reports, other similar material handling systems operational feedback	**oper_fma.clm[3,4]; serv_prcd.clm[3,4]**
	protype_rep.evd4	report on prototype experiment	**oper_fma.clm[3,4]; serv_prcd.clm[3,4]**

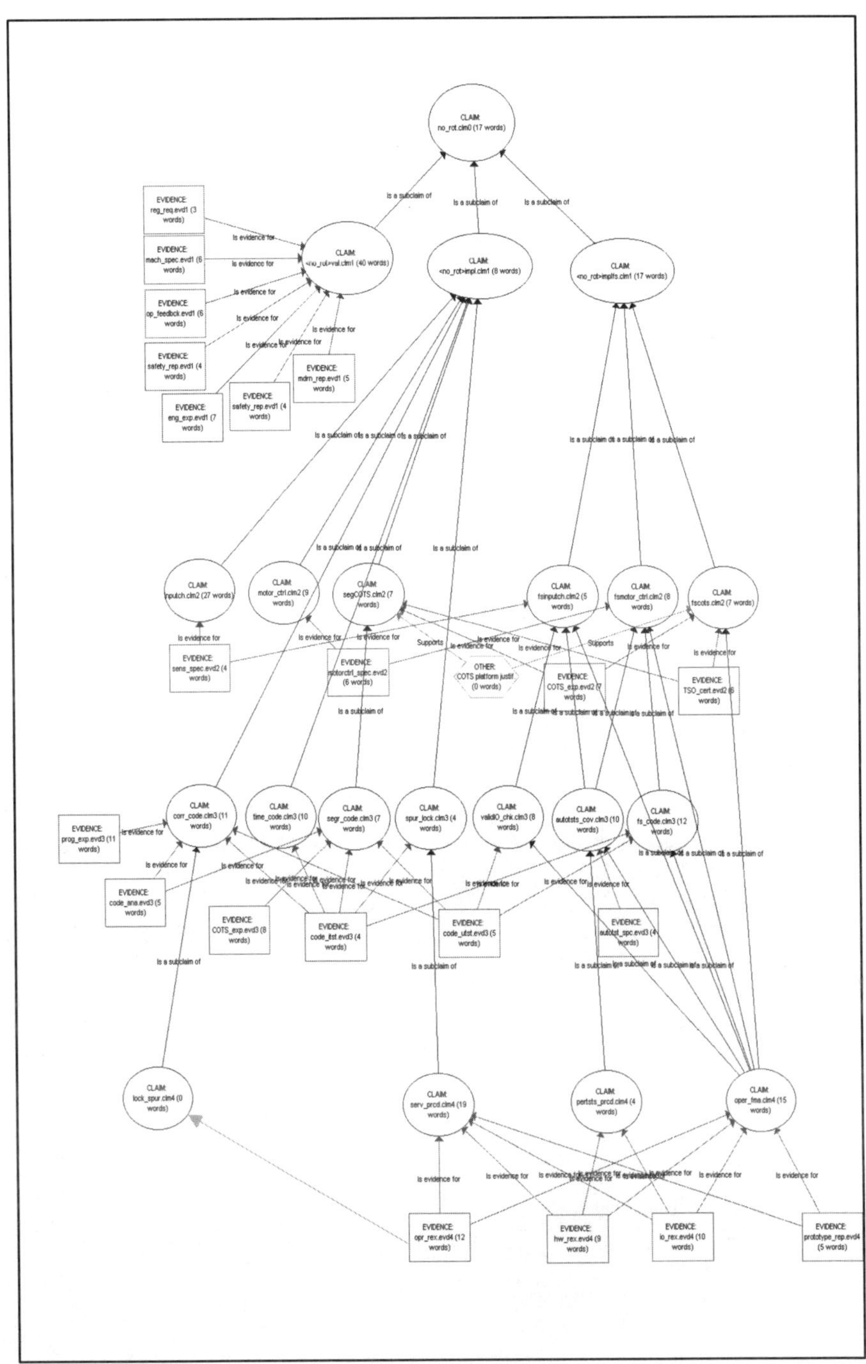

Figure B.4. Expansion of the claim **no_rot.clm1** produced with the graphical Adelard editor ASCE[TM]

References

1. Alexander C (1964) Notes on the Synthesis of Form. Harvard Univ. Press, Cambridge, Massachusetts

2. Atkinson C *et al.* (eds) (2005) Component-Based Software Development. Lecture Notes in Computer Science. vol 3778. Springer-Verlag Berlin Heidelberg

3. Alvesson M, Sköldberg K (2000) Reflexive Methodology. Sage Publications, London

4. Aschenbrenner JF, Colling JM, Raimondo E (1988) Reactor Instrumentation and Control for the French N4 Nuclear Power Plants Units. Proc. IAEA Specialists' Meeting on Microprocessors in Systems Important to the Safety of Nuclear Power Plants, London, UK. International Atomic Energy Agency, Vienna.

5. Barwise J (1977) Handbook of Mathematical Logic. North Holland, Amsterdam NewYork Oxford

6. Bell J, Machover M (1986) A Course in Mathematical Logic, 2nd edn. North Holland, Amsterdam New York Oxford Tokyo

7. Bishop PG, Bloomfield RE (1998) A Methodology for Safety Case Development. Safety-Critical Systems Symposium (SSS 98), Birmingham, UK

8. Bishop PG, Bloomfield RE, van der Meulen MJP (2004) Public Domain Case Study. An Example Application of the CEMSIS Guidance. EC CEMSIS project FIS5-1999-00355 deliverable D5.6. Downloadable from http://www.cemsis.org/

9. Blanc P, Guesnier G (1984) Conception du Contrôle – Commande des Tranches REP 1300MW d'Electricité de France. Proc. IAEA Specialists' Meeting on the Use of Digital Computing Devices in Systems Important to Safety. Saclay, France

10. Boehm B (2008) Making a Difference in the Software Century. IEEE Computer, 41, (3) 32–38

11. Chang CC, Keisler HJ (1990) Model Theory, 3rd edn. North Holland, Amsterdam NewYork Oxford Tokyo

12. Courtois P-J (1977) Decomposability. Queuing and Computer System Applications. Academic Press, New-York San Francisco London

13. Courtois P-J, Littlewood B, Strigini L, Wright D, Fenton N, Neil M (2000) Bayesian Belief Networks for Safety Assessment of Computer-based Systems. System

Performance Evaluation: Methodologies and Applications, (E. Gelenbe, ed.), 349–363, CRC Press

14. Courtois P-J (2001) Semantic Structures and Logic Properties of Computer-Based System Dependability Cases. Nuclear Engineering and Design 203, 87–106

15. Courtois P-J, Delsarte P (2006) On the Optimal Scheduling of Periodic Tests and Maintenance for Reliable Redundant Components. Reliability Eng. and System Safety, Elsevier, 91(1), 66–72

16. Courtois P-J, Parnas DL (1993) Documentation for Safety Critical Software. Proc. 15th International Conference on Software Engineering, Baltimore, May 1993, 315–323

17. Dahll G (2000) Combining Disparate Sources of Information in the Safety Assessment of Software Based Systems. Nuclear Engineering and Design 195, 307–319

18. de Bruijn NG (1983) Computer Program Semantics in Space and Time. Selected Papers on Automath Series – Studies in Logic and the Foundations of Mathematics, Ed: Nederpelt RP, Geuvers JH, de Vrijer RC, 113, North-Holland, Amsterdam NewYork Oxford

19. DeMillo RA, Lipton RJ, Perlis AJ (1977) Social Processes and Proofs of Theorems and Programs. POPL 1977, 206–214

20. Descartes R (1637) Discours de la Méthode. Présentation, Notes, Dossier, Bibliographie et Chronologie par L. Renault. GF Flammarion, Paris, 2000

21. Deetz S (1992) Democracy in an Age of Corporate Colonization: Developments in Communication and the Politics of Everyday Life. State University of New-York Press, Albany

22. DeVa (1998) European Esprit Long Term Research Project 20072, Third Year Report. Computing Science Laboratory, Univ. of Newcastle upon Tyne.
Available from www.newcastle.research.ec.org/deva/trs/index.html

23. Dijkstra EW (1968) The Structure of the "THE" Multiprogramming System. Communications of the ACM, 11 (5). ACM Press, New-York

24. Dijkstra EW (1969) Complexity Controlled by Hierarchical Ordering of Function and Variability. Software Engineering. Scientific Affairs Division, NATO, Brussels, 181–185

25. DOD 178 (1992) Software Considerations in Airborne Systems and Equipment Certification. RTCA Secretariat, Washington D.C. 20036

26. Dunn M (1985) Relevance Logic and Entailment. In [31], Chapter III.3, 117–224.

27. Dupuy JP (2002) Quand l'Impossible est Certain. Le Seuil, France

28. EUR 19265 (2000) Common Position of European Nuclear Regulators for the Licensing of Safety Critical Software for Nuclear Reactors. European Commission – DG for the Environment, Unit C.2 – Nuclear Safety, Regulation and Radioactive Waste. Luxemburg

29. Fenton NE, Littlewood B, Neil M, Strigini L, Sutcliffe A, Wright D (1999) Assessing Dependability of Safety Critical Systems Using Diverse Evidence. IEE Proceedings Software Engineering, 145(1), 35–39

30. Freeman P A (2007) Risks Are Your Responsibility. Comm. of the ACM, 50(6), 104

31. Gabbay DM, Guenthner F (eds) (1985) Handbook of Philosophical Logic – vol III: Alternatives to Classical Logic, vol 166 of Syntheses Library. D. Reidel Publishing Company, Dordrecht, Holland

32. Gallier JH (1986) Logic for Computer Science – Foundations of Automatic Theorem Proving. Harper & Row Publishers, New York

33. Grigory M (1992) A short Introduction to Modal Logic. vol 30 of CSLI Lecture Notes. Centre for the Study of Language and Information, Stanford, CA

34. Guidelines for Using Programmable Electronic Systems (PES) in Nuclear Safety and Nuclear Safety-Related Applications (2001) Issue 2 Draft D, British Energy Generation Ltd. East Kilbride, G74 5PG, UK

35. Habermas J (1971) Towards a Rational Society. Heinemann, London.

36. Habermas J (1984) The Theory of Communicative Action, vol 1. Beacon Press, Boston

37. Haapanen P, Helminen A (2002) Failure Mode and Effects Analysis of Software-based Automation Systems. STUK-YTO-TR 190, Radiation and Nuclear Safety Authority, Helsinki, Finland. ISBN 951-712-584-4

38. Heineman GT *et al.* (eds) (2005) Component-Based Software Engineering. Lecture Notes in Computer Science, vol 3489. Springer-Verlag Berlin Heidelberg

39. Heitmeyer C, Bharadwaj R (2000) Applying the SCR Requirements Method to the Light Control Case Study. Journal of Universal Computer Science (JUCS), 6(7), 650–678

40. Hester SD, Parnas DL, Utter DF (1981) Using Documentation as a Software Design Medium, Bell System Tech. J. 60(8), 1941–1977

41. Houck MM (2006) CSI: Reality. Scientific American 295(7), 67–71

42. Hoffman DM, Weiss M, Ed. (2001) Software Fundamentals. Collected Papers by David L. Parnas. Addison-Wesley, Boston San Francisco New-York

43. HSE Technical Assessment Guide (2002) Guidance on the Purpose, Scope and Content of Nuclear Safety Cases. Guide T/AST/051. Health & Safety Executive,UK. Downloadable from www.hse.gov.uk/foi/internalops/nsd/tech asst guides/tast051.pdf

44. Huang S, Lappa D, Chiao T, Parrish C, Carlson R, Lewis J, Shikany D, Woo H (1997) Real-time Software Use in Nuclear Materials Handling Criticality Safety Control. Report UCRL-ID 127772. Lawrence Livermore National Laboratory, Livermore

45. Hudak P (1989) Conception, Evolution and Application of Functional Programming Languages. ACM Computing Surveys 21(3), 359–411. ACM Press, New-York

46. International Atomic Energy Agency (1998) Modernization of Instrumentation and Control in Nuclear Power Plants. TECDOC-1016, IAEA, Vienna, Austria

47. International Atomic Energy Agency (2006) Storage of Radioactive Waste. IAEA Safety Standards No. WS-G- 6.1, IAEA, Vienna, Austria

48. IEC 60880 Ed2 (2006) Nuclear Power Plants – Instrumentation & Control - Software for computers in the safety systems. International Electrotechnical Commission

49. IEC 61226 (1993) Nuclear Power Plants – Instrumentation and control systems important for safety – Classification. International Electrotechnical Commission. CH-1211 Geneva 20, Switzerland

50. IEC 62096 (2001) Nuclear Power Plants – Instrumentation and Control – Guidance for the Decision on Modernization (committee draft 2001-11-03). International Electrotechnical Commission. CH-1211 Geneva 20, Switzerland

51. Jackson D, Thomas M, Millette LI (2007) Software for Dependable Systems: Sufficient Evidence? National Research Council. The National Academies Press, Washington D.C. Available from http://www.nap.edu/catalog/11923.htlm

52. Johnston WM, Hanna JRP, Millar RJ (2004) Advances in Dataflow Programming Languages. ACM Computing Surveys, 36 (1), 1–34. ACM Press, New-York

53. Jones CB (1993) Reasoning about Interference in an Object-based Design Method. FME'93: Industrial-strength Formal methods. Lecture Notes in Computer Science, vol 670. Springer-Verlag Berlin Heidelberg

54. Karp MR (1972) Reducibility Among Combinatorial Problems. In Miller RE and Thatcher JW (eds): Complexity of Computer Computations. Plenum, New York, 85–103

55. Keisler HJ (1977) Fundamentals of Model Theory. In Barwise J (ed): Handbook of Mathematical Logic. North Holland, Amsterdam New York Oxford Tokyo

56. Klein E (2007) La Naissance de la Physique Quantique. Les Dossiers de la Recherche, 29, 6–12, SES publication, Paris, France

57. Laprie J-C (1995) Guide de la Sûreté de Fonctionnement. Cépaduès Editions, Toulouse, France

58. Lardner D (1834) Babbage's Calculating Engine. Edinburgh Review, July 1834, No. CXX. Reprinted pp.163 and ff. in "Charles Babbage and his Calculating Engines. Selected Writings by Charles Babbage and Others" P. and E. Morrison (eds) Dover Publications, Inc. New York, 1961

59. Lepage C, Guery F (2001) La Politique de Précaution. Presses Universitaires de France (PUF)

60. Lions JL (Chairman) (1996) Ariane 501 Failure: Report by the Inquiry Board. European Space Agency, 19 July 1996. Available from http://sunnyday.mit.edu/accidents/Ariane5accidentreport.html

61. Littlewood B, Wright D (1995) A Bayesian Model that Combines Disparate Evidence for the Quantitative Assessment of System Dependability. Proc 14th International Conference on Computer Safety, 173–188. Springer, Berlin Heidelberg

62. Littlewood B, Popov P, Strigini L (2001) Modeling Software Design Diversity: a Review. ACM Computing Surveys, 33(2), 177–208. ACM Press, New-York

63. Littlewood B, Wright D (1997) Some Conservative Stopping Rules for the Operational Testing of Safety-Critical Software. IEEE Trans Software Engineering, 23(11) 673–683

64. Lyotard JF (1984) The Postmodern Condition: A Report on Knowledge. University of Minnesota Press, Minneapolis

65. Maibaum T, Wassyng A (2008) A Product-Focused Approach to Software Certification. IEEE Computer, 41(2), 91–93

66. McDermid JA (1994) Support for Safety Cases and Safety Argument Using SAM, Reliability Eng. and System Safety, 43(2) 111–127

67. Mendelson E (1987) Introduction to Mathematicl Logic, 3rd edn. Wadsworth & Brooks/Cole, Monterey, California

68. Mercuri R, Lipsio V, Feehan B (2006) COTS and Other Electronic Voting Backdoors. Comm. of the ACM 49(11),112. ACM Press, New York

69. Mills HD (1975) The New Math of Computer Programming. Comm. of the ACM, 18(1), 43–48. ACM Press, New York

70. Mosleh A, Rasmuson DM, Marshall FM (1998) Guidelines on Modeling Common-Cause Failures in Probabilistic Risk Assessment. NUREG/CR-5485, U.S. Nuclear Regulatory Commission, Washington, DC

71. Neumann P (2007) Widespread Network Failures. Comm. of the ACM, 50(6)104. ACM Press, New York

72. Oh Y, Yoo J, Cha S, Seong Son H (2005) Software Safety Analysis of Function Block Diagrams Using Fault Trees. Reliability Eng. and System Safety, 88, 215–228

73. Parnas DL (1972) On the Criteria to be Used in Decomposing Systems into Modules. Comm. of the ACM, 15 (2), 1053–1058. ACM Press, New York

74. Parnas DL (1974) On a Buzzword: Hierarchical Structures. IFIP Congress 74, North Holland, 336–339. Reprinted in [42], 161–168

75. Parnas DL (1978) Another View of the Dijkstra-DMLP Controversy. ACM Software Engineering Notes, 3(4) 20–21. ACM Press, New York

76. Parnas DL, Weiss DM (1985) Active designs reviews: Principles and practices. Proc.8th Int. Conf. Software Eng., London. Reprinted in [42]

77. Parnas DL, Clements PC (1986) A Rational Design Process: How and Why to Fake It. IEEE Trans. Software Eng. SE-12 (2), 251–257

78. Parnas DL, Madey J (1990) Functional Documentation for Computer Systems Engineering. Technical Report 90-287, Queen's University, C&IS, Telecommunications Research Institute of Ontario (TRIO)

79. Parnas DL, Asmis GJK, Madey J (1991) Assessment of Safety Critical Software in Nuclear Power Plants. Nuclear Safety, 32 (2)

80. Parnas DL (1994) Mathematical Description and Specification of Software. Proceedings IFIP World Congress 94, North Holland, 354–359. Reprinted in [42], 93–105

81. Parnas DL (2007) Use The Simplest Model but Not Too Simple. Communications of the ACM, Forum, 50 (6), 7. ACM Press, New York

82. Parnas DL (2007) Monopoly Versus Diversity. Communications of the ACM, 50(7). ACM Press, New York

83. Pavey DJ, Winsborrow LA (1993) Demonstrating Equivalence of Source Code and PROM Contents. The Computer Journal, 36 (7), 654–667

84. Pavey DJ, Winsborrow LA (1995) Formal Demonstration of Equivalence of Source Code and PROM Contents. In Mitchell C and Stavridou V (eds): Mathematics of Dependable Systems (Proc. of the IMA Conf on Mathematics of Dependable Systems, Royal Holloway College, Sept 1993, Clarendon Press, Oxford

85. Pavey DJ (2003) CEMSIS Cost Effective Modernization of Systems Important to Safety. FISA Conf. Proc. Downloadable from http://www.cemsis.org/

86. Pearl J (2000) Causality – Models, Reasoning, and Inference. Cambridge University Press. Cambridge, UK

87. Perry W (1995) Effective Methods for Software Testing. J. Wiley & Sons Inc. New York

88. Randell B, Laprie J-C, Kopetz K, Littlewood B (eds) (1995) Predictably Dependable Computing Systems. European publication N° EUR 16256EN, Springer-Verlag, Berlin Heidelberg New York

89. Rechtin E (1991) Systems Architecting. Creating & Building Complex Systems. PTR Prentice Hall, Englewood Cliffs, New Jersey

90. Ristord L, Esmenjaud C (2002) FMEA Performed on the SPINLINE3 Operational System Software as part of the TIHANGE 1 NIS Refurbishment Safety Case. Proc. OECD CNRA/CSNI Workshop on Licensing and Operating Experience of Computer-based I&C Systems. NEA/CSNI/R (2002)1, vol 2, 37–50

91. Saglietti F (2004) Licensing Reliable Embedded Software for Safety-Critical Applications. Real-Time Systems, 28, 217–236

92. Schopenhauer A (1864) Die Kunst, Recht zu behalten. In Arthur Schopenhauers Handschriftlicher Nachlaß von Julius Frauenstädt, Brockhaus, Leipzig

93. Simon HA (1953) Causal Ordering and Identifiability. In W.C. Hood and T.C. Koopmans (eds): Studies in Econometric Methods. Wiley, New-York

94. Simon HA (1969) The Sciences of the Artificial. MIT Press, Cambridge

95. SQUALE (1999) Dependability Assessment Criteria. European Commission Research Project ACTS95/AC097. Available from LAAS-CNRS, Toulouse, France

96. Tanenbaum AS (1992) Modern Operating Systems. Prentice Hall Inc., New Jersey

97. Tarski A, Vaught R (1957) Arithmetical Extensions of Relational Systems. Comp. Math. 18, 81–102

98. Taylor RP, Mills S, Chen S, El-Saadany S (1991) Reliability Assessment for Safety Critical Systems. AECB Report. Canadian Nuclear Safety Commission (CNSC), Ottawa, Ontario K1P 5S9, Canada

99. Theofanous TG (1996) On the Proper Formulation of Safety Goals and Assessment of Safety Margins for Rare and High-consequence Hazards. Reliability Eng. and System Safety, 54, 243–257

100. Toulmin S (1958) The Uses of Argument. Cambridge University Press, Cambridge, UK

101. Updike J (1960) Rabbit, Run. Alfred A. Knopf USA, 1st edn. Penguin Books UK (1962)

102. U.S. Nuclear Regulatory Commission (2002) HuMan–System Interface Design Review Guidelines. NUREG-0700 Rev. 2, U.S. Government Printing Office, Washington DC 20402-001. Available from http://bookstore.gpo.gov/

103. van Schouwen AJ (1990) The A-7 Requirements Model: Re-examination for Real-Time Systems and an Application to Monitoring Systems. Technical Report 90-276, Queen's university, Kingston, Ontario K7L 3N6.

104. Voas JM, Laplante PA (2007) Standards. Confusion and Harmonization. IEEE Computer, 7, 94–96.

105. Voges U (ed) (1988) Software Diversity in Computerized Control Systems. Springer-Verlag, Wien NewYork.

Index[*]

---

[*] Page numbers in bold refer to definitions.

Detectability (of events), 94, 134, 166, 197, 226
Deterministic
    assessment, 14, 139
    program behaviour, 212
DeVa project, 285
Dewey notation, 275
Displays, 218
Dissensus, 64
Diversity
    architecture, 169, 171
    design (software), **201**, 203, 205
    evidence, 275
    functional, **136**, 138, 169, 200, 288
    pre-existing component, 266
Documentation
    level-0 (environment), 122
    level-1, 150
    level-2, 185
    level-3, 215
    level-4 (dependability case), 237
Drift, 155

E

$\exists_r$, 176
$\varepsilon_3(m_i(t))$, 194
Elicitation (of claims), 26, 272
    see also justification ordering
Elicitation theorems, 95, 100
Encapsulation, 263, 265
Enlightened catastrophism, 27
Environment
    assumptions, 126
    constraints, 25, 122, 125
    (safe) states, 25, 124
Epistemic, *see* uncertainty
Equivalence class, 232
Ergonometry, 236
Error, **119**
Event
    external/internal, **118, 119**
    Postulated Initiating (PIE), 25 **119**, 122, 298
    random, 189, 196, 200
    rare, 284

undesired (UE), **120**, 156, 165, 196, 220, 293
Evidence, 33, **38**, (valid) **92**, 274
    components, 305
    disparity, 33
    levels, 34
    naming, type, 275
    weight, 18, **62**
    see also plausibility
Exceptions, 190, 201, 203, 205
Expansion
    of claim, **49**, 54, 270, 271, 299, 301, 307
    of language, **86**, 107
    level-0, 129
    level-1, 144, 273, 300
    level-2, 157
    level-3, 192
    level-4, 235
Expansion structure, 97
Expansion theorem, 99, 106
Expert judgment, 123
    see consensus

F

$\mathcal{F}^1$, 127
$f(t)$, **140**, 177
$F_1(x,t)$, **140**, 207
$\bar{F}(t_1,t_2)$, **141**
$\underline{f}_k(t)$, **177**
$\bar{F}_k(t_1,t_2)$, **178**
F_{kmax}, **179**
FS(t), **124**, 225, 254
$FS^s(t)$, **253**, 254
$fsw_k(t)$, 207
Fsw_{kmax}, 208
FFP method, 278
Failure, **119**
    communication, 267
    human, 219, 226
Failure Mode Analysis (FMECA), 47, 61, 213, 214, 292
Fail-safe control, 224, 292
Faults, **119**
    transitory, Byzantine, 267
Feasibility
    architecture, 159

Printing: Krips bv, Meppel, The Netherlands
Binding: Stürtz, Würzburg, Germany